Rising Above: A Practical Guide to Overcoming Climate Anxiety and Finding Hope in a Changing World

Table of Contents

Also by John Shoufler 4

Part I: Understanding Climate Anxiety 15

    Chapter 1: Naming the Feeling 15

    Chapter 2: Why Are We So Anxious? 32

    Chapter 3: How It Affects Us Day-to-Day 50

    Chapter 4: Context from the Past and Around the World 70

Part II: Finding Emotional Ground 89

    Chapter 5: Immediate Relief When Fear Spikes 89

    Chapter 6: Reframing Your Thoughts 106

    Chapter 7: Building Lasting Resilience 124

Part III: Strength in Numbers—Connecting with Others 162

    Chapter 9: Breaking Isolation Through Community 162

    Chapter 10: Honest, Compassionate Communication 182

    Chapter 11: Turning Anxiety into Collective Action 197

Part IV: Sustaining Hope and Engagement Over Time 217

    Chapter 12: Recognizing Progress and Celebrating Wins 217

    Chapter 13: Hope as a Practice, Not a Fantasy 239

    Chapter 14: Learning from Nature's Resilience 256

    Chapter 15: A Lifelong Journey of Caring 273

Epilogue: Embracing Complexity, Choosing Courage 287

Appendices 299

    Worksheets & Prompts: Tools for Reframing Anxious Thoughts, Tracking Emotional States, and Planning Next Steps 299

        Worksheet A: Reframing Anxious Thoughts 300

        Worksheet B: Tracking Emotional States Over Time 301

        Worksheet D: Embracing Complexity and Uncertainty 303

        Worksheet E: Mentoring and Intergenerational Wisdom Exchange 304

        Worksheet F: Updating Your Coping Toolkit Regularly 305

        Organizations and Initiatives for Continued Engagement 306

        Books That Offer Moral Depth, Complexity, and Hope 307

        Podcasts for Regular Inspiration and Nuanced Updates 308

        Online Communities and Discussion Forums 309

Approaches for Using These Resources Mindfully 310
Tables and Charts to Customize Your Resource Use 311
Chart: Information-Action Cycle 312
Glossary: Key Terms for Quick Reference 314

Copyright Disclaimer

© 2024 by John Shoufler. All rights reserved.

No part of this publication may be reproduced, distributed, or transmitted in any form or by any means, including photocopying, recording, or other electronic or mechanical methods, without the prior written permission of the publisher, except in the case of brief quotations embodied in critical reviews and certain other non-commercial uses permitted by copyright law.

For permission requests, please contact John Shoufler at johnshoufler@sarapalooza.com.

The information in this book is intended to be a general resource for career development and job search strategies. The author and publisher disclaim any liability in connection with the use of this information. This book is not intended as legal, financial, or professional advice, and individuals should consult with appropriate professionals for guidance in their specific circumstances.

All company names, products, logos, and trademarks mentioned in this book are the property of their respective owners and are used for identification purposes only. Use of these names, logos, or trademarks does not imply endorsement.

Also by John Shoufler

You can find more books by John Shoufler, available in various formats, listed below:

- Defend Yourself From Scammers: The Ultimate Guide to Identifying and Avoiding Scams: Protect Your Finances, Privacy, and Peace of Mind in a Digital World
    - Paperback ISBN: 979-8346296911
    - Hardcover ISBN: 979-8346296072
    - Kindle Edition ASIN: B0DMR2S58Y
- Bully-Proof Your Child: The Ultimate Parent's Guide to Ending School Bullying
    - Paperback ISBN: 979-8346113010
    - Hardcover ISBN: 979-8346113690
    - Kindle Edition ASIN: B0DMLQFJ87
- Digital Balance: Reclaiming Your Life Through Mindful Technology Use: Achieving Harmony in a Digital World: Practical Steps for Balanced Technology Use
    - Paperback ISBN: 979-8345898819
    - Hardcover ISBN: 979-8345897522
    - Kindle Edition ASIN: B0DJ3K9KDK
- Mid-Career Transitions: A Comprehensive Guide - Volume One: Navigating the Path from Stability to Opportunity in a Changing Workforce (The Mid-Career Transformation Collection Book 1)
    - Paperback ISBN: 979-8344570846
    - Hardcover ISBN: 979-8344569734
    - Kindle Edition ASIN: B0DL2FMFZ7
- Mid-Career Transitions: A Comprehensive Guide - Volume Two: Advanced Strategies for Navigating Change, Overcoming Challenges, and Sustaining Career Growth ... Transformation Collection Book 2)
    - Paperback ISBN: 979-8344572789
    - Hardcover ISBN: 979-8344571980
    - Kindle Edition ASIN: B0DL2N9Q1P

- Six-Figure Career Mastery: Your Guide to High-Paying Jobs, Strategic Networking, and Career Growth: Unlocking Your Path to Leadership, Financial Success, and Lifelong Career Growth
    - Paperback ISBN: 979-8344019994
    - Hardcover ISBN: 979-8344015705
    - Kindle Edition ASIN: B0DKGCZ8DV
- Freelance Writing Mastery: The Complete Guide to Building and Scaling a Six-Figure Writing Business
    - Paperback ISBN: 979-8340686701
    - Hardcover ISBN: 979-8340686718
    - Kindle Edition ASIN: B0DJ6YZ2GJ
- Mastering Technical Writing: A Comprehensive Guide to Effective Documentation and Communication
    - Paperback ISBN: 979-8350914562
    - Hardcover ISBN: 979-8350914579
    - Kindle Edition ASIN: B0DJ7MXTQR
- Flutter & Rhyme: Whimsical Tales for Little Minds
    - Paperback ISBN: 979-8350914586
    - Hardcover ISBN: 979-8350914593
    - Kindle Edition ASIN: B0DJN2J14C
- Mastering the Online Tutoring Business: Build, Scale, and Succeed in the New Era of Digital Education
    - Paperback ISBN: 979-8340987129
    - Hardcover ISBN: 979-8340987136
    - Kindle Edition ASIN: B0DJN2J14C
- Virtual Assistant Success: A Step-by-Step Guide to Starting, Scaling, and Profiting from Your VA Business
    - Paperback ISBN: 9798350914562
    - Hardcover ISBN: 9798350914579
    - Kindle Edition ASIN: B0DGRNRHP8

- Dropshipping and eCommerce Secrets: A Comprehensive Guide to Creating a Six-Figure Business
    - Paperback ISBN: 9798339218425
    - Hardcover ISBN: 9798339218432
    - Kindle Edition ASIN: B0DH19H62M
- Affiliate Marketing: Promoting Other Companies' Products and Earning Commissions for Each Sale Made
    - Paperback ISBN: 9798340987143
    - Hardcover ISBN: 9798340987150
    - Kindle Edition ASIN: B0DJ6YZ2GJ
- How to Build an Inexpensive Chicken Feeder
    - Paperback ISBN: 9798340987167
    - Hardcover ISBN: 9798340987174
    - Kindle Edition ASIN: B00QUCBDR

Introduction

We live in a time when the climate—once easily relegated to science reports, policy debates, or distant headlines—has entered our kitchens, our gardens, our workplaces, and our dreams. It's no longer possible to think of environmental change as an abstract concept concerning polar bears, remote rainforests, or some distant point in the future. Instead, it shapes what we eat, how we move, where we settle, and the prospects we imagine for ourselves and future generations. You might feel a subtle shift in your mood when you see daffodils blooming weeks earlier than expected, or a pang of worry when an unusually fierce storm ravages a community far from home. Perhaps you've caught yourself lingering over your morning coffee, not just enjoying its warmth but wondering whether erratic rainfall patterns or depleted soils threaten the beans that made it into your mug.

It's natural that this new reality has begun to stir powerful emotions. You may find yourself uneasy in ways you can't quite name—an intermittent tension in your chest, a heaviness behind your eyes when reading the news, or a nagging sense that something, somewhere, is off-kilter. You might discover that what once seemed like mere background information—the melting of ice sheets, the bleaching of coral reefs, the creeping deserts—now affects how you think about career choices, long-term plans, or even the moral responsibility of having children. There is a name that has emerged to describe this cluster of feelings: **climate anxiety** (sometimes called eco-anxiety).

Validating Our Emotional Responses

Climate anxiety is not, as of now, a formal psychiatric diagnosis. It's not something you'll necessarily see in a medical manual. But it is very real and increasingly recognized. Around the world, people are reporting worry, dread, sadness, anger, grief, guilt, and confusion tied to our changing planet. These emotions are responses to genuine threats and uncertainties. They arise because we understand, at some level, that the very conditions that have allowed human life and civilization to flourish are shifting beneath our feet. Recognizing these feelings is the first step toward understanding them. There is no shame in feeling unsettled. On the contrary, it often means you're paying attention.

A few decades ago, open discussions about environmental feelings were rare. People might have worried privately, activists certainly expressed concern, and Indigenous communities around the world sounded warnings, but broad public acknowledgement that ecological upheaval could produce widespread emotional distress was not common. Now we understand that these inner responses matter—not only because they influence how we cope, but also because they reflect our moral and ethical sensibilities. Climate anxiety often signals that we care deeply: about each other, about future generations, about other species, about places we cherish, and about sustaining a livable planet.

This introduction is an invitation into a conversation that treats your feelings about climate change as relevant, legitimate, and valuable. Instead of pushing these emotions aside, we will examine them openly. Instead of ignoring the dread or dismissing it as irrational, we'll explore why it's emerging and what it tells us about ourselves and our world. This book acknowledges that while technical solutions—new policies, clean energy, sustainable agriculture—are crucial, they aren't the whole story. We must also tend to our inner landscapes. Understanding why we feel anxious and how to manage that anxiety is an essential piece of navigating these uncertain times.

Why Acknowledging Climate Anxiety Matters

Understanding and addressing climate anxiety helps prevent it from becoming paralyzing or destructive. When left unchecked, anxiety can lead to despair, burnout, denial, or apathy—all states that diminish our capacity to respond constructively. Conversely, when we learn to recognize and work with these feelings, they can be transformed into clarity, motivation, courage, and a sense of purpose. Far from a weakness, acknowledging climate anxiety can become a form of strength, a catalyst for engagement rather than withdrawal.

It takes moral courage to face the facts. If you feel uneasy, it's likely because you realize that the balance of life on Earth is shifting. This includes not only melting glaciers or endangered species halfway around the world, but also changes that affect your daily life: food availability, housing security, economic stability, public health, cultural traditions, and more. The crisis is both global and personal, both vast and intimate. Recognizing the emotional toll this complexity takes is an act of honesty. It allows you to move beyond denial or numbness, stepping into a more authentic relationship with reality.

Different Perspectives, Shared Concerns

Not everyone experiences climate anxiety in the same way. Some people, especially those on the frontlines of climate disruption—coastal communities facing rising seas, farmers contending with failing crops—may feel immediate fear and grief for their livelihoods. Others, living in regions so far spared from extreme weather, may feel a more diffuse, abstract worry, a subtle sense that something is "off" without always connecting it to the broader ecological picture. Age, cultural background, economic status, and personal history influence how we perceive these changes. A teenager who grew up with dire predictions finds it harder to imagine what "normal" used to mean. An older adult who recalls stable seasons might feel profound loss and confusion as long-held expectations unravel.

Cultural stories also matter. In some societies, environmental stewardship and living harmoniously with nature are longstanding values, providing moral frameworks that can guide emotional responses. In other cultures, where growth and technological mastery have been celebrated as hallmarks of progress, acknowledging limits and vulnerabilities may feel disorienting or threatening. Recognizing these differences helps us approach the conversation with empathy. Climate anxiety is a shared human predicament even if it manifests differently across communities.

Moral Dimensions and Justice Issues

Climate change isn't just a technical puzzle of greenhouse gases and melting ice; it's also a moral and ethical challenge. Those who have contributed the least to the crisis often face its severest impacts—an injustice that stirs feelings of anger, guilt, and sorrow. As we understand these moral dimensions, anxiety sometimes intensifies. We realize that this is not only about temperature graphs and emissions curves. It's about fairness, responsibility, and what we owe to each other and the future.

This understanding invites us to see climate anxiety not as an overreaction, but as an appropriate response to real moral stakes. If you feel uneasy, it may be because deep down you recognize that ignoring these issues would be a profound ethical failing. Naming this moral conflict helps clarify why we feel anxious. We're not just "sensitive" or "fragile"; we're grappling with a situation where high-stakes decisions—by governments, businesses, communities, and individuals—will shape the world for centuries to come.

Our Minds' Limited Tools for Long-Term, Global Threats

Part of what makes climate anxiety so challenging is the way our minds are wired. Human beings evolved to respond to immediate dangers—predators lurking nearby, sudden resource shortages—not slow-moving, complex, and globally dispersed threats. Climate change defies easy categorization. It unfolds over decades and centuries, affects every corner of the globe, and has no single villain to confront or single solution to implement.

This evolutionary mismatch helps explain why some respond with denial, numbness, or despair. Our brains struggle to process diffuse hazards that lack a clear endpoint. Understanding this can foster compassion for ourselves. Instead of blaming ourselves for feeling overwhelmed, we can say: "Of course this is hard to grasp. Of course it feels unreal even though I know it's real. My mind wasn't built for this scenario, and that's okay." This insight can free us from self-criticism and allow us to focus on finding constructive ways to cope.

From Fear to Active Engagement

This conversation is not about dwelling on fear. Rather, it's about equipping you with understanding and tools. Just as a person with a health concern benefits from learning about their condition to make informed choices, someone feeling climate anxiety benefits from recognizing the signs, identifying triggers, and practicing coping strategies. These strategies might be as simple as focusing on your breath when headlines overwhelm you, taking a mindful walk in a local park, writing down your worries to see them more clearly, or seeking community support.

The goal is not to pretend everything will turn out perfectly, but to replace panic or resignation with "active hope." Active hope acknowledges the seriousness of the problem while affirming our agency. It does not guarantee a happy ending, but it insists that our choices matter. Active hope reminds us that small steps—reducing personal carbon footprints, supporting sustainable businesses, voting for climate-conscious policies, engaging in activism—can add up. Collective action magnifies these efforts. There's strength in knowing you're not alone, that others share your concern and are willing to work together toward solutions.

Balancing Individual and Collective Responsibility

It's true that personal efforts alone won't solve the crisis. Climate anxiety often flares up when we feel that powerful institutions or leaders are failing to act. Corporate interests might stall meaningful reforms, or political gridlock may block urgent measures. Acknowledging these systemic barriers can initially heighten anxiety. But it can also motivate us to seek allies, join movements, and push for change at broader levels. Realizing that we are part of a larger human endeavor can alleviate feelings of isolation. We can share our burdens and aspirations, inspiring each other to keep going despite setbacks.

By recognizing that climate anxiety is rooted in both personal and systemic factors, we move away from self-blame. Instead of feeling solely responsible for fixing everything, we understand that we are one thread in a vast tapestry of concerned citizens, policymakers, entrepreneurs, scientists, farmers, and educators. This perspective can ease some of the emotional load. You do not carry the future of the planet alone. None of us does. But each of us plays a meaningful role.

Drawing on Our Shared Human Heritage

While the scale and complexity of today's climate crisis are unprecedented, humanity has a long history of confronting adversity. Across millennia, people have faced floods, droughts, plagues, wars, and collapses of societies. While these challenges were often localized or regional, they taught us that humans possess resilience, creativity, moral imagination, and the capacity to adapt and cooperate. Knowing this can offer solace. Though we cannot rely on old solutions to entirely fix new problems, we can learn from our past and trust in our ability to find new pathways.

This is not to minimize the severity of our current predicament. Rather, it's a reminder that we are not the first generation to face profound uncertainty. By acknowledging this lineage, we can approach climate anxiety with a sense of belonging to a long tradition of human resilience. Our task is enormous, but not hopeless. We have tools, knowledge, and ethical frameworks that our ancestors could not have imagined. With every scientific breakthrough, every grassroots initiative, every international treaty negotiated, we demonstrate our capacity to respond, learn, and evolve.

What This Book Offers

This book aims to serve as a companion on your journey of understanding and managing climate anxiety. It does so by moving through four main arcs: understanding, coping, connecting, and sustaining hope.

1. **Understanding Climate Anxiety:**
   We begin by clarifying what climate anxiety is and why it arises. We'll explore its psychological roots, cultural influences, and moral dimensions. Understanding helps normalize your feelings, showing that you're not alone and not "overreacting." It provides context and language, turning vague unease into something you can name and reflect upon.

2. **Coping with Emotional Distress:**
   Once we understand these emotions, we'll delve into practical strategies to cope when anxiety flares up. These might include simple breathwork, journaling, reframing negative thought patterns, seeking professional help if needed, or setting healthy boundaries with media consumption. Coping doesn't mean ignoring the problem—it means equipping yourself to remain grounded, alert, and mentally strong so that you can engage rather than retreat.
3. **Connecting with Others:**
   Climate anxiety often feels isolating. Yet millions share these concerns. We'll look at how connecting with communities—whether online forums, local sustainability groups, faith-based initiatives, or advocacy organizations—can alleviate despair. Sharing stories, learning from others, and working together can transform solitary worry into collective empowerment. Relationships become anchors in a changing world, allowing you to draw on shared wisdom, skills, and moral support.
4. **Sustaining Hope Over the Long Term:**
   Addressing the climate crisis is not a sprint; it's a marathon with no fixed finish line. Sustaining hope means learning to live with uncertainty without giving up on the possibility of better outcomes. We'll consider how to maintain motivation when progress is slow or uneven, how to celebrate incremental improvements, and how to find inspiration in the resilience of nature and human communities. Hope is not naïve optimism; it's a deliberate stance that acknowledges difficulty while affirming that our actions and values still count.

These four arcs are not rigid steps. They interweave and loop back on each other. Some days you might revisit understanding because new information challenges your previous assumptions. Other times coping strategies will come to the forefront when a particularly grim headline rattles you. When feeling isolated, you'll turn to connecting. When doubt creeps in, you'll practice sustaining hope.

Integrating Complexity Without Overwhelm

A key challenge in writing about climate anxiety is the risk of overwhelming readers. After all, the climate crisis itself is overwhelming. But this book's structure and approach aim to prevent spirals of despair. By first normalizing your feelings, we defuse shame and confusion. By offering coping tools, we ensure that knowledge doesn't lead to panic but to managed resilience. By encouraging connection, we show that you're not alone. By focusing on sustained hope, we remind you that while outcomes aren't guaranteed, possibilities remain open.

The complexity of environmental changes also means this book touches on multiple spheres—mental health, ethics, politics, culture, technology, spirituality, and community life. None of these realms can solve the crisis in isolation, but together they can help us navigate it. Emotions bridge these domains, providing signals about what matters. By legitimizing our emotional responses, we allow moral insights to surface, helping us discern priorities, values, and avenues for action.

Embracing Emotional Nuance

In discussing climate anxiety, it's important to recognize the range of emotions that might arise. You might feel anger at political inaction, sadness at the loss of familiar landscapes, guilt over personal consumption patterns, or admiration for those working tirelessly to restore ecosystems. These emotions can coexist, contradict, and evolve. Embrace this emotional nuance. It reflects the complexity of what we face. Rather than seeking a single "correct" feeling, welcome the interplay of moods as part of a meaningful engagement with reality.

This openness to emotional nuance also helps reduce polarization. Climate debates sometimes descend into factions—some champion technological solutions, others call for systemic overhaul, some focus on personal behavior change, others emphasize global policies. Emotions remind us that these perspectives are not merely intellectual stances. They emerge from deep values, fears, and hopes. By starting conversations from what we feel, rather than just what we think, we can find common ground and empathy even when we disagree on specifics.

Practical and Personal Applications

As you read on, you'll encounter concrete examples, stories, exercises, and reflections. For instance, you might learn a breathing exercise to steady yourself after reading dire headlines, or discover how a community supported each other through a sudden climate-induced disaster, drawing lessons on solidarity and adaptation. You might encounter philosophical frameworks that expand your sense of what's possible, or spiritual traditions that offer comfort and guidance in times of uncertainty.

Not every strategy will resonate with you, and that's okay. Consider this book a toolkit, not a strict program. Feel free to try some approaches and discard others. Climate anxiety, like any complex emotional state, is personal. Different individuals find comfort and motivation in different places. One might be inspired by technological innovation, another by the regenerative cycles of nature, another by policy wins at local or international levels, and yet another by art, literature, or spiritual practice. Adapt what you find here to your needs.

Building a Relationship with Your Emotions

Ultimately, this book aims to help you develop a more skillful relationship with climate anxiety. Instead of seeing it as an enemy, you can learn to view it as a messenger. Anxiety signals that something you care about is at risk. If you listen, you can discover what values lie behind your worry—perhaps a love of biodiversity, a commitment to fairness, a desire to protect future generations. By understanding these underlying values, you can transform anxiety from a state of paralysis into a driver of meaningful engagement.

This does not mean romanticizing anxiety or glossing over pain. Emotional distress is still distress. But acknowledging it as natural and meaningful can lighten the load. You move from feeling at the mercy of anxiety to recognizing it as one element of your emotional landscape. With practice, you'll become better at catching unhelpful patterns—like catastrophizing or self-blame—early on, guiding yourself back to a balanced state of concern mixed with action-oriented thinking.

Recognizing That the Conversation Continues

It's important to understand that the conversation about climate anxiety doesn't end with this book. Climate change is ongoing. Our emotional responses will evolve as conditions shift, new data emerges, solutions are tested, and policies come and go. You may find yourself revisiting these feelings many times over the years, each time with a slightly different perspective and set of experiences.

The aim here is not to provide a final, one-size-fits-all solution to emotional turmoil. Instead, it's about equipping you with resources—intellectual, psychological, relational—that you can draw upon when needed. By normalizing and exploring these emotions now, you'll be better prepared to handle the emotional ups and downs that the future may hold.

Welcome to This Community of Concern

As we set forth, imagine that you're not reading these pages alone. Picture a circle of diverse readers—teenagers, parents, elders, professionals, students, farmers, city dwellers—each with their own stories. Some have witnessed climate disasters firsthand, others have gleaned their understanding from studies, documentaries, or conversations with friends. All are asking: How do we live wisely, ethically, and fully in uncertain times?

No single reader or author has all the answers. But together, we can learn from each other, support one another, and find strength in our collective efforts. Your voice matters. Your emotions matter. Your search for meaning and purpose matters. By acknowledging climate anxiety and learning to navigate it, you become part of a community that values honesty, courage, empathy, and a willingness to engage deeply with the world as it is—and as it might yet become.

An Invitation to Move Forward

As you finish reading this introduction, take a moment to notice how you feel. Perhaps some of your worries have been affirmed; perhaps you sense relief in knowing these feelings are shared and valid. Maybe you remain skeptical or cautious, unsure how much a book can help with such a vast problem. Whatever you feel is welcome. The simple act of acknowledging your emotions is a step toward understanding them.

In the chapters ahead, you will find frameworks, stories, and practical suggestions. Not all will fit your situation, but many may help you navigate the landscapes of climate anxiety. This is not a prescription for how you must feel. It's a set of tools, a companion, a guide you can consult as you chart your own course through this era of transformation.

Welcome to this conversation. Welcome to acknowledging that fear, grief, anger, and hope can coexist. Welcome to understanding that these emotions are signals of engagement, not weakness.

Welcome to exploring techniques to cope, to connect, and to sustain hope. May the journey ahead empower you to be informed, resilient, hopeful, and ready to contribute to healing and protecting the planet we call home.

Part I: Understanding Climate Anxiety

Chapter 1: Naming the Feeling

Defining Climate (Eco)Anxiety and Its Signs

Climate anxiety, sometimes referred to as eco-anxiety, is an emerging term describing the uneasy, distressed, or fearful feelings people experience when confronted with the reality of climate change and its consequences. While not yet recognized as a formal psychiatric diagnosis, it has gained significant attention in recent years because it articulates a widespread emotional response that many previously struggled to name. Understanding what climate anxiety is, and how it manifests, is key to validating our feelings and developing strategies to manage them. Without this understanding, these emotions can feel nebulous, isolating, or even shameful. By defining climate anxiety and recognizing its signs, we can begin to respond more compassionately to ourselves and others who share these concerns.

Climate anxiety arises in response to a multi-layered problem that defies simple categorization. Climate change isn't just another issue on the news; it is a pervasive challenge that influences the foundations of life on Earth: weather patterns, food production, biodiversity, economic stability, cultural continuity, and public health. This complexity means that our emotional reactions are rarely straightforward. Someone experiencing climate anxiety may feel fear over anticipated disasters—rising sea levels, mega-droughts, and extreme storms. At the same time, they might also grieve losses already unfolding: species extinctions, coral reef bleaching, the disappearance of ancestral lands, or the erosion of cultural traditions tied to the rhythms of nature. This emotional mixture can also include anger at the slow pace of policy change or frustration with perceived inaction by leaders and institutions. Taken together, these feelings form a psychological landscape where anxiety is both a signal and a symptom.

Unlike a discrete phobia triggered by a well-defined threat, climate anxiety encompasses a spectrum of emotional responses to a diffuse and ongoing crisis. For instance, you might feel vaguely uneasy after reading an article about melting glaciers. You may carry a subtle tension in your shoulders, an increased heartbeat, or a sinking feeling in your stomach. Another person might experience more pronounced, acute moments of panic when confronted with grim climate projections, while yet another feels persistent low-grade worry that something is amiss. In all these scenarios, the underlying theme is a worry about the future viability of life as we know it, coupled with uncertainty about the efficacy of responses—both personal and collective.

It's useful to distinguish climate anxiety from garden-variety stress or general anxiety disorders, though they can overlap. Generalized anxiety disorder involves excessive worry about a range of issues—health, finances, relationships—and is not necessarily tied to climate conditions. Climate anxiety, by contrast, focuses on environmental changes and their socio-ecological ramifications. Though some people with underlying anxiety disorders may find that climate-related fears latch onto their existing worries, climate anxiety is more specifically rooted in our understanding of environmental instability. Recognizing this specificity can help people seeking support find resources and communities that understand the unique nature of these concerns.

One reason climate anxiety has become more visible is that climate change has shifted from an abstract, future-oriented problem to one with tangible, present-day impacts. Not too long ago, public discussions about climate were often couched in scientific models predicting changes by 2050 or 2100. Now, people witness abnormal seasons, unprecedented storms, shifting agricultural zones, and supply chain disruptions. With the crisis at the doorstep, it's easier to draw a line between what we read in reports and what we observe outside our windows. This direct exposure can intensify anxiety, as individuals realize that what once seemed like a distant challenge is now affecting their own lives—or soon will.

Cultural narratives also shape how climate anxiety emerges. In societies that have long cherished a myth of endless progress—one where technology and economic growth inevitably lead to better living conditions—the recognition that nature has limits can be disorienting. The idea that not everything can be fixed by ingenuity or market solutions shakes deeply held beliefs. For people raised with an expectation of linear improvement, climate anxiety can manifest as a sense of betrayal or confusion: Why isn't the future going to be as bright as promised? This disappointment can fuel feelings of unease, anger, or disillusionment.

By contrast, communities with traditions that emphasize living in harmony with nature might frame climate anxiety differently. For them, the crisis may confirm long-held warnings and beliefs about ecological respect. While the anxiety might still be painful, it's grounded in a worldview that has long acknowledged environmental fragility, making the experience of eco-anxiety less about sudden shock and more about mourning what they have always known could happen if humans disrespect the Earth's boundaries. In this context, climate anxiety might blend with cultural sadness, grief, and moral outrage at those who failed to heed ancestral wisdom.

Identifying the signs of climate anxiety begins with acknowledging that it can appear in diverse forms. Some of the most common signs include an overarching sense of worry about the future. You might catch yourself feeling uneasy whenever the topic of climate arises in conversation, or you may find yourself drawn compulsively to news about extreme weather events, political failures on climate agreements, or new scientific reports that paint an increasingly dire picture. This information can trigger a range of bodily responses: restlessness, difficulty sleeping, tightness in the chest, heart palpitations, or a feeling of heaviness in the limbs. Emotionally, one might feel sadness, anger, hopelessness, or a kind of anticipatory grief for losses that may occur in the future—or are happening right now somewhere else on the planet.

Another sign is a lingering sense of moral conflict. Climate anxiety often comes tethered to guilt and shame. Individuals aware of their personal carbon footprints—traveling by plane, driving a car, consuming products with large ecological footprints—may feel complicit in the problem. While personal responsibility should be contextualized within larger systemic issues, these emotions can still weigh heavily, intensifying anxiety. You might feel torn between the life you lead and the life you believe you should lead, constantly judging your choices against a backdrop of impending environmental crises. Over time, this moral tension can lead to frustration, cynicism, or self-criticism.

People experiencing climate anxiety might also notice changes in their behavior or decision-making patterns. Perhaps you hesitate before planning a long-distance vacation, worried about the environmental toll. Maybe you feel uneasy investing in a home near a coastline when you know sea levels may rise. Or you might find yourself drawn to sustainability initiatives or ecological activism, partly as a way to channel worry into meaningful action. On the flip side, climate anxiety can also lead to avoidance behaviors—some people, overwhelmed by fear, might tune out environmental news entirely, trying to escape the emotional burden. Recognizing such behavioral signs is important because they highlight the tangible influence of climate anxiety on how we live day-to-day.

It's equally important to note that climate anxiety is not always front and center. For some, it remains a background hum, a low-level stress that flares up occasionally. For others, it can become pervasive, coloring their worldview and impacting their mental health in more pronounced ways. In extreme cases, individuals might experience panic attacks triggered by climate-related stories or descend into despairing states where they question the point of making long-term plans. Such intense reactions underscore the need for support systems, whether through community networks, professional counseling, or group discussions focused on eco-distress.

Another sign of climate anxiety involves recurring thoughts or nightmares. People report dreaming about flooded cities, scorched landscapes, or barren fields where once-lush forests stood. Such dreams, though symbolic, reveal the psyche's attempt to grapple with forces that feel overwhelming or uncontrollable. Recurrent imagery of environmental devastation in one's mind—whether asleep or awake—signals a deep emotional processing. If you find yourself frequently imagining worst-case scenarios or feeling a constant sense of impending doom, it may be time to acknowledge that climate anxiety is at play.

For younger generations, climate anxiety can take a particularly acute form. Many young people grew up hearing about climate change as a defining issue of their era. They understand that the decisions made today will shape their entire adult lives and the lives of their children. This foreknowledge can lead to intense anxiety about personal futures—career paths in unstable economies, where to settle in an increasingly unpredictable environment, whether to have children at all. Young activists who speak publicly about these fears highlight how climate anxiety is not just an emotional inconvenience but a factor shaping the trajectories of their lives.

Cultural differences also mean that the signs of climate anxiety may vary depending on what people consider "normal" or "appropriate" emotional expression. In some cultures, openly expressing fear or sadness about environmental issues might be encouraged, leading to community gatherings, rituals, or artistic expressions that help process these feelings. In others, where stoicism or optimism is prized, climate anxiety may manifest more subtly, maybe through irritability, changes in appetite, or a general sense of dissatisfaction without clearly identifying the cause. Understanding these cultural nuances helps us see that climate anxiety is not a one-size-fits-all experience, and its signs must be interpreted within social contexts.

A critical aspect of defining climate anxiety is recognizing that it often involves a loss of predictability and stability. Humans have long relied on stable seasons and ecological patterns to guide agricultural practices, cultural events, and economic planning. Climate change erodes these patterns, making it harder to rely on historical knowledge. When previously fertile land becomes dry or winter festivals lack the reliable snow they once celebrated, people lose not just resources but symbols of meaning and continuity. This loss can trigger anxiety as we realize the ground rules we took for granted are shifting, and we must adapt to conditions we cannot fully anticipate or control.

People might also exhibit climate anxiety by withdrawing from conversations about the future or feeling uneasy when others speak about long-term plans. You might catch yourself hesitating before discussing retirement ideas, imagining your children's adulthood, or envisioning a stable community decades from now. The inability to imagine a secure future can induce anxiety, as the mind constantly grapples with uncertainty. This can lead to a sense of powerlessness, another hallmark sign. Feeling powerless—like no matter what you do, larger forces remain unresponsive—magnifies anxiety because it suggests that even informed and concerned individuals lack a meaningful role in shaping outcomes.

On the other hand, climate anxiety can sometimes inspire intense engagement. Not all signs are negative in the sense of withdrawal or despair. Some people, in response to these fears, dive deep into research, activism, or community work. They attend climate rallies, join environmental organizations, adopt zero-waste lifestyles, or attempt to influence policy changes. While such engagement can be productive and meaningful, it can also be driven by a sense of urgent anxiety. When activism itself becomes a coping mechanism, it's important to remain aware of burnout risks. Anxiety-fueled activism can lead to exhaustion if one's emotional state isn't also tended to. Recognizing this balance is essential for sustaining long-term involvement without sacrificing mental well-being.

In families and communities, climate anxiety might appear as generational tension. Older family members may remember more stable times and dismiss younger relatives' concerns as overblown, while younger people insist that the world has fundamentally changed. Disagreements on how serious the situation is, or what actions are justified, can strain relationships. If you notice recurrent conflicts at the dinner table about environmental issues, or if conversations about weather or future plans regularly ignite emotional reactions, these interpersonal dynamics can be viewed as signs that climate anxiety is seeping into family and social networks.

Media consumption habits can also reveal climate anxiety. Some may find themselves "doomscrolling"—endlessly reading negative news about climate disasters, biodiversity loss, and political inertia. This behavior, while intended to stay informed, can feed anxiety and create a feedback loop of despair. Conversely, complete avoidance of environmental news might signal an overwhelming fear that leaves the person unable to cope with updates. Being aware of how your media consumption affects your mood and thoughts can help identify if climate anxiety is shaping your habits.

Spiritual or existential reflections can also surface as signs of climate anxiety. People who once drew comfort from religious or spiritual worldviews that promise stability and order might find themselves questioning these narratives in the face of climate upheaval. Or, conversely, individuals might turn more deeply to spiritual practices seeking solace, guidance, and ethical frameworks for responding. If you notice yourself grappling with existential questions—what is humanity's role on Earth, what do we owe future generations, what is the purpose of striving if the future is uncertain—these are often signs that climate anxiety has prompted deeper philosophical inquiries.

It's worth mentioning that while climate anxiety is deeply uncomfortable, it can also be a sign of empathy and moral engagement. Feeling anxious often means you care about the well-being of others, human and non-human. You're concerned about justice, fairness, and long-term flourishing. This moral dimension highlights that climate anxiety is not just a personal ailment; it's interwoven with ethical considerations. Recognizing this can help reframe anxiety as evidence of your capacity to perceive injustice and feel compassion. In this sense, the sign of climate anxiety—an emotional alert that something important is at risk—can also be a clue to your values and priorities.

However, if anxiety becomes debilitating—if you find yourself unable to focus on daily tasks, plagued by insomnia, experiencing persistent panic attacks, or losing interest in activities you once enjoyed—it may be time to seek professional help. Therapists, counselors, or support groups specializing in eco-anxiety have begun to emerge. These professionals can help you navigate intense emotions, distinguishing between rational concern and catastrophic thinking, and equip you with coping tools that restore your psychological equilibrium. Recognizing this tipping point is crucial. While anxiety can motivate action, it can also paralyze if not addressed.

Identifying the line between healthy, motivating concern and overwhelming anxiety isn't always easy. One approach is to ask yourself whether your climate-related worries leave room for constructive engagement. Can you still appreciate small joys, build meaningful relationships, and find purpose in your daily life despite awareness of environmental challenges? Or do your fears consume every thought, leaving you stuck in despair and dread? If the latter, it's a sign that you might need extra support to find balance. It's also a reminder that climate anxiety, like other mental health challenges, deserves compassion, not self-blame.

Another subtle sign of climate anxiety involves changes in your sense of identity. As the reality of climate change sinks in, some people redefine who they are in response. Maybe you now see yourself as a "climate guardian," "environmental steward," or a more conscientious consumer. Others may struggle with identity crises, feeling that the narratives they once embraced—about endless growth, or simple solutions—no longer hold. This identity shift might feel unsettling, but it can also open doors to new forms of meaning, deeper connections with the natural world, and a clearer sense of purpose. Noticing these shifts in how you view yourself can be part of understanding the psychological impact of climate change.

One often overlooked sign is the spontaneous emotional reaction to seemingly mundane triggers. For example, a simple conversation about seasonal fruit availability at a grocery store might spark anxiety if it reminds you that changing weather patterns affect crop yields. Hearing about a distant wildfire might evoke a disproportionate emotional response, not because you are personally threatened, but because it taps into the broader anxiety about a warming world. These triggers highlight how climate-related fears permeate everyday life. You might realize that what once were ordinary observations—like the timing of a flower's bloom or the dryness of the local reservoir—now carry emotional weight.

In professional settings, climate anxiety can show up as difficulty concentrating, lower motivation, or questioning the relevance of your work. If your job seems disconnected from building a sustainable future, you may feel restless or dissatisfied. Conversely, those working directly on environmental issues—scientists, activists, policymakers—might feel intense anxiety about the gap between what their research shows and the slow pace of change. Recognizing that these professional struggles might be signs of climate anxiety can help you approach them more compassionately and seek appropriate adjustments, whether that's finding more purpose-driven work or setting emotional boundaries to prevent burnout.

As you observe these various signs—emotional, physical, behavioral, cultural, moral, existential—it's clear that climate anxiety isn't a simple, one-dimensional experience. It's a web of reactions woven into our evolving relationship with the Earth. Defining climate anxiety, then, is not merely about assigning a label. It's about understanding that our emotional responses to environmental instability are part of being human in this era. We worry because we sense that something precious is at risk, and we cannot simply shrug off that knowledge.

Ultimately, defining climate anxiety and recognizing its signs is the first step in a longer journey. Once we know what we're dealing with, we can name it out loud. We can talk to friends and family, join support groups, engage in community resilience projects, or seek professional help. We can practice coping strategies that help regulate our emotions, keeping anxiety at manageable levels so it fuels engagement rather than despair. We can transform fear into a catalyst for learning, adaptation, and moral growth.

It's also important to remember that acknowledging these signs doesn't mean accepting that anxiety is our permanent state. Emotions are fluid. They evolve as circumstances change and as we develop new skills. We can learn to hold the complexity of climate change without collapsing. We can cultivate inner resilience that allows us to face uncertainty with steadiness and courage. By defining climate anxiety, we claim the power to understand and shape our emotional responses, ensuring they lead us not toward paralysis, but toward compassionate action, creative problem-solving, and enduring hope.

In sum, climate (eco)anxiety is an umbrella term capturing a wide range of emotional, cognitive, and physical reactions to environmental change and its looming consequences. The signs manifest in subtle and overt ways—unsettling thoughts, bodily tension, altered behavior, moral conflicts, identity shifts, changes in media consumption, strained relationships, and existential questions. Each sign points to the central truth that we are living in unprecedented times, times that challenge our assumptions and invite us to rethink how we relate to the natural world. By recognizing these signs, we position ourselves to respond thoughtfully to climate anxiety, turning emotional distress into an impetus for meaningful engagement and purposeful living.

A Shared Emotional Landscape

It's easy to feel alone in the swirl of emotions prompted by a changing climate. When the future seems uncertain and the present feels increasingly volatile, you might assume that the tension you carry in your shoulders, that uneasy flutter in your stomach, or the heaviness in your heart is yours alone. You might wonder whether you're overreacting, imagining problems that are too big, too distant, or too complicated. Yet when we step back and look more broadly, a different picture emerges. As it turns out, a growing number of people, across generations and geographies, share similar feelings—feelings of worry, sadness, grief, anger, or moral unease in response to environmental disruptions. This collective emotional current reveals that you are not navigating these turbulent waters by yourself.

In neighborhoods and towns scattered around the world, individuals and communities are grappling with shifts that challenge previously stable rhythms. Some are adjusting to more erratic weather patterns, others are reading about distant disasters with unsettling frequency, and still more are witnessing long-familiar ecosystems transform in real-time. Whether it's farmers wondering if their crops can still thrive, coastal residents contemplating relocation, or urban dwellers noticing a steady uptick in extreme heat waves, countless people now face scenarios they once considered hypothetical. In these circumstances, feelings of worry and apprehension arise not because one person is uniquely sensitive, but because the ground beneath many feet is shifting.

These emotions spread beyond any single cultural setting. In urban centers, people feel a subtle but persistent stress as they track headlines about melting ice sheets or drying rivers. In rural communities, discussions over kitchen tables touch on erratic growing seasons or the disappearance of once-common species. Indigenous communities, long attuned to the land's fluctuations, find ancestral knowledge colliding with unprecedented environmental change. Across continents, students raised with stories of progress and boundless opportunity confront the possibility that not all futures are equally bright. Elders who recall stable climates now wonder how to guide younger generations through territory they themselves never had to navigate. This convergence of emotional responses, visible in different languages, customs, and belief systems, suggests that these feelings are not rooted in personal failings. Instead, they reflect a shared human response to a rapidly transforming planet.

Conversations—whether whispered between friends, debated in online forums, or raised in public meetings—often touch on this pervasive sense of uncertainty. People who might never have considered themselves "environmentalists" find themselves disturbed by the changes they see: the timing of blossoms, the odd patterns of migrating birds, or the eerie quiet that follows a vanished pollinator population. Even those who approach the topic from a purely pragmatic angle—concerned about local economies, infrastructure, or public health—encounter a similar unease. This widespread feeling underscores that our emotional reactions are not esoteric or marginal; they are part of the mainstream experience in a world where once-stable systems are becoming less predictable.

Cultural narratives further illustrate how these emotional responses are woven into the fabric of our era. Films, novels, art installations, and music increasingly reflect themes of environmental vulnerability and uncertainty. Creators tap into an undercurrent of shared feeling, not manufacturing it. They give voice to what many already sense: that something fundamental is at stake, and we cannot pretend otherwise without cost. The resonance of such works suggests that the emotions surrounding climate change strike a chord with wide audiences. People recognize these themes in themselves, find solace in seeing them expressed, and realize they are not outliers for feeling unsettled—they are part of a collective story unfolding around the globe.

This sense of a shared emotional landscape becomes even clearer when we consider the diversity of those experiencing it. The feelings are not confined to a single demographic. Wealthy, poor, young, old, urban, rural—climate-related emotions cross these boundaries. True, the intensity or focus of these feelings may differ. Those directly suffering from drought, floods, or rising seas may feel more immediate fear and grief than those who know these events mainly through headlines. Still, a baseline awareness, a stirring sense that the world is changing in disquieting ways, emerges in countless conversations. It's as if we are all standing in different corners of a vast hall, hearing echoes of each other's murmurs, even if we can't see everyone clearly.

Far from being an obscure phenomenon, this landscape of shared emotions is increasingly discussed by journalists, researchers, mental health professionals, and community leaders. They note the uptick in stories about "eco-anxiety" and related terms—evidence that a once-unspoken feeling now has a public vocabulary. Environmental psychologists and therapists observe a rising demand for support from clients worried about what lies ahead. Community organizers report that climate concerns surface frequently in local forums, even if the main topic is water management, public safety, or food security. These patterns confirm that our emotions are not outliers. They are part of a broader tapestry of human response to global change.

For many, simply recognizing that others feel similarly is a powerful step. Realizing that a knot in your stomach at news of a faraway wildfire is not a sign of personal fragility but rather echoes the feelings of countless others can be comforting. It suggests you are responding as a caring individual, alive to the world's challenges and empathic toward those affected. This recognition fosters empathy and reduces feelings of isolation or self-doubt. Instead of asking, "What's wrong with me?" you might ask, "What is it about these changes that rightfully troubles so many of us?" Such reframing can open pathways toward more constructive engagement.

One might also consider that this shared emotional landscape extends beyond national borders. Environmental shifts defy political boundaries; the atmosphere and oceans connect us all. Thus, it makes sense that people on opposite sides of the planet, who may never meet or even speak the same language, nonetheless harbor comparable worries. They, too, wonder if their children will know certain animals only through pictures, or if cherished hometowns will remain livable. Even as climate impacts vary, the underlying sense of vulnerability and responsibility links communities across distant terrains.

The universality of these feelings can, paradoxically, encourage action rather than despair. Recognizing that our emotional responses are not private flaws but collective signals can shift the narrative. When emotions are personal quirks, we might try to repress them. But when they are widely shared, they gain legitimacy. They hint that addressing climate change isn't just about data or policies—it's also about attending to our shared moral compass, our collective sense of what is right and fair, and our common hope to sustain a world worth inheriting.

Within this emotional landscape, you are free to locate yourself. Maybe you lean more toward sadness or grief, feeling the weight of what's already lost. Perhaps you hover in uncertainty, worried about what might be lost tomorrow. Or you might find yourself stirred to anger at inaction, or energized to join hands with others to bring about change. Every emotional stance is part of the spectrum that many others occupy. No single feeling disqualifies you from being part of this community of concern. Whether you oscillate between dread and determination or hold steady in cautious optimism, your responses have parallels in countless hearts and minds.

This communal aspect also implies that support and understanding are available. Just as climate anxiety doesn't exist in a vacuum—untouched by the social and cultural context—neither must we face it alone. Friends, family members, colleagues, local groups, faith communities, and online forums can offer listening ears and shared strategies for coping. Recognizing that your feelings are not bizarre or unwarranted may give you the courage to speak up and find that others nod in recognition. From such exchanges, solidarity grows. Solutions to environmental challenges come not only from technology and policy but also from communities that support each other emotionally, cultivating resilience together.

In literature and journalism, first-person narratives of climate anxiety have surfaced, adding personal voices to the dialogue. Whether it's a farmer penning an essay about uncertain harvests or a coastal resident describing the gradual encroachment of the sea, these accounts resonate widely. Each testimony reminds us that emotions connected to environmental changes are not confined to private thoughts—they are part of public discourse. People read these accounts and exhale in relief, realizing their own feelings aren't isolated incidents. The chorus of voices suggests that the emotional dimensions of climate change are woven into our collective story, as integral as any scientific statistic.

The shared nature of these feelings also paves the way for an expanded notion of empathy. If people everywhere feel anxious about climate conditions, then we can empathize not only with those who experience the same localized problems as ourselves but also with distant others. Empathy travels along emotional currents, allowing us to care about places we may never visit and communities we may never know personally. This broadened empathy can be harnessed in support of equitable solutions, guiding us toward policies and actions that honor the global community and future generations.

By recognizing that we inhabit a shared emotional landscape, we position ourselves to respond more constructively. Instead of pathologizing our concerns—treating them as eccentricities or private burdens—we can view them as a natural human response to widespread environmental signals. This shift in perspective doesn't trivialize the distress. On the contrary, it acknowledges that our emotions are meaningful, reflecting legitimate apprehensions and moral instincts. Understanding that others feel similarly can provide a sense of collective purpose. Our emotional responses become data points, evidence that we are attuned to changes that cannot be ignored.

This communal aspect also suggests that strategies for coping, adapting, and finding hope might emerge from shared knowledge. If many people are grappling with these feelings, many solutions—both emotional and practical—are being tested and refined. By sharing stories, tools, and experiences, we can discover what helps to keep anxiety in check. Some might find comfort in focusing on achievable local projects, while others draw strength from engaging in political advocacy or supporting international agreements. Some turn to spiritual or cultural practices that provide continuity and meaning. In a shared emotional landscape, no one must invent coping mechanisms from scratch; we can learn from each other.

As you reflect on this landscape, take a moment to consider the implications. If these feelings are widespread, then the burden of "just dealing with it" alone diminishes. You can take a step back and appreciate that anxiety, fear, and concern are part of a collective alert system. They inform us that something in our environment requires attention. Instead of wrestling with doubts about whether you "should" feel this way, recognize that your feelings align with a broader, human response to environmental strain. They are a testament to your sensitivity, awareness, and care.

In time, we may look back and see this era not only as one of environmental upheaval but also as a period of profound emotional reckoning. The fact that these feelings are not confined to a few individuals but are emerging simultaneously across diverse societies underscores that we are living through a global shift in consciousness. That shift can be uncomfortable—who enjoys feeling anxious?—but it can also be transformative. By realizing we share this emotional terrain, we open the door to collective growth, sustained effort, and a more empathetic approach to navigating an uncertain future.

This shared emotional landscape, then, is a place of recognition. In it, you find not only your own anxieties but also countless echoes. As you listen closely, you may hear these echoes forming harmonies of mutual understanding. The discomfort doesn't vanish, but it becomes part of a chorus. And a chorus, unlike a solitary voice, has the power to resonate more deeply, to shift perspectives more effectively, and perhaps, to inspire the kind of collective willpower and compassion that is so urgently needed in these changing times.

The Ties That Shape Our Climate Fears: Culture, Morality, and Generations

Our emotional responses to climate change do not arise in a vacuum. They are influenced—often subtly, sometimes profoundly—by the cultural stories we inherit, the moral frameworks we embrace, and the generational vantage points from which we observe the unfolding crisis. Culture shapes the narratives we tell about progress, nature, and human destiny; morality informs what we consider right or just; and generational context sets our baseline expectations for what is "normal" and how quickly and dramatically we expect the world to change. Understanding these layers can illuminate why our fears take the forms they do, why some concerns resonate more strongly in certain communities, and how we might find common ground amid differences.

Cultural Narratives and the Lens They Provide

Culture can be understood as the shared values, beliefs, traditions, and practices that give a group of people a sense of identity and meaning. Every culture carries stories about humanity's place in the world: are humans masters over nature, or stewards within it? Does prosperity come from controlling resources, or from living in balance with natural rhythms? Are we destined for continuous technological advancement, or must we tread carefully to avoid upsetting delicate ecosystems?

These cultural stories impact how we perceive environmental threats and, by extension, how we feel about them. Consider a society steeped in a narrative of endless progress: for generations, it might have believed that innovation, industry, and human ingenuity would always push the world forward, increasing comfort, security, and abundance. In such a cultural context, confronting climate change—a problem that challenges the assumption of limitless growth—can trigger deep disillusionment and anxiety. It's not just that the climate is changing; it's that the foundational story of guaranteed improvement is crumbling. This can translate into feelings of betrayal or confusion, intensifying fear because the future no longer seems reliably brighter.

By contrast, consider cultures or communities where there has long been an understanding that nature sets limits and that human actions must align with ecological constraints. Many Indigenous peoples, for example, have traditions emphasizing reciprocity, humility, and respect for non-human life. For them, the climate crisis might not be surprising; it may confirm long-held warnings about unsustainable practices. Yet this does not necessarily ease their anxiety. If anything, it can produce grief and frustration, as what they've feared for decades or centuries now manifests more broadly. This cultural orientation might also foster moral outrage toward those who ignored warnings and contributed most heavily to environmental damage.

Cultural differences extend to how much uncertainty people are comfortable with. Some societies value stability and continuity, anchored in long traditions and stable livelihoods tied to the land. For them, climate disruptions that erode predictable seasons—affecting planting, harvesting, and ritual calendars—strike at the core of cultural identity. A community might feel fear not only for physical survival but also for the survival of its cultural fabric. If a festival timed to a certain season no longer matches the climate reality, the loss goes beyond inconvenience; it's a disruption in how people understand themselves and pass knowledge forward. Thus, cultural narratives shape the contours of climate anxiety, turning a scientific phenomenon into a deeply personal and communal concern.

Moral Frameworks and the Weight of Responsibility

Alongside culture, moral beliefs profoundly influence how we experience climate-related fears. Morality here refers to our sense of right and wrong, what we owe one another, and how we judge actions or policies. Climate change, by its nature, has an undeniable moral dimension. Those who have contributed least to greenhouse gas emissions often suffer the most severe consequences. Future generations, who have no say in current decisions, will inherit the results. Non-human species and ecosystems, which cannot advocate for themselves, bear immense harm. Such injustices stir moral emotions—anger, guilt, sorrow, and indignation.

If your moral framework places a high value on fairness and protection of the vulnerable, witnessing climate injustices can intensify anxiety and frustration. Imagine someone raised with strong ethical principles about equality and stewardship suddenly realizing that entire communities face displacement, famine, or disasters due to actions taken by distant industries or affluent nations. This moral clash can transform what might have been mild concern into urgent fear and resolve, or into despair if no remedy seems forthcoming. The anxiety is not just about the climate's instability; it's about the moral violation of allowing preventable suffering and about the burden of knowing that inaction or insufficient action contributes to continued harm.

Moral beliefs also influence how we perceive personal responsibility. In societies that celebrate individual freedom and consumption, climate awareness can trigger guilt—am I contributing to the problem through my lifestyle? Should I bear moral shame for my part, however small, in supporting carbon-intensive systems? These questions add layers to climate anxiety, blending worry about environmental outcomes with self-critique and moral tension. On the other hand, in cultures with collective moral orientations—where community well-being and shared accountability are emphasized—individuals might feel supported by communal efforts to reduce impact, thereby mitigating personal guilt but raising concern about whether the community's collective actions are sufficient.

Religious and spiritual moral frameworks further shape these fears. For those whose faith traditions teach care for creation, climate disruption might feel like a profound moral offense, intensifying anxiety over humanity's failure to honor a sacred trust. Conversely, some religious narratives might downplay human responsibility or encourage faith that divine intervention will restore balance, affecting how believers emotionally engage with the crisis. Whether faith intensifies anxiety by highlighting moral failure or tempers it by offering hope can depend on how religious teachings interpret environmental stewardship.

In moral landscapes, climate anxiety also surfaces when questioning what it means to lead a good life. If one once believed that success meant accumulating wealth or rising in a corporate hierarchy, learning that these pursuits may indirectly contribute to environmental degradation can feel deeply unsettling. Moral frameworks that highlight simplicity, moderation, or interdependence can either soothe anxiety (by providing a clear ethical path forward) or heighten it (if they make one acutely aware of systemic moral shortcomings). Thus, morality weaves itself into every emotional thread of climate fear, guiding how we weigh responsibilities, distribute blame, and envision justice.

Generational Perspectives and Shifting Norms

Generational factors influence climate anxiety by setting our baseline assumptions about what is normal and possible. Each generation grows up with particular cultural reference points. For older adults who recall a time when seasons were more predictable, certain species more abundant, and environmental problems less pronounced, today's changes can spark anxiety rooted in memory. They feel the loss of what once was: stable weather patterns, reliable harvests, or ecosystems they took for granted. Their fear may center on the tragic erosion of familiar worlds and the shame or sadness that their generation did not prevent it.

Younger generations face a different set of emotional triggers. They inherit a world where climate instability is a given, not a hypothesis. Many young people cannot remember a time without warnings of global warming, rising seas, or mass extinctions. This generational reality shapes their fears differently. Instead of mourning lost stability, they fear that they may never experience it. Their anxiety might revolve around questions like: Should I have children in an uncertain climate? Will my career be defined by disaster response and adaptation? Are my dreams for the future realistic if ecosystems collapse or politics remain deadlocked?

For adolescents and young adults, climate anxiety can meld with the pressures of forging identity and independence. They face a dual challenge: establishing their place in a society that may not fully acknowledge their concerns and planning futures in a context that feels shaky. This generational layer can make climate anxiety more intense, as young people see the clock ticking on the planet and on their personal aspirations. They might also resent older generations for perceived inaction, intensifying moral outrage and intergenerational conflict.

Meanwhile, middle-aged adults might feel caught between care for the elderly, who recall a more stable past, and concern for children, who anticipate a fraught future. Their climate anxiety may arise from a sense of protective responsibility. They want to provide a better world for their kids and honor the legacy of their parents, but climate instability complicates this mission. This generational tension can magnify fear, as they juggle multiple moral and emotional commitments simultaneously.

It's important to note that generational factors do not operate in a vacuum. They intersect with culture and morality. Consider a young person in a culture that has long emphasized harmony with nature. Their fear might be framed less as personal frustration and more as a continuation of their community's longstanding reverence for the land. Conversely, an older adult in a high-consumption culture might feel moral guilt and anxiety over having enjoyed the fruits of an unsustainable economy, leaving a mess for the younger generation to clean up. Such intersections show that the lines between cultural, moral, and generational influences are fluid, each informing the other.

Intersecting Identities and Climate Anxiety

Beyond these broad categories, it's crucial to acknowledge that individuals hold multiple identities influenced by culture, morality, and generation simultaneously. For example, an Indigenous youth may experience climate anxiety through all these lenses at once. Culturally, they may have inherited a worldview that stresses environmental reciprocity. Morally, they might feel a duty to protect ancestral lands and honor traditions that have sustained their people for centuries. Generationally, they see themselves as stewards for a future that demands radical adaptation. The resulting climate anxiety is not a simple formula; it is a dynamic interplay of multiple narratives and responsibilities.

Similarly, a recent immigrant to a new country might draw upon moral beliefs from their homeland, cultural narratives from their upbringing, and generational outlook shaped by the unique historical moment they occupy. If their new home faces rising sea levels, they might feel anxiety compounded by cultural displacement and moral imperatives to secure a safe environment for their children.

These examples illustrate that understanding climate anxiety requires examining how multiple influences stack and interact. No single factor fully explains why one person's fear manifests as sorrow while another's manifests as anger. Culture, morality, and generational context blend to produce unique emotional tapestries.

Changing Cultural Scripts and Moral Evolution

Cultural narratives and moral frameworks are not static. They evolve as societies encounter new challenges. Climate change, as a destabilizing force, may prompt communities to rewrite their stories. What happens when a culture that believed in infinite growth begins to question that assumption publicly? Books, films, community workshops, and educational curricula may start emphasizing resilience, sustainability, and limits. Over time, this cultural shift could mitigate some forms of climate anxiety by offering a new story: humans can learn, adapt, and find balance if they choose. Similarly, moral frameworks can evolve to emphasize intergenerational justice and the rights of non-human life, guiding policies that reduce anxiety by showing that collective action is possible and meaningful.

We've begun to see such shifts. Terms like "just transition," "intergenerational justice," and "decolonizing environmental policy" reflect moral evolutions. They acknowledge that addressing climate change isn't merely technical—it's ethical. As these ideas enter mainstream discourse, they can help individuals cope. Knowing that one's moral outrage aligns with emerging ethical standards can transform helpless anxiety into purposeful engagement. Recognizing that cultural scripts are changing can ease the sense of disorientation. Instead of feeling trapped by outdated narratives, people may find hope in the idea that new stories are being written.

Generational changes also play a role in this cultural and moral evolution. Younger generations often lead the charge in challenging old assumptions. They raise their voices at international summits, organize school strikes, and use social media to highlight inequities. Their efforts contribute to reshaping the moral consensus. As older generations witness these movements, some adapt their views, acknowledging that responding to climate threats may require abandoning cherished narratives of unfettered growth. This intergenerational dialogue—sometimes tense, sometimes inspiring—fuels cultural shifts that can help reduce collective anxiety by replacing denial or apathy with a sense of direction and accountability.

The Role of Communication and Education

How do we navigate these cultural, moral, and generational influences productively? Communication and education are key. Open dialogues within families can help bridge generational gaps, allowing older adults to share memories of more stable times and younger ones to articulate their visions for a sustainable future. Such conversations can ease anxiety by creating understanding: the older generation learns that younger people's fear is not hysteria but a response to real threats, and the younger generation sees that previous norms weren't always perfect, nor were earlier generations fully aware of the damage accumulating over decades.

In educational settings, incorporating multiple cultural perspectives and moral dimensions of environmental issues can help students understand that climate change is not solely about atmospheric chemistry or policy debates. It's about ethical responsibilities, cultural continuities, and generational stewardship. This broadened understanding can reduce anxiety by providing a more holistic view, affirming that many paths to engagement exist—community leadership, activism, science, art, policy, faith-based initiatives, and more.

Culturally sensitive mental health support can also play a role. Therapists, counselors, and support groups familiar with environmental distress can tailor their approaches to a client's cultural background and moral beliefs. They can help individuals reframe their anxieties, understanding them as natural responses to cultural narratives in flux or moral compasses reacting to injustice. By validating these influences, mental health professionals can guide people to find meaning and purpose amidst fear, rather than feeling paralyzed or alienated.

Practical Examples of Cultural and Moral Intersections

To ground these abstract ideas in concrete terms, consider a few scenarios:

1. **Coastal Fishing Community:**
   In a village where fishing traditions have passed from generation to generation, climate change alters migration patterns of fish and raises sea levels. The community's cultural identity is tied to these practices, and moral obligations to maintain these traditions run deep. Elders feel anxiety because they sense that their cultural legacy might end with them. Young fishers, meanwhile, worry if they can carry on the trade at all. The moral imperative to preserve a way of life collides with environmental reality, intensifying collective fear. Yet, this shared anxiety may lead to community adaptation—investing in sustainable methods, diversifying livelihoods, or advocating for marine protections. A new cultural narrative might emerge: resilience and adaptation become as much a part of identity as the old traditions.
2. **Urban Professionals in a High-Consumption Society:**
   In a wealthy city known for consumerist values, well-educated professionals start feeling uneasy about their carbon footprints. Cultural narratives of success once meant upward mobility and material comfort. Now, these narratives clash with moral understanding that limitless consumption fuels climate harm. Anxiety arises from the tension: Should they continue to strive for bigger houses and luxury cars, or shift toward low-impact lifestyles? The moral dilemma intensifies worry about personal complicity. However, as these professionals discuss their feelings with peers, cultural norms may begin to shift. Green policies, corporate responsibility measures, and new social standards of "eco-conscious success" might emerge. Over time, the community's anxiety recedes as it finds congruence between moral values and cultural aspirations.
3. **Intergenerational Climate Advocacy Group:**
   Imagine a coalition that includes elders who remember pristine landscapes, adults who work in industries now seen as problematic, and youth activists demanding rapid change. Their meetings are fraught with emotional tension. Elders mourn losses, adults feel defensive or guilty, and youth are angry at perceived delays. All share climate anxiety, but it manifests differently. Through dialogue, they learn that each perspective contributes pieces of a larger puzzle. The elders' memories provide motivation, the adults' experience aids in crafting pragmatic solutions, and the youth's passion pushes innovation. As they collaborate, cultural stories evolve to include all generational voices, and moral frameworks expand to emphasize collective responsibility. Anxiety becomes more manageable when the group acknowledges and bridges these differences.

Looking Forward: Transforming Anxiety Through Understanding

Recognizing how cultural, moral, and generational factors influence our fears can be liberating. Instead of seeing anxiety as a mysterious, personal failing, we understand it as a product of intersecting influences. By examining these roots, we gain tools to address them. Realizing that your fear may stem from cultural narratives of endless growth can inspire you to explore alternative stories of sustainability. Noticing moral tensions in your worry can push you toward actions that align better with your values, reducing internal conflict. Understanding generational differences can help you empathize with others' responses, building solidarity and reducing the isolation that exacerbates anxiety.

This process is not about dismissing fear but about placing it in context. When we see fear as connected to cultural transitions, moral reckonings, and generational shifts, it no longer seems arbitrary. It becomes meaningful feedback—a sign that we are collectively grappling with what it means to be human in a rapidly changing world. Such understanding can replace helpless dread with a sense of participation in an unfolding narrative. We are, in a way, co-authors of the next chapters of human culture, ethics, and practice.

It's also worth noting that not all cultures, moral viewpoints, or generations will respond identically. Diversity in responses is valuable. Different perspectives can complement each other, leading to more robust strategies. For example, a culture with deep ecological wisdom can inform policy debates in places that have historically prioritized economic expansion over environmental balance. Moral philosophies emphasizing future generations' rights can influence cultures that never considered unborn lives in their calculus. Youth-led movements can challenge outdated moral complacencies and spark essential cultural reforms. Diversity becomes a strength, not a source of confusion, when we learn to appreciate how it enriches our collective emotional and intellectual response to climate challenges.

Empathy and Dialogue as Pathways to Healing

Ultimately, moving through these cultural, moral, and generational dimensions can help us find empathy for others and for ourselves. When we understand that the older generation's reluctance to change might come from a cultural narrative that promised them stability, we feel less judgmental. When we grasp that a young activist's despair stems from the moral imperative to protect future life, we respect their stance even if it feels confrontational. When we see that a rural community's fear is intertwined with cultural traditions tied to the land, we honor their anxiety as rooted in identity and value rather than dismissing it as misguided.

Dialogue fosters this empathy. Conversations that explore why we feel what we feel, rather than just stating facts and figures, can lead to breakthroughs in understanding. Such dialogues humanize the crisis. Instead of viewing climate change as an external problem "out there," we begin to see how it infiltrates our collective psyche, pushing us to adapt culturally, ethically, and across generations.

From Fear to Informed Engagement

Recognizing these influences does not eliminate fear, but it can transform it. When fear is contextualized, it may feel more manageable. Knowing that your anxiety is partly a response to cultural narratives in flux might prompt you to seek new narratives—through art, education, spirituality, or political movements. Understanding the moral roots of your worry can guide you toward advocacy or lifestyle changes that align with your ethics, turning fear into motivation rather than paralysis. Appreciating generational differences may encourage you to support intergenerational dialogues and initiatives, building coalitions that are stronger than any single generation's voice.

In this sense, cultural, moral, and generational factors are not just sources of anxiety; they are also resources. They remind us that we possess moral compasses that can guide us toward justice, cultural adaptability that can foster resilience, and generational interplay that can spur innovation. Each source of fear also holds potential for growth. A culture can reinvent itself, moral beliefs can expand, and generations can learn from one another.

As we navigate the climate crisis, acknowledging these deeper dimensions of our fears offers a richer understanding of ourselves and our communities. Far from being alone in anxiety, we participate in a widespread, historically significant transition. The emotions we feel are signals that old norms no longer suffice, that moral boundaries are being tested, and that generational handoffs are fraught with complexity. Interpreting these signals with compassion and curiosity can help us chart a path forward—one that honors cultural inheritance, upholds moral responsibility, bridges generational divides, and ultimately transforms anxiety into constructive, collective action.

Recognizing how cultural, moral, and generational factors influence our fears grounds our understanding of climate anxiety in a rich, multidimensional context. It shows that our emotions reflect more than personal sensitivity; they emerge from a nexus of shifting values, contested ethics, and evolving expectations. By situating our fears within these broader influences, we gain insights that can help us respond more wisely, fostering empathy, dialogue, and a renewed sense of purpose.

Chapter 2: Why Are We So Anxious?

Evolutionary roots: Why our minds struggle with slow-moving, global threats

Human beings stand at an unusual juncture in history. On one hand, we possess unprecedented knowledge about the Earth's climate systems, gleaned from satellites, advanced models, and centuries of scientific inquiry. On the other, we find ourselves strangely unmoved—or not as moved as one might expect—by what that knowledge tells us. Even as researchers warn of rising temperatures, melting ice sheets, and intensified storms, many individuals and communities struggle to feel the urgent alarm that such information arguably warrants. This puzzling emotional gap between understanding the gravity of climate change intellectually and responding with decisive emotional engagement can be traced, in part, to our evolutionary heritage.

For the vast majority of human history—counting back tens of thousands of years rather than the blink of a few centuries—our ancestors inhabited environments fundamentally different from the globally interconnected, resource-intensive world we know today. Early humans lived in relatively small, close-knit groups, often nomadic or semi-nomadic, responding to immediate hazards like predators, famine, disease outbreaks, and territorial disputes. In such settings, threats were local, visible, and required prompt action. The ability to detect, prioritize, and respond quickly to immediate, tangible risks provided a clear survival advantage.

Over countless generations, natural selection favored cognitive and emotional mechanisms well-suited for dealing with short-term dangers. The human brain became adept at recognizing direct cues of threat—like the sound of a predator approaching in the brush, the smell of smoke indicating fire, or the sudden scarcity of a known food source. Responses to these cues were swift and visceral. A rush of adrenaline, heightened alertness, and a quick calculation of fight, flight, or concealment strategies would unfold almost automatically. In these ancestral contexts, those who hesitated or who overfocused on distant potential dangers might have found themselves at a disadvantage, as immediate survival challenges always took precedence.

Fast-forward to the present. The climate crisis unfolds at a pace and scale that sharply contrast with the conditions that shaped our ancestors' minds. Instead of a lion lurking behind a nearby tree, we have rising levels of carbon dioxide in the atmosphere—an invisible, odorless gas that cannot be directly sensed. Instead of a known rival clan encroaching on our hunting grounds next week, we face complex changes in global temperature averages, shifting precipitation patterns, and ocean acidification that intensify over decades. The feedback loops and tipping points warned of by climate scientists occur at spatial and temporal scales unimaginable to early humans. A one-degree Celsius increase in global average temperatures might sound negligible to a mind attuned to immediate perils, but it can have profound long-term consequences for agriculture, weather extremes, sea levels, and ecosystems.

Our inherited cognitive biases push us to discount the future and prioritize the present. For our forebears, focusing on immediate challenges was sensible. Securing today's meal or ensuring tonight's shelter often mattered far more than planning decades ahead. Even stable seasonal cycles were managed through cultural knowledge rather than intricate long-term forecasts, because life expectancies and living conditions seldom demanded extensive multi-decade planning. This evolutionary legacy means that distant threats—like what might happen to polar ice caps in 2100—fail to trigger the same alarm bells as a sudden shortage of local game would have in a hunter-gatherer society. Our brains, simply put, are not primed to respond viscerally to data-driven projections about events that seem remote in time and space.

The complexity and abstraction of climate change further muddy the waters. Humans excel at understanding linear, direct cause-and-effect scenarios. If you eat a bitter berry and feel ill immediately afterward, the lesson is clear. If prey disappears from your hunting ground, you understand that the environment has changed in a visible, comprehensible way, prompting you to migrate elsewhere or adopt new hunting strategies. Climate change, on the other hand, involves non-linear causality. Burning fossil fuels in one part of the world leads to increased concentrations of greenhouse gases. These gases trap more heat in the Earth's atmosphere, altering global circulation patterns and gradually shifting climate zones. The results—droughts in some regions, floods in others, bleaching coral reefs thousands of miles away—don't line up neatly with any one identifiable action or moment in time. This invisibility and complexity of causation stymie our intuitive alarm systems.

Moreover, humans evolved as social creatures highly responsive to group norms and cues. In small tribes, if everyone displayed alarm, that often meant a real, immediate danger was present. Calmness among group members signaled a lack of threat. Today's climate predicament involves billions of people and complex institutions. Many leaders, media outlets, and social groups do not consistently convey the sense of emergency that climate scientists feel. Without a unified social signal, people's ancient instincts fail to register crisis-level urgency. If those around you act as though things are normal, or if the political discourse remains contentious and divided, the ancestral mental wiring interprets this lack of collective alarm as an indication that danger might not be imminent after all.

Another factor is the abstractness of the victims and beneficiaries of climate action. Early humans cared for their kin, the immediate tribe whose members they knew personally. Climate change affects distant populations—people in low-lying islands or arid agricultural zones that we will never meet—and involves future generations not yet born. Our empathetic capabilities, honed to respond strongly to the plight of visible, present individuals, struggle to extend the same emotional intensity to hypothetical descendants or strangers halfway across the globe. Without direct emotional resonance, moral outrage or protective instincts may fail to ignite, allowing complacency or denial to persist.

Denial itself can be viewed as a psychological defense mechanism rooted in evolutionary logic. If a threat is too complex, too overwhelming, or too distant to address meaningfully in the short term, it might have been adaptive for ancestors to ignore it in favor of immediate problems that they could solve. Chronic stress from worrying about uncertain and long-term hazards might have reduced day-to-day survival efficiency. Thus, minimizing or rationalizing away the importance of a slow-burning threat could, in certain contexts, have helped maintain mental stability and focus on solvable problems. While this reasoning no longer serves us in the face of climate change—where acknowledging and acting on the threat is crucial—it helps explain why some people instinctively downplay the crisis.

Cognitive biases such as optimism bias and system justification bias also emerge from these evolutionary roots. Optimism bias—the tendency to assume that things will generally turn out alright—may have encouraged resilience and perseverance in challenging environments. But now it can lead people to discount dire climate projections, hoping that technology, markets, or future innovations will magically solve the problem without major sacrifices. System justification bias, which leads people to defend the status quo and resist changes to established social or economic orders, also played a stabilizing role in ancestral communities. Radical upheavals could threaten social cohesion, so a mental nudge to trust existing structures often kept tribes cohesive. In the modern era, however, defending outdated energy systems and consumption patterns in the face of clear environmental harm becomes maladaptive.

The scale mismatch between what we evolved to handle and what climate change demands is perhaps the largest barrier. Our ancestors rarely contemplated global phenomena; their worlds were local and immediate. Today, preventing catastrophic climate outcomes requires global coordination, emission reductions in multiple sectors, international treaties, and shifts in cultural values. None of these steps are intuitive. They challenge the "tribal mind," which is better at rallying around immediate threats—like defending a village against attackers—than addressing a diffuse problem that affects humanity as a whole. The ancient wiring that encourages loyalty to a small group and suspicion of outsiders can even hinder global cooperation on climate issues, as nationalist or isolationist attitudes prompt people to protect local interests and ignore global consequences.

Yet, while these evolutionary influences partially explain why our emotional responses lag behind intellectual understanding, they do not condemn us to inaction. One hallmark of humanity is our capacity for learning, cultural innovation, and moral imagination. We may not have evolved instincts to handle global, long-term threats, but we have language, reason, and empathy that can be extended beyond their original evolutionary scope. Awareness of these mismatches empowers us to design interventions that overcome our inherited blind spots.

For instance, since invisible gases and distant impacts fail to trigger alarm, we can use storytelling and imagery to bring climate change closer to home. Personal narratives from communities already suffering from rising seas or more frequent storms can bridge the empathy gap. Instead of relying on abstract statistics, educators and activists can highlight tangible changes—disrupted seasons affecting local farmers, unusual heatwaves stressing urban populations—so that minds accustomed to immediate sensory cues can feel the problem more directly.

Focusing on short-term co-benefits of climate solutions can help counter our aversion to distant payoffs. Investing in renewable energy not only mitigates future warming but also cleans the air now, improving health and reducing hospital costs. Planting trees or restoring wetlands provides immediate ecosystem services: cooler neighborhoods, better flood management, habitats for wildlife. These immediate rewards align with our evolutionary bias toward the present, making climate action feel sensible and beneficial, even without the distant threat fully internalized.

Shaping social norms also becomes crucial. If ancestral minds rely on group cues, then visible consensus and leadership can create a sense of collective alarm, not through fear-mongering but through honest communication of stakes. When communities, faith leaders, celebrities, and trusted authorities display genuine concern and take meaningful steps, it sends signals our tribal brains recognize: a respected ally is on high alert, so perhaps we should be, too. Such shifts can gradually rewire our emotional responses, making climate action a culturally endorsed priority rather than an abstract debate.

Moral and philosophical frameworks can help expand our circle of empathy and concern beyond immediate kin, tribes, or present generations. Religious teachings, spiritual practices, ethical philosophies, and humanist ideals can foster a sense of stewardship and responsibility for distant strangers and future lives. By encouraging identification with humanity as a whole, rather than a small group, and emphasizing interdependence with non-human species, these frameworks align moral reasoning with the needs of the planet. Although our ancestors evolved empathy primarily for known individuals in their immediate circles, we have shown throughout history that we can extend moral considerability to broader groups—different races, ethnicities, and now, potentially, all living beings.

Cultural evolution can work alongside these strategies. Cultures are not static; they develop new stories and norms as circumstances change. As societies integrate environmental education into schools, produce art that grapples with climate themes, and celebrate role models who champion sustainability, the cultural narrative shifts. Gradually, the perception of climate change can move closer to something emotionally resonant, blending long-term ethical considerations with the sense of immediacy once reserved for visible predators or seasonal scarcities.

Technological and communicative innovations also offer tools. Virtual reality can let people experience what future sea-level rise might mean for their hometowns. Interactive simulations can show how today's emissions influence tomorrow's extreme weather events. These techniques speak to the senses, creating vivid impressions that approximate the kind of immediate sensory input our ancestors used to gauge threats. By making the distant future temporarily feel present, we hack our primal alarm systems into caring about long-term instability.

We should recognize that these evolutionary mismatches do not operate in isolation. They interact with socio-political factors such as corporate lobbying, media fragmentation, and ideological polarization. While these factors are not evolutionarily derived, they exploit our cognitive biases, making it even harder to rally an effective emotional response. Understanding the evolutionary backdrop helps us see that certain forms of denial or indifference aren't just stubbornness or ignorance; they are also manifestations of ancient mental patterns inadvertently hijacked by modern complexities.

In adapting our responses, we must acknowledge that our minds will never find climate change as straightforward to respond to as a direct, immediate threat. However, we can train ourselves to think differently. Education that emphasizes systems thinking, long-term planning exercises, and critical reflection on cognitive biases can gradually shift mindsets. Just as literacy and numeracy were once not universal but have now become widespread, we can cultivate a form of "climate literacy" that normalizes considering decades-long timelines, global interconnections, and moral obligations beyond immediate interests.

Over time, humanity has repeatedly shown the capacity to surpass our evolutionary constraints. We have embraced agriculture, creating food surpluses and planning harvests months in advance. We built cities, requiring some foresight and cooperation beyond the small tribe. We established religions, philosophies, and laws that extended moral consideration beyond direct family ties. We ended slavery in many societies, gave women the right to vote, recognized human rights, and protected endangered species—each step expanding the circle of care and planning horizons. Climate action may be another stage in this moral and cognitive evolution, requiring that we acknowledge our limitations and transcend them.

If we refuse to acknowledge these evolutionary roots, we risk misunderstanding why climate communication sometimes fails to spark the urgency we expect. Without self-awareness, we might grow frustrated with people who don't share our sense of alarm, labeling them ignorant or selfish. Recognizing the built-in cognitive hurdles opens a path to empathy: we see that everyone struggles to some degree with engaging a threat so unlike what our ancestors faced. This compassionate perspective encourages patience, creativity, and humility in how we discuss and address climate issues.

Though the evolutionary mismatch is daunting, it is not insurmountable. Our cultural and intellectual tools are flexible enough to reorient priorities, highlight moral duties, and develop emotional resonance with the well-being of future generations and distant ecosystems. By designing stories, policies, and social norms that leverage our known biases—focusing on present co-benefits, encouraging group consensus, using vivid imagery—we can partially align the slow-moving global threat with the triggers our minds respond to naturally.

Embracing complexity is part of this adaptation. Our ancestors dealt with simpler cause-and-effect patterns, but we have the capacity for abstract thought, scientific reasoning, and moral philosophy. By teaching systems thinking, we encourage the recognition that invisible greenhouse gases and incremental changes can yield massive results over time. This not only helps us rationally comprehend the crisis but, if communicated well, can also instill a sense of awe and responsibility powerful enough to motivate action.

None of these strategies guarantee that we will collectively muster the emotional energy needed to address climate change fully. Yet without understanding our evolutionary hurdles, we risk repeating the same communication errors: presenting climate as a distant, uncertain future problem or flooding audiences with data that never quite hits home. The evolutionary perspective pushes us to ask how we can make climate information feel as urgent as an approaching storm—how to replicate, in a way, the immediate emotional resonance that once saved our ancestors' lives.

This reframing doesn't mean resorting to panic or despair. It means acknowledging that a measured sense of urgency is crucial and that our minds need help bridging the gap between statistical abstraction and heartfelt concern. It suggests that creative storytelling, moral leadership, visual dramatization of climate impacts, and emphasizing here-and-now benefits can all be tools to engage the old neural hardware for new ends.

As we implement these approaches, we do so with an understanding that change may be gradual. Just as literacy and scientific thinking took generations to become widespread, so may the capacity to treat climate threats as emotionally real and pressing. Cultural evolution, if steered wisely, can produce societies where the default mindset incorporates long-term stewardship and global empathy.

While our ancestors may not have faced anything like climate change, their legacy lives on in how we think and feel. Recognizing the evolutionary roots of our struggles clarifies that we are not flawed for finding global, slow-moving threats emotionally elusive. We are simply operating with mental tools forged in a different arena. Now, through conscious effort, cultural innovation, and moral reasoning, we can retool these mental frameworks to meet the challenge before us. Our capacity to learn, reflect, and adapt intellectually and emotionally may yet prove the key to overcoming evolutionary constraints and rallying effective action for the sake of the world we will leave behind.

The emotional weight of uncertainty and the loss of a stable future vision

For much of recorded human history, individuals and societies have relied on certain bedrock assumptions about the future. Seasonal cycles might have varied slightly, but farmers generally knew when to plant and harvest. Economic conditions ebbed and flowed, but people had a reasonable sense of what it meant to provide for themselves and their children. Cultural traditions persisted across generations, reinforcing the idea that, while some aspects of life might shift, the underlying patterns of nature and society would remain intelligible and relatively steady. Even amid wars, epidemics, and other disruptions, many believed that, in the long run, human life would continue along a familiar trajectory, improving in some respects, remaining constant in others, and offering at least a stable framework of expectations.

Climate change, however, has introduced a profound new element of uncertainty that undermines these long-standing assumptions. Instead of trusting that tomorrow's conditions will resemble today's (give or take some predictable fluctuations), we now face an unpredictable environmental future. Weather patterns that once held steady, guiding agriculture and migration, grow erratic. Landscapes that communities considered permanent fixtures—coastlines, forests, glaciers—are visibly transforming. Vital ecosystems strain under stressors whose outcomes are hard to forecast with confidence. The knowledge that these large-scale environmental shifts are ongoing and may accelerate places a psychological weight on individuals who no longer know what kind of world they will pass on to their descendants, what local environments will look like a generation from now, or what baseline of normality society can assume.

This uncertainty bears emotional consequences that run deep. When people lack a stable vision of the future, they struggle to plan their lives, craft their identities, and anchor their moral commitments. Instead of trusting that retirement plans, education pathways, or family decisions fit into a relatively predictable societal fabric, they must consider that the world could be drastically altered by the time these plans come to fruition. The feeling that one's personal trajectory—and that of one's children, community, or culture—is clouded by environmental instability can evoke anxiety, sadness, frustration, anger, guilt, and grief. These emotions do not stem merely from practical concerns but from something more existential: a recognition that the foundational narrative of incremental progress and reliable continuity may no longer hold.

Uncertainty itself is a potent psychological stressor. The human mind craves predictability and structure, not merely for comfort but to make informed decisions. If we believed that rainfall would always come at roughly the same time each year, we could build routines around that. When that assumption falters, what was once automatic becomes fraught with doubt. Should we invest in certain crops if droughts might intensify? Should we live near a coastline if storm surges grow more frequent? Such questions, once uncommon or peripheral, become central and unnerving. Compounding this is that climate change effects often unfold gradually or episodically, interspersed with periods that feel deceptively "normal," making it challenging to know when to change course or how drastically to adapt. The resulting state—oscillating between moments that still feel familiar and others that hint at looming upheaval—intensifies the emotional burden. It's one thing to adapt to a crisis that arrives suddenly and definitively; it's quite another to dwell in a twilight zone of indefinite uncertainty.

Culturally, the loss of a stable future vision undermines narratives of progress. Many modern societies grew up on stories of continuous improvement: technological advancements improving living standards, policies ensuring more rights and security, and economies growing steadily. Environmental stability underpinned these stories. Now, climate disruption challenges the narrative that each generation will automatically inherit a better world. If future generations face harsher climates, scarcer resources, and more frequent natural disasters, the moral architecture that praised perpetual progress comes into question. People may feel a sense of betrayal: they were told that by working hard, saving money, and following societal rules, they would secure a better tomorrow for themselves and their children. Instead, they grapple with the possibility that even the most diligent planning cannot ensure stability in a climate-altered world. This mismatch between cultural expectations and ecological reality feeds disillusionment and moral outrage.

A significant aspect of this uncertainty involves anticipatory grief. While grief usually accompanies tangible losses—like the death of a loved one, the extinction of a known species, or the transformation of a cherished landscape—climate change introduces a subtler form of mourning. People grieve not only what is lost today but also what might never come to be. Parents grieve the possibility that their children may not enjoy the same seasonal pleasures or natural wonders they once took for granted. Communities grieve traditions tied to environmental cycles that may no longer align with shifting patterns. Even those who have not yet suffered direct impacts feel the weight of what could vanish: coral reefs fading before children ever get to see their color and vitality, stable coastlines receding so that iconic beaches or ancestral lands disappear. This preemptive sadness over possible futures that now seem compromised underscores the emotional complexity of uncertainty. It is not just fear of what might happen, but sorrow at the threatened loss of once-assumed inheritances.

Moral and intergenerational dimensions add further depth. If we accept that present actions—such as greenhouse gas emissions—set the stage for future conditions, the uncertainty can feel like an ethical burden. We sense responsibility not only for ourselves but also for people not yet born. If we cannot guarantee them a stable environment, what does it mean to live ethically now? This moral unease amplifies anxiety. Without clarity about future scenarios, how can we assess the morality of current decisions? The question, "Are we doing enough to prevent catastrophes for future generations?" hovers ominously. The uncertainty extends beyond personal or cultural identity crises into the realm of moral responsibility, heightening the emotional toll.

Generational tensions emerge from these uncertain visions as well. Younger individuals, who've grown up hearing dire predictions, may feel existential dread more acutely because they know no baseline of security. For them, uncertainty is the norm. They may resent older generations who once seemed confident about future trajectories, suspecting that complacency or short-term gains took precedence over long-term stewardship. Older individuals, on the other hand, may feel guilt, shame, or despair, wondering if their complacency or lack of foresight contributed to the destabilization younger generations now face. This interplay of blame, regret, and frustration exacerbates anxiety, as does the difficulty of forging intergenerational solidarity in a context where no one can promise what tomorrow holds.

Media and cultural discourse play a role, too. Reporting on climate topics often swings between alarmist predictions and hopeful stories of innovation and adaptation. While balanced reporting tries to offer realism and solutions, the net effect can still leave audiences uncertain about just how bad things could get or how effectively humans can respond. Without a stable baseline for interpreting these stories, individuals may find themselves overwhelmed by mixed messages. Are we on the brink of irreversible disaster, or are green technologies poised to save us? This ambiguity further erodes confidence and intensifies the emotional strain of uncertainty. The mind craves clarity to form coherent responses. Deprived of it, confusion and stress mount.

In social contexts, uncertainty can strain relationships and community cohesion. People might disagree on whether to take immediate, drastic measures or adopt a wait-and-see approach. Divergent responses to uncertainty can fracture social units. Some might withdraw into denial, insisting that the future remains bright despite evidence of instability, while others become activists, pushing for radical changes. Polarization can arise not only from ideological differences but also from diverging emotional strategies for coping with uncertainty. This social friction adds yet another emotional layer, as anxiety is compounded by conflicts with family, friends, or neighbors who interpret the uncertain future differently.

The marketplace of solutions—techno-fixes, policy reforms, personal lifestyle changes—also contributes to uncertainty. None of these approaches guarantee success, and each comes with trade-offs, unknown side effects, and complexity. People aware of these caveats find themselves asking: Which path is truly safe? Should they invest in renewable energy now, or might breakthroughs in carbon capture technology arrive later and change the landscape? Should governments prioritize adaptation (like building seawalls) or focus on rapid mitigation (reducing emissions at the source)? The array of partial, uncertain solutions means anxiety persists, since no single strategy offers a surefire return to a stable vision of the future. Instead, people must tolerate a landscape of possibilities and risks, a scenario that challenges the desire for firm, actionable guarantees.

Coping with this emotional weight of uncertainty involves various psychological strategies. Some individuals attempt to compartmentalize their worries, focusing on daily tasks and personal joys to avoid being paralyzed by long-term doubts. Others immerse themselves in learning, hoping that deeper understanding might yield clarity and thus reduce anxiety. Still others find solace in spiritual or philosophical traditions that emphasize impermanence and the necessity of adapting to change. By reframing uncertainty as an intrinsic part of existence, one might mitigate the shock of losing stable future visions. This does not remove the underlying problem, but it can ease emotional suffering and foster resilience.

Community-level responses to uncertainty often involve creating support networks. Local groups discussing climate worries, sharing coping techniques, and volunteering for environmental projects can transform individual anxiety into collective empowerment. By acknowledging together that no one has all the answers, communities can reinforce each other emotionally, sustaining a sense of purpose even when certainty is elusive. Mutual validation and empathy go a long way in preventing isolation. Recognizing that others share similar fears and uncertainties can alleviate feelings of personal inadequacy or irrationality. Instead, it frames these emotions as logical responses to unprecedented conditions.

Art, literature, and storytelling also offer pathways through uncertainty. When authors, poets, and filmmakers explore the tension between known pasts and unknown futures, they articulate what many struggle to express. By naming and depicting these emotions, cultural creators help audiences process them indirectly. Such works can remind people that uncertainty, while painful, can also inspire creativity, moral reflection, and a search for meaning. In this sense, the arts become a bridge between intellectual knowledge of climate issues and the deep emotional landscapes they conjure. Through metaphor, narrative arcs, and character struggles, the arts help us grapple with complexity and nuance.

In academic and policy circles, there is ongoing recognition that uncertainty is not simply a communication challenge but a psychological and moral one. Traditional risk assessments, cost-benefit analyses, and scenario planning struggle to convey emotional resonance. Emerging interdisciplinary approaches combine climate science with behavioral economics, psychology, philosophy, and cultural studies to better understand how to guide people through uncertain futures. These new frameworks acknowledge that uncertainty itself demands careful emotional navigation, not just more data or better models. Policymakers and scientists increasingly realize that fostering public engagement requires addressing emotional needs, providing transparent reasoning about uncertainties, and setting realistic expectations while maintaining hope and solidarity.

The moral dimension of uncertainty cannot be overstated. If the future were guaranteed to be stable, people might feel confident that incremental improvements in technology, policy, and culture would ensure ongoing progress. Without that guarantee, moral responsibility weighs more heavily. Uncertainty raises the stakes of today's choices, because the outcomes remain ambiguous. People must act without the comfort of knowing their efforts will suffice, investing energy and resources into mitigation, adaptation, and equity measures whose full benefits may remain unknown for decades. Courage and moral integrity thus gain a new importance: stepping into the unknown, striving to do right, without a scripted outcome, requires a moral fortitude that differs from the confidence of past eras.

For younger generations, the emotional toll of uncertainty can be particularly acute. With their whole lives ahead, they cannot confidently map out careers, family life, or long-term goals without factoring in climate instability. Education and career guidance do not traditionally prepare young people for a world where regional weather patterns may shift drastically, altering which industries thrive, which regions remain habitable, or which skill sets become valuable. This uncertainty can trigger existential questions: Is it fair to bring children into such a world? How do you invest in a home or business if you're unsure the local climate will remain supportive? These dilemmas feed anxiety and can hinder the sense of agency and forward momentum that young adults typically seek.

Older adults, too, bear emotional burdens. Those who remember stable climates and abundant natural resources might struggle to reconcile their memories with the shifting reality. They may feel regret—could they have done more when evidence of climate change first emerged? Could previous generations have set better policies or lessened dependencies on fossil fuels? This retrospective guilt or sadness, coupled with uncertainty about whether current late-stage efforts will mitigate future harm, makes emotional processing complex. Many older individuals feel protective of the world they imagined passing on, grieved to see that continuity jeopardized, and helpless about how to restore confidence in the future.

The interplay of knowledge and uncertainty heightens emotional pressure. We now know far more about the Earth's systems than any previous generation. Satellite data, ice core analyses, and computer models give detailed projections. Yet the nature of these projections—often probabilistic and scenario-based—does not yield the crisp certainty people crave. Instead, it multiplies conditional statements: if emissions decrease by this amount, we might limit warming to that degree; if certain technologies scale up rapidly, some outcomes may be averted. Instead of comforting reassurance, this conditional complexity can feel like an emotional labyrinth with no guaranteed exit.

Social justice and inequality further complicate the emotional landscape of uncertainty. The knowledge that vulnerable communities—often low-income, indigenous, or otherwise marginalized—will bear the brunt of climate impacts, even though they contributed least to the problem, adds moral frustration. If we cannot predict exactly how or when these impacts will manifest, alleviating suffering becomes harder. Uncertainty here not only spawns anxiety but also moral indignation, as people worry that inequities will deepen when instability reigns. Emotions combine: fear for the disadvantaged, anger at perceived injustice, shame at collective inaction, and sorrow over lost fairness. Uncertainty magnifies these tensions because no one can definitively say what reforms will prevent or mitigate such harms in time.

Yet uncertainty also invites reevaluation of priorities and values. If we cannot rely on stable external conditions, we might turn inward, seeking stable internal guides. Ethical principles, compassion, community support, and intellectual humility may gain new importance. People may invest more in social bonds, local resilience projects, or spiritual and philosophical inquiry to find equilibrium amid external unpredictability. This inward shift acknowledges that emotional stability can sometimes be cultivated even when environmental stability cannot be assured.

Some respond to uncertainty by pushing for better early warning systems, more robust scenario planning, and flexible policies designed to handle multiple possible futures. While this approach does not erase the emotional strain, it can provide some psychological relief by suggesting pathways for adaptive resilience. If people see institutions preparing for a range of outcomes, they might feel less helpless. Flexibility and adaptability become virtues, replacing the old ideal of long-term predictability. Re-envisioning stability as a dynamic capacity to respond effectively to change, rather than an absence of change, helps individuals cope emotionally. It encourages viewing uncertainty as a challenge that can be met with creativity, learning, and moral courage.

In addition, the emotional weight of uncertainty can stimulate deeper dialogues about meaning and purpose. Without a guaranteed stable future, what anchors our sense of why we strive, build, and care? Some find meaning in the act of caring itself, in protecting what can be protected, in stewarding whatever remains of ecological richness, or in ensuring that even uncertain futures contain possibilities for human dignity and ecological flourishing. The emotional burden can thus push people to forge new narratives of meaning that do not rely on the promise of continuity but on the resolve to act ethically regardless of outcome.

These narratives might highlight heroic efforts to restore degraded ecosystems, invest in sustainable technologies, or safeguard cultural diversity and knowledge for future generations—whoever they are and whatever conditions they face. By focusing on deeds rather than assured results, individuals and communities channel uncertainty's emotional energy into constructive endeavors. Instead of despair, they cultivate a determined acceptance that while the future is unclear, our present actions matter deeply in shaping it.

In the end, the emotional weight of uncertainty and the loss of a stable future vision pervade multiple layers of human experience: personal identity, cultural narratives, moral commitments, generational relations, and global justice. Each aspect compounds the others, weaving a complex emotional tapestry that includes anxiety, grief, frustration, and, at times, a quiet resolve. Understanding this multidimensional stress can lead to greater empathy for ourselves and others as we navigate a world where the old signposts vanish and new ones remain elusive. It can inspire interdisciplinary efforts to communicate climate risks more effectively, foster resilience in communities, support mental health resources, and develop ethical norms that honor obligations to future life under uncertain conditions.

By grappling honestly with uncertainty, people may discover reserves of emotional strength they did not know they possessed. They learn to live with moral complexity, to care fiercely about future lives despite no guarantees, to pursue sustainability and equity not because success is certain but because striving is morally right. Far from trivial, these emotional adaptations represent an evolution in how humans conceptualize their place on Earth. In this uncertain age, cultivating emotional resilience, humility, and solidarity can help societies maintain coherence and purpose even as the stable future visions they once relied upon recede into the past.

Moral dilemmas and the tension between personal responsibility and systemic issues

As the climate crisis intensifies, we find ourselves confronting a tangled web of moral dilemmas. At the heart of these is a deep tension between what we, as individuals, can and should do to address environmental degradation, and what is required at broader societal, economic, and political levels to meaningfully alter the course of climate change. The complexity of this tension challenges our understanding of morality, fairness, agency, and accountability. It urges us to ask where personal responsibility ends and systemic responsibility begins, or whether it makes sense to separate the two at all.

On one hand, moral intuition often suggests that each person should do their part. The logic is appealingly simple: if everyone makes sustainable choices—reducing waste, conserving energy, biking instead of driving, avoiding excess consumption—collective action would sum to significant change. Behavioral recommendations abound: eat less meat, switch to renewable energy where possible, reduce single-use plastics, vote for environmentally conscious policies. These actions, we are told, contribute to lower carbon footprints and more sustainable communities. And indeed, there's moral resonance in the notion that no one is exempt from caring or from adjusting their lifestyle to ease environmental stress.

Yet, as individuals try to adopt greener habits, they soon encounter systemic barriers and countervailing forces. Corporations continue extracting and burning fossil fuels at massive scales. Governments subsidize polluting industries or fail to enact regulations stringent enough to rein in emissions. Global supply chains rely on cheap, carbon-intensive transportation. Even those who carefully recycle their plastic may learn that much of it ends up incinerated or in landfills because the recycling infrastructure and markets are not robust. Buying organic produce might reduce pesticide use locally, but what of the global commodity markets that shape land-use patterns or the deforestation driven by distant economic incentives? A household installing solar panels must still contend with a grid powered largely by coal or gas, and their personal emissions cut may be dwarfed by a single industrial plant's output.

Such realizations breed frustration and moral confusion. Is it fair or rational to place heavy moral burdens on individuals when systemic factors—policies, industries, infrastructures—constrain the effectiveness of personal actions? This tension can generate feelings of guilt, helplessness, or resentment. Some individuals respond by trying even harder to live sustainably, hoping their personal purity or minimal footprint can stand as an example, or at least ease their conscience. Others respond by disengaging, arguing that personal sacrifices seem pointless if major polluters go unchecked. Both approaches reflect moral struggles: the former striving for moral consistency, the latter conceding the futility of moral striving in isolation.

Moral philosophers and ethicists grapple with these questions. The scale of climate change is unlike a typical moral scenario—there is no single villain to blame, no direct line from one individual's actions to a specific harm easily seen. Instead, myriad actors contribute incrementally to global emissions. Emissions from driving your car may not directly cause a flood in a distant coastal community, but cumulatively, billions of such actions contribute to greenhouse gas accumulation. The diffuse nature of causation complicates how we assign moral responsibility. Traditional moral frameworks often assume a more tangible cause and effect, making it challenging to determine the ethical obligations of any single person in a systemic crisis.

Some moral thinkers argue that because the problem is collective, so too must the solution be. From this viewpoint, individual moral behavior cannot be properly separated from collective arrangements. The need is not just for personal virtue but for systemic changes—new laws, international treaties, market interventions, and cultural shifts. This perspective suggests that while personal efforts are commendable and can raise awareness, they cannot substitute for institutional accountability. After all, how can one seriously claim that by recycling diligently and reducing meat consumption alone, one is taking on a fair share of responsibility, when entire energy infrastructures or trade networks remain unchanged?

At the same time, another set of moral arguments defends personal responsibility. They point out that systemic injustices and destructive policies do not arise from thin air; they are sustained by social norms, market demands, and political leaders who reflect societal values. Personal choices, when aggregated, can shift cultural expectations. If enough people reduce their meat intake, markets respond, potentially lowering deforestation for cattle ranching. If enough communities embrace renewable energy, political pressures mount to support these sectors. Personal actions, while limited in isolation, become morally meaningful steps towards building momentum for systemic transformation.

This interplay fosters a moral dilemma: should individuals focus on personal lifestyle changes or on pushing for systemic reforms via activism, political engagement, and collective action? Many activists suggest that we need both: personal moral integrity and systemic advocacy. Adopting a greener lifestyle isn't just about lowering one's personal carbon footprint; it can also serve as a form of moral signaling, reinforcing the idea that society is ready to move beyond a high-carbon status quo. The combination of personal virtue and political pressure might create a synergy—personal changes normalizing sustainable living, collective action demanding institutional change, and institutional change making personal efforts more effective.

There's also an emotional dimension to this moral tension. People can feel distressed when realizing that their best personal efforts are dwarfed by systemic inertia. A sense of unfairness arises: "I'm doing my part, why aren't the big players doing theirs?" This emotional friction sometimes leads to moral fatigue or cynicism. People question the point of their sacrifices if multinational corporations continue extracting oil or governments stall on climate agreements. On the flip side, moral motivations can prompt individuals to join grassroots movements, attend protests, sign petitions, and build networks that can influence policy and create structural shifts. By participating in collective endeavors, individuals find moral solidarity and reduce the feeling of isolation that often accompanies personal environmental choices.

Inequality further complicates these moral dilemmas. Some individuals have greater financial means, educational backgrounds, or geographical advantages that make sustainable choices easier—installing solar panels, affording an electric car, buying locally sourced organic produce. Others, constrained by economic hardship or living in food deserts and poorly served neighborhoods, find eco-friendly options too expensive or inaccessible. Should the less privileged be morally judged for not "doing their part" if systemic factors have priced them out of low-impact lifestyles? This raises questions about moral responsibility's intersection with social and economic justice. If systemic issues create conditions that make sustainable choices a luxury, then the moral burden falls disproportionately and unfairly on those least capable of carrying it. Such scenarios highlight that climate moralities cannot be divorced from broader questions of equity and justice.

Moreover, cultural differences and historical responsibilities loom large. Climate change has roots in centuries of industrialization, colonial exploitation, and uneven development. Certain countries and corporations bear a historically greater share of emissions and environmental damage. Does this mean individuals in historically high-emitting nations owe more morally? Should their personal efforts to reduce emissions count as partial compensation for past and present injustices? This moral calculus can be emotionally fraught. It suggests that not all personal carbon footprints carry the same moral weight, and that systemic corrections—like reparations, international climate finance, technology transfer—are needed to address moral debts embedded in the global system.

The moral tension also extends to the realm of activism and leadership. A climate activist, for instance, might renounce flying to reduce personal emissions, setting an example of moral consistency. Yet what if this sacrifices their ability to travel to important conferences where they could influence policymakers to enact emissions cuts orders of magnitude greater than any they could achieve personally? Balancing personal moral purity against strategic systemic influence is no small task. At what point does the pursuit of personal ethical consistency hinder broader impact? Conversely, can one justify carbon-intensive actions by claiming it's for the greater systemic good? Such questions highlight that moral reasoning must navigate trade-offs, strategic thinking, and an honest reckoning with outcomes rather than just principles.

Additionally, moral frameworks differ. Some adhere to a deontological view—arguing that individuals should do what's right regardless of outcomes. From this stance, living sustainably is a duty no matter what others do. Even if personal actions seem insignificant relative to global emissions, the moral agent remains responsible for not contributing unnecessarily to harm. Others take a consequentialist view, focusing on outcomes: if personal lifestyle changes barely shift the needle on global emissions, the moral imperative might lean more towards political advocacy, collective mobilization, or pressuring major emitters. The tension here is between moral purity—doing what feels right on an individual level—and moral effectiveness—channeling efforts into areas likely to yield meaningful results. The interplay of these ethical philosophies can cause internal moral conflict, as people weigh the comfort of personal righteousness against the strategic need for broader systemic engagement.

Cultural narratives further influence how individuals resolve these moral dilemmas. In some societies, personal responsibility and individual action are highly prized, leading citizens to embrace lifestyle changes as the primary route to moral rectitude. In others, there is a stronger tradition of collective problem-solving, where moral responsibility is understood in the context of community action and political advocacy. Likewise, religious or spiritual traditions may shape perspectives, encouraging believers to see stewardship of the Earth as a moral obligation. Such frameworks might reduce the tension by offering moral guidelines that emphasize communal duties and the necessity of structural transformations. On the other hand, if a faith tradition stresses personal virtue without equally highlighting systemic reforms, adherents may struggle to reconcile their moral sense with the need for larger-scale changes.

The role of corporations and markets intensifies these moral questions. Corporate leaders, shareholders, and advertisers shape consumer behavior and choices. If major companies push high-carbon products as the default, is it fair to blame individual consumers who lack viable alternatives? Here, moral tension arises from a recognition that systemic actors—those controlling production lines, supply chains, and marketing strategies—wield disproportionate influence. Individuals might wonder if they should boycott certain products, support ethical businesses, or invest in sustainable funds. While these personal market-based choices can signal preferences, if fundamental policy frameworks remain unchanged, the transformation may be slow or negligible. This highlights that moral efforts to reshape demand must be complemented by policies that reshape supply. Without the latter, individuals run into the frustration of feeling morally compelled to choose the "least bad" options in a marketplace structured for environmental harm.

Another dimension lies in the storytelling and media representations of responsibility. Environmental campaigns often focus on personal solutions—using metal straws instead of plastic ones, bringing reusable bags to the store—because these actions are visible, concrete, and empower the individual to feel part of the solution. But such messages can inadvertently reinforce the idea that if we fail, it's because individuals didn't try hard enough, obscuring systemic inertia and the need for political overhaul. Individuals internalize this narrative and may feel crippling guilt or shame when confronted with climate data, forgetting that no personal action alone can offset the emissions of an entire industrial sector. Conversely, if media downplays personal action and focuses solely on systemic failures, people might feel powerless, believing their choices are irrelevant. Balancing these narratives is crucial to maintain moral motivation without oversimplifying the complexity.

Understanding these moral dilemmas also involves acknowledging emotional labor. The emotional toll of trying to be "good" in a system that encourages harm—through normalized consumption patterns, fossil-fuel dependencies, and cultural habits—is substantial. People who diligently track their carbon footprints, avoid air travel, minimize plastic, and forego certain pleasures often feel weighed down by the effort. They may resent the ignorance or apathy of others who don't share their moral concerns. They might also feel lonely, alienated in social contexts where sustainable habits stand out as unusual or inconvenient. This emotional weight can lead to burnout, cynicism, or withdrawal from climate engagement. Recognizing that moral dilemmas generate emotional strain can help communities support each other emotionally, offering empathy and understanding instead of judgment.

If moral dilemmas stem from the interplay of individual and systemic responsibilities, solutions likely involve cultivating moral maturity that embraces complexity. Instead of framing the issue as either personal virtue or systemic reform, a nuanced moral approach can hold both elements together. One might accept that personal actions are morally significant symbols and stepping stones, even if insufficient alone, and that such actions gain meaning when linked to collective pressures, political engagement, and social movements. Conversely, when pushing for systemic change, individuals can recall that institutions and policies ultimately reflect collective values and choices. Bringing these perspectives into harmony reduces the moral tension by showing that the dichotomy—personal vs. systemic—is partly artificial. Both levels reinforce and influence one another.

Educational and institutional efforts can address these moral dilemmas by teaching systemic thinking. By helping people see how personal consumption and systemic structures interlock, education can clarify that moral responsibility exists in multiple layers. If well-implemented, this understanding prevents moral frustration. Instead of oscillating between guilt-laden personal scrutiny and helpless outrage at inaction in high places, people can find constructive paths: joining local climate action groups, advocating for fair and ambitious policies, voting for candidates who promise systemic reform, and supporting businesses transitioning to sustainable models. The moral sense shifts from "either I fix this alone or nothing changes" to "my actions contribute to shifting cultural norms, which influence policy and corporate behavior, and my political and civic engagement pushes institutions toward accountability."

Another strategy involves highlighting success stories where moral dilemmas have been partially resolved. Communities that transition to renewable energy cooperatives, cities that redesign public transport for low emissions, or nations that enact just transition policies for workers in high-carbon industries offer glimpses of moral coherence. In such scenarios, personal efforts align with systemic improvements. Individuals find low-carbon options readily available, policies encourage sustainable behavior, and markets respond to ethical demands. These examples prove that it's not futile to make personal sacrifices; done in a conducive environment, they become part of a larger moral mosaic.

In religious or spiritual contexts, faith leaders might emphasize that personal stewardship is essential not just because of the sum of individual actions but because it prepares the moral ground for demanding more from leaders and institutions. Rituals, prayers, and teachings can frame the tension as a test of collective moral character, calling on believers to support policies that protect creation. By blending personal piety with communal advocacy, religious traditions might resolve some of the moral tension by showing that individual virtue is a prerequisite for collective righteousness, not a distraction from it.

Similarly, artistic and cultural expressions can depict characters or communities navigating these dilemmas, offering empathy for those who feel small in the face of global problems and admiration for those who strive to influence structural change. This cultural reflection helps normalize the complexity of moral reasoning in climate crises. People can identify with fictional narratives that show flawed but earnest protagonists learning to direct their moral impulses beyond personal purity and into collective struggles. Art can thus alleviate the moral pressure by reminding audiences that no one is morally perfect, that embracing complexity and incremental progress is itself a moral stance.

In economic theory and policy debates, acknowledging moral dilemmas might encourage designing policies that reduce individual burdens. Carbon pricing, if equitable, makes sustainable choices more accessible and rewarding, aligning individual incentives with systemic goals. Regulations on polluting industries relieve individuals of the guilt of benefiting from harmful systems, as the system itself evolves to offer cleaner options. Public investments in renewable infrastructure or public transportation ensure that personal moral aspirations meet supportive structural frameworks. Such policies minimize moral conflict by making virtuous choices easier, cheaper, and more natural, gradually lessening the sense of being caught between personal integrity and systemic inertia.

Critical thinking and open dialogue remain key. By openly discussing these moral tensions without resorting to shaming or cynicism, societies can cultivate a more mature environmental ethic. People might voice frustrations: "Why must I give up conveniences while big polluters remain unchecked?" and receive informed responses that highlight ongoing efforts to regulate emissions, petitions for corporate responsibility, and success stories from local initiatives. Such conversations can steer moral energy towards productive channels: encouraging more people to engage in activism, vote in elections influenced by climate policy, and support organizations that hold powerful actors accountable. The moral dilemma evolves from a paralyzing contradiction into a motivating call to integrate personal ethics with structural efforts.

As individuals acknowledge that no single approach suffices, moral humility emerges. Accepting that the problem is too large for personal action alone prevents despair because it clarifies that moral responsibility is relational and collective. One doesn't have to shoulder everything individually. Instead, one can find purpose in contributing part of the solution—living more sustainably, supporting political candidates who promise real climate action, aligning investments with ethical principles—and trusting that others will do their parts as well. This shared moral endeavor dilutes the tension, replacing isolated moral burdens with a sense of participation in a moral community striving for systemic shifts.

While none of these reflections conclude the moral dilemma neatly or resolve the tension once and for all, they illuminate paths through it. Recognizing the complexity might ironically provide emotional relief: it is not a personal failing that leaves one feeling conflicted; it is the nature of the problem itself. If climate change were simple, an individual could solve it by just recycling more or driving less. The fact that the crisis demands both personal virtue and massive structural change testifies to its scale and significance.

In facing this moral complexity, people learn to tolerate ambiguity. They realize that doing their best does not guarantee a fixed outcome. Instead, moral action becomes a statement of identity and values, a contribution to changing norms, and a building block in the pressure for systemic reform. Even without certainty of success, moral commitments gained through personal responsibility and systemic engagement guide people towards hope and integrity. And while the tension may never vanish, understanding its roots and possible responses ensures it does not paralyze moral will, but rather motivates it in more nuanced, collaborative, and determined ways.

It becomes clear that the struggle to fully grasp and emotionally respond to climate change does not arise from apathy, ignorance, or a lack of moral fiber, but rather from ancient mental patterns formed in environments unlike our own. Instead of labeling ourselves as flawed for not feeling constant alarm, we can recognize that our instincts have simply not caught up to the unprecedented complexity and timescale of global climate disruption. This recognition need not inspire resignation. Rather, it highlights the adaptability that has always defined human societies. We can build upon our inherited capacities for empathy, reason, and culture to forge new frameworks that transcend old cognitive limitations. By telling vivid stories that resonate with our senses, emphasizing immediate co-benefits of climate action, shaping collective norms that treat long-term stewardship as common sense, and drawing on ethical and spiritual traditions that broaden our circle of care, we can cultivate a mindset that, over time, becomes as naturally attuned to the subtle rise of ocean acidity or the slow retreat of ice as our ancestors were to the snap of a twig. There is nothing inevitable about emotional indifference to distant, systemic threats. With conscious effort, cultural innovation, and moral imagination, we can attune our hearts and minds to the realities of this century, allowing truly global, long-term responsibility to feel as intuitively necessary and emotionally compelling as any immediate danger faced in eras long past.

Rising Above: A Practical Guide to Overcoming Climate Anxiety and Finding Hope in a Changing World

Chapter 3: How It Affects Us Day-to-Day

Physical, Emotional, and Social Impacts of Climate Anxiety

Human lives unfold within a tapestry of environmental cues so consistent, so deeply woven into daily experience, that many people seldom stop to acknowledge them. Yet these cues—predictable seasons, stable coastlines, abundant wildlife, reliable food supplies—form a psychological backdrop supporting our emotional equilibrium. When these once-reliable markers begin to shift, as they do under the pressure of climate change, minds and bodies respond. This response may not roar into consciousness immediately. Instead, it often trickles in subtly: a lingering tension in the jaw after reading another record-breaking temperature headline, a vague sense of restlessness that colors a drive through a drought-stricken landscape, or the flutter of anxiety arising when deciding whether to invest in a home near a coastline subject to rising seas. These small, scattered reactions reflect something larger—an emotional process, both personal and collective, set in motion by an era of environmental instability.

On a physical level, climate anxiety often manifests in symptoms associated with stress and uncertainty. Stress hormones like cortisol may spike when contemplating dire projections of future climate scenarios, even if these scenarios remain decades away. Individuals may find their sleep disturbed—tossing and turning at night, waking abruptly with concerns that linger at the edge of dreams. Over time, chronic tension might settle into shoulders, necks, or backs, as the body braces for an undefined threat it can't confront head-on. Headaches might become more frequent; appetite patterns could shift. While none of these symptoms are unique to climate-related worries, their presence in this context underscores how intangible fears can become tangible burdens on the body. Consider someone who reads daily reports of irreversible melting in polar regions. They might lie awake, muscles clenched, breathing shallowly, even if the arctic meltdown doesn't personally threaten them tomorrow. The body's alarm system, honed by millennia of responding to immediate dangers, now grapples with a global, slow-moving hazard it cannot easily parse.

Emotional responses vary widely and can be complex. A subtle background dread may pervade the day, low-level and persistent, never erupting into panic but quietly sapping joy from activities once taken for granted. Others might experience sharper bouts of fear—moments of acute distress triggered by encountering alarming news about intensifying storms or new data on biodiversity loss. Sadness often intertwines with fear: sadness for species fading from existence, sadness for the idea that cherished childhood landscapes may no longer greet future generations with the same abundance. Anger also emerges, directed at political leaders who fail to act decisively, at corporations prioritizing profit over planetary well-being, or at a societal inertia that seems incongruent with the stakes. Some feel guilt, knowing their lifestyle contributes to emissions. Even if trying to reduce their carbon footprint, they might still rely on cars or imported goods, recognizing that personal actions are insufficient to halt global warming. Shame can creep in, too—the uneasy sense of participating in systems that push ecosystems toward collapse.

These emotional states can mix, blend, and shift over time. A single individual might experience fear when reading an IPCC report in the morning, anger after watching a documentary on deforestation later in the afternoon, and sadness that evening upon recalling childhood memories of more bountiful nature. This emotional kaleidoscope can feel overwhelming. People might struggle to articulate what they feel, finding it hard to explain why headlines about coral bleaching leave them more irritable or withdrawn. Yet these feelings arise logically from the cognitive dissonance between the stable, predictable world many grew up imagining and the uncertain, dynamic reality scientists describe. In this way, emotions serve as a barometer, registering that something fundamental is amiss.

Socially, climate anxiety does not remain confined within one's own mind. It often surfaces in conversations, community gatherings, and relationships. Friends might find themselves disagreeing on the severity of climate risks, straining ties when one person dismisses worries as overblown while another feels deeply concerned. Family members discussing future plans—like where to buy a home or whether to have children—may encounter unexpected tension if some hold onto assumptions of environmental stability while others question whether those assumptions still hold. Communities grappling with tangible climate impacts—like recurrent floods or heatwaves—experience collective unease. Where once festivals or seasonal traditions proceeded according to known environmental rhythms, shifts in weather can disrupt these events, leaving a sense of cultural disorientation. Even workplace interactions might be affected; colleagues trying to introduce sustainability policies could face resistance from others who see no immediate need for change.

Within these social dynamics, feelings of isolation and misunderstanding can arise. Someone who internalizes a great deal of climate anxiety may feel lonely if their peers seem carefree or indifferent. This gap in emotional investment can generate resentment or confusion. "Why don't they care?" one might wonder. Alternatively, a person seeking to avoid the discomfort of acknowledging climate instability may downplay or mock concerns expressed by others, fragmenting trust in the process. Over time, repeated friction on such topics may erode the sense of belonging that is vital for emotional well-being. The very relationships that provide support during stressful times become sites of discord, paradoxically intensifying anxiety.

Yet not all social consequences are negative. As awareness grows, many find solace in new communities formed precisely to address eco-anxiety. Support groups dedicated to climate worries have emerged, both online and in person. Environmental organizations, religious congregations emphasizing stewardship, and social movements pushing for climate justice can offer a sense of shared mission, validating that these fears are not personal neuroses but collective responses to real threats. In such supportive contexts, individuals transform isolated anxiety into communal energy, forging bonds that help carry emotional loads together. The process of naming and acknowledging these feelings publicly can alleviate shame and loneliness. Realizing that anxiety is a reasonable response to extraordinary circumstances makes it easier to seek help, empathy, and constructive dialogue.

Physical, emotional, and social impacts often interact. Take the example of someone feeling persistent low-level dread about future water shortages in their region. This emotional state might lead to restless sleep (physical impact) and, as a result, reduced patience and irritability in daily interactions. That irritability might cause friction in their household (social impact), further exacerbating tension and perpetuating the cycle. Conversely, finding emotional support in a community group might reduce that person's stress, improve their sleep, and restore harmony in their relationships. The interplay shows that these impacts do not occur in isolation. They feed into one another, shaping and reshaping the individual's daily reality.

Gender, age, socioeconomic status, and cultural background can influence how these impacts manifest. Younger people, aware they'll live longer into the climate-altered future, may feel more intense anxiety or anger. They might question whether to have children, given the uncertain legacy their offspring might inherit. This line of worry introduces emotional strain into relationships and family planning discussions, linking personal existential dilemmas to broader social discourses. Those with fewer resources might experience heightened anxiety simply because they know adaptation—like cooling homes in extreme heat or relocating from flood-prone areas—is costlier and harder for them. Environmental inequality thus magnifies emotional and physical health burdens on vulnerable communities, while social stratification complicates efforts to find common ground.

Cultural narratives influence the intensity and expression of these impacts. Communities with strong traditions emphasizing harmony with nature may feel a different kind of grief and moral outrage, rooted in a sense that sacred balances are being upended. Others, raised on stories of endless progress and human dominion over nature, might struggle with the dissonance that arises when confronted by stark evidence that nature sets limits humans cannot simply evade. In the first scenario, cultural lenses intensify emotional responses by framing the climate crisis as a spiritual or moral violation, while in the second, anxiety may stem from the collapse of cherished myths of control and infinite growth. Both examples show how cultural frameworks shape the emotional register of eco-anxiety.

These impacts also shift over time. Early encounters with climate anxiety may produce mild unease that people brush aside. As headlines accumulate, ecosystems degrade, and future scenarios become more dire and visible, anxiety can deepen. What begins as subtle tension may evolve into persistent worry or moments of panic. Alternatively, some adapt to the emotional load by numbing themselves. They might notice, ironically, a drop in conscious anxiety not because the problem has improved but because their minds have resorted to denial or emotional withdrawal to protect well-being. This adaptation, however, comes at a cost, potentially dulling empathy and moral responsiveness and complicating efforts to mobilize effective solutions.

Even small instances of anxiety reflect larger phenomena at play. For example, a spike of irritation when encountering a neighbor's wasteful habits—perhaps seeing them water their lawn excessively during a drought—may seem trivial. Yet that irritation is rooted in a recognition that local actions accumulate global consequences. Similarly, feeling disheartened after reading about coral bleaching on a reef you've never visited is not random sadness; it's a manifestation of understanding that what happens in one part of the world can ripple out, affecting global ecological integrity and, by extension, human futures. Thus, the emotional resonance of even minor triggers reveals underlying awareness that we're all entwined in a planetary web whose stability no longer seems guaranteed.

Awareness of these impacts can lead to proactive responses. Individuals who recognize their growing tension might seek healthier coping mechanisms: exercise routines to reduce stress hormones, mindfulness practices to calm racing thoughts, or therapy sessions with professionals familiar with eco-anxiety. Socially, acknowledging the pressure climate anxiety exerts on relationships may prompt more compassionate communication strategies, encouraging family and friends to discuss these topics openly and supportively. Emotionally, understanding that these anxieties are shared can alleviate self-blame—one may realize that feeling anxious about climate change isn't a personal failing, but a natural response to unprecedented conditions.

Media organizations, educators, and policymakers can also help mitigate the negative impacts. Communicating climate information in ways that are neither alarmist nor dismissive can create more productive emotional responses. Presenting realistic but hopeful narratives, highlighting successful adaptation projects or policy reforms, can restore some balance to emotional landscapes. Schools that introduce climate literacy and coping strategies at an early age can help younger generations handle eco-anxiety more gracefully, reducing fear and uncertainty.

Still, no single approach can fully eliminate the physical, emotional, and social toll of climate anxiety. The complexity of the crisis ensures that people will feel its weight in diverse ways. Some might lean into activism, channeling their fears into political engagement and community initiatives that foster solidarity and purpose. Others may prefer quieter forms of resilience, like gardening or preserving local habitats, finding solace in constructive action that links personal well-being with environmental stewardship. Such activities can have positive feedback loops: improving one's immediate environment eases anxiety, which, in turn, encourages further engagement and emotional healing.

Addressing climate anxiety's impacts does not mean underplaying the gravity of the crisis. Rather, it acknowledges that fear, sadness, and anger are valid responses to real threats. Individuals experiencing these emotions are not being irrational; they are responding to shifts that challenge ancestral assumptions about stability and predictability. By validating these feelings, communities can foster emotional resilience instead of shame. Recognizing that many share similar fears encourages mutual support and normalizes seeking help, whether through professional counseling, community workshops, or spiritual gatherings.

There are also nuanced differences in how climate anxiety influences day-to-day life depending on a person's sense of agency. Those who believe they can influence outcomes—through voting, activism, scientific research, or technological innovation—may find that anxiety, while uncomfortable, motivates engagement. Their physical symptoms might prompt them to run for office or volunteer in an environmental nonprofit, converting stress into action. Conversely, those who feel powerless might experience their anxiety as paralyzing. Without seeing a path from inner turmoil to outer change, their worry can stagnate, contributing to despair or apathy.

Social narratives can help shift this dynamic. If communities openly discuss the connection between emotional distress and the climate crisis, highlighting avenues for collective action, they may empower even the powerless to discover small but meaningful contributions. For example, a neighborhood initiative to plant climate-resilient trees could provide tangible local improvement, relieving some social and emotional strain. A company adopting sustainable practices might ease employees' anxiety by showing that their workplace is part of the solution rather than indifferent.

Physical health professionals may begin to recognize climate anxiety as a factor in stress-related conditions—hypertension, insomnia, chronic fatigue—enabling more holistic treatment approaches. Instead of treating sleeplessness in isolation, doctors might consider a patient's environmental concerns. Similarly, mental health professionals may incorporate discussions of eco-anxiety into therapy sessions, normalizing these feelings and guiding patients toward acceptance, coping strategies, and balanced perspective.

At the intersection of personal health and public policy, some forward-thinking strategies might emerge: adding green spaces in urban areas not only mitigates some climate risks but also offers havens for emotional recovery, reducing stress and fostering a sense of stability amid changing conditions. Encouraging community dialogues can transform anxiety into a shared impetus to advocate for local climate adaptation measures that protect everyone's interests. In these ways, acknowledging the physical, emotional, and social impacts of climate anxiety can catalyze preventive and adaptive solutions.

The presence of these impacts also reveals how interconnected human life is with the planet's well-being. Historically, environmental changes were often perceived as external challenges—droughts, floods, plagues—that communities adapted to physically. Now, we understand that such changes also infiltrate our emotional worlds and social relationships. The intimate connection between planetary health and human mental health underscores that climate instability is not just an ecological or economic issue, but a profound psychological and cultural one as well.

As time advances, we might notice generational shifts in how these impacts are managed. Younger cohorts raised amid ongoing climate discourse may develop more nuanced emotional resilience or innovative community models to face uncertainty. Cultural narratives may evolve to incorporate climate anxiety as a recognized part of life, offering rituals, art forms, and educational curricula aimed at helping people process these feelings constructively. By integrating the emotional dimension of climate change into mainstream understanding, societies can prevent individuals from feeling alone or defective for experiencing stress about the planet's trajectory.

Physical manifestations can also become cues for intervention. If an individual notices recurring tension headaches after reading climate reports, they might learn to pace their media consumption or take breaks to engage in calming activities. These self-regulation techniques, once learned by a subset of environmentally conscious individuals, might spread throughout populations as normal life skills, as important as learning to budget finances or maintain a balanced diet. A new form of emotional literacy might emerge, one that includes understanding how large-scale environmental factors influence personal well-being and relationships.

On the global stage, acknowledging these impacts could influence international cooperation. If leaders realize that climate inaction not only threatens future resources but also imposes present emotional and psychological burdens on citizens, it might add urgency to climate negotiations. Emotional well-being, typically seen as a downstream effect of social and economic conditions, becomes part of the moral calculus. Protecting mental health by ensuring environmental stability might offer another compelling argument for ambitious climate policies. While intangible, the collective mental strain is real and can erode quality of life, productivity, and social cohesion.

As for research and scholarly inquiry, the link between climate anxiety and physical, emotional, and social impacts may spur interdisciplinary studies. Psychologists, neuroscientists, anthropologists, and climate scientists could collaborate to understand how exactly environmental instability triggers these reactions and how best to mitigate them. The result might be more informed recommendations for governments, educators, and health professionals. Research could identify which coping strategies yield the best outcomes, which community interventions prevent polarization, and how cultural narratives can evolve to nurture resilience.

At the interpersonal level, simply having a name for what people feel—climate anxiety—can reduce confusion and self-judgment. Instead of dismissing their tension as inexplicable, individuals can say, "I'm feeling eco-anxiety today," and look for ways to address it. Partners can support each other through difficult conversations, parents can reassure children that their worries are understandable, and neighbors can collaborate on small sustainability projects that offer emotional relief along with ecological benefits.

The complex interplay of physical symptoms, emotional reactions, and social ramifications forms a feedback loop that shapes not only how people perceive the climate crisis but also how they respond to it. If the net effect becomes too distressing, some might retreat into denial or fatalism. Others, however, when fully aware of what these feelings represent, may lean into action—lobbying for policy changes, planting community gardens, reducing personal waste, helping to design local adaptation plans, or joining environmental education efforts. Every such action, in turn, can ease emotional strain, as people see tangible results and realize they are not powerless.

While no single approach completely alleviates the weight of these impacts, understanding their nature, acknowledging their validity, and seeking supportive networks can allow individuals and communities to navigate them more gracefully. Recognizing that climate anxiety stems from genuine ecological instability rather than personal weakness reframes the experience. Instead of feeling victimized by unexplained stress, one can interpret these emotions as signals that prompt engagement, reflection, and empathy. In this light, even uncomfortable feelings have a role to play, nudging societies toward collective reflection and moral recalibration.

By integrating these insights, individuals and groups might construct environments—both literal and metaphorical—that foster emotional resilience. Schools could incorporate lessons that teach students how to process eco-anxiety, families might develop open communication channels about fears and hopes for the future, and communities can hold gatherings where scientific updates are presented alongside discussions on coping and adaptation strategies. Cultural narratives that once assumed environmental predictability may shift toward celebrating adaptability, moral courage, and cooperation in the face of changing conditions.

In considering these physical, emotional, and social impacts, it's clear that even seemingly minor feelings or bodily reactions can be meaningful indicators. They reveal how climate change, while often perceived as abstract and distant, already shapes our internal worlds and interpersonal dynamics. Each headache, knot of tension, or anxious dream hints that people are absorbing the world's instability at a visceral level. Each uneasy conversation or interpersonal conflict about environmental issues shows that these changes are not staying in the realm of theory; they are infiltrating daily life.

By paying attention to these signs, individuals and communities can move from a state of passive distress to one of informed awareness. Instead of feeling caught off-guard by stress, they can recognize it as an invitation to learn, adapt, and respond. Even small steps, like setting boundaries on how much environmental news one consumes at a time, can help maintain emotional balance. Finding a healthy rhythm between staying informed and protecting mental well-being allows people to remain engaged without becoming overwhelmed.

The interplay of physical, emotional, and social dimensions underscores that climate change is not a problem external to humanity—it is also internal, reshaping how we feel, how we relate to each other, and how we envision the future. This realization can motivate more holistic approaches that address the crisis not just technologically or politically, but also psychologically and culturally. As research and experience deepen, communities can develop best practices for mitigating emotional harm, ensuring that the struggle against climate change includes compassion, understanding, and care for people's mental health and relationships.

In this evolving era, reading bodily stress signals, acknowledging emotional responses, and mending social strains become as crucial as devising renewable energy plans or drafting emission reduction targets. Environmental solutions that ignore human emotional well-being risk failing to garner the consistent support needed for long-term success. Conversely, by taking emotional and social factors seriously, societies can foster an engaged, resilient population better equipped to handle the uncertainties and changes ahead.

Identifying Subtle Signs in Our Bodies, Relationships, and Choices

In the modern world, people often think of anxiety as something that manifests clearly and unmistakably—racing heartbeats, clammy palms, restlessness, and nervous pacing. Yet when it comes to eco-anxiety, the feelings that arise in response to climate change often flow in quietly, like a gentle but persistent tide. They may not present as dramatic panic attacks or overwhelming dread, at least not initially. Instead, the signs can be subtle, woven into the fabric of daily life. A slight tension in the shoulders whenever the evening news covers another unusual weather pattern, a minor hesitation before planning a vacation that involves air travel, a fleeting sadness triggered by noticing a certain bird species less frequently—these small, almost imperceptible signals reveal the ways climate anxiety infiltrates everyday existence. Understanding these subtle signs can be a powerful step toward acknowledging the influence of environmental uncertainty on emotional landscapes and making more deliberate choices in how we respond.

The human body is often the first place these quiet stirrings of distress appear. The body is an instrument tuned by evolution to detect dangers and respond with stress hormones and physical readiness. Yet, climate change does not announce itself with the roar of a predator or the crash of immediate calamity; it broadcasts through abstract data points, scientific projections, and complex feedback loops. Even so, the body can internalize a sense of unease. Individuals may notice a slight tightness in their chest or shallowness in their breathing when confronted with a difficult environmental story. Perhaps they scroll through an article about vanishing coral reefs and find their jaw clenched. These bodily cues might seem unrelated to the topic at hand—maybe the person attributes the tension to a poor night's sleep or too many hours at the computer. Over time, though, patterns emerge. Each new climate-related stressor leaves a physical imprint, a mild discomfort or restlessness that accumulates.

Muscular tension is one common manifestation. People might hold their shoulders higher than usual, tense their forearms, or feel mild headaches that seem to come and go without a clear cause. Stomach discomfort or subtle changes in digestion can also appear. While none of these signs scream "climate anxiety," they become meaningful when correlated with exposure to environmental news or personal reflections on uncertain futures. Even sleep patterns can shift. Insomnia or troubled sleep might not follow a frightening news segment immediately, but over weeks or months, as environmental concerns settle into the subconscious, a person may find their rest less restorative. Dreams may grow restless, populated by vague worries, shifting landscapes, or unplaceable concerns that mirror ecological instability.

These bodily signs are subtle partly because climate threats are slow-moving and dispersed. The body does not flood with adrenaline in response to an abstract scenario decades down the line. Instead, it produces a chronic, low-level stress response, akin to background radiation of worry that never fully dissipates. This low-level tension can leave individuals more susceptible to other stressors, more easily annoyed at minor inconveniences, or more inclined to avoid situations that remind them of environmental uncertainty. By paying attention to these nuanced bodily cues, people can gain insight into how deeply climate issues have penetrated their psyche, even if they haven't consciously acknowledged feeling anxious.

Emotional subtlety is just as prevalent. Climate anxiety often masquerades as mild irritability, a vague sadness that arises at unexpected moments, or a heightened sense of vulnerability that's difficult to pinpoint. For example, someone might notice they've become more sensitive to certain phrases or jokes about the weather that once seemed harmless. Another might feel an inexplicable gloom after reading about a species facing extinction, even if they've never encountered that animal in person. These emotional undercurrents may not be dramatic, but they influence mood and can alter the tone of everyday interactions. The shift might be as slight as feeling less enthusiastic about future plans or noticing a pang of guilt when engaging in habits previously taken for granted—like driving a short distance instead of walking, or buying a product packaged in excessive plastic.

This emotional subtlety is partly due to the abstract nature of the climate crisis. Unlike immediate threats, which trigger fear and adrenaline, climate anxiety often induces a quieter form of distress. It might lean toward sorrow rather than panic, or manifest as a kind of moral unease—an internal tug-of-war between what one knows should be done for the planet and what seems feasible or convenient in daily life. Over time, these emotional flickers can accumulate. The occasional twinge of guilt, when repeated across multiple decisions—buying imported foods, taking an extra flight, ignoring opportunities to reduce waste—gradually intensifies. Eventually, a person may find themselves feeling generally uneasy about their lifestyle without clearly linking this unease to climate concerns. Recognizing the root of these feelings can empower individuals to address them constructively, whether by making small changes that align with their values or by seeking out supportive communities to discuss their worries.

Social realms—our relationships, interactions, and communal practices—are another area where subtle signs of climate anxiety emerge. Consider how conversations around future plans, homeownership, career choices, or having children are subtly shifting. Ten or twenty years ago, many people discussed such plans under the assumption of relative environmental stability. Now, subtle hesitations creep in. A couple might pause for just a moment longer when talking about the future: "Should we really settle near the coast?" "Is it wise to plan that long overseas trip if extreme weather events are becoming more frequent?" These hesitations may not spark immediate arguments or overt distress, but they thread uncertainty into everyday dialogues, indicating that climate concerns are shaping priorities and diminishing some of the carefree optimism once associated with planning ahead.

Friendships and family relationships can also pick up on these subtle tensions. Picture a group of friends planning a summer vacation. One person suggests a destination known for its natural beauty, but another hesitates, recalling recent reports of bleaching corals or drought-stressed forests in that region. They might feel discomfort voicing their concern, worried about being perceived as alarmist or spoiling the fun. They might say nothing, but their reluctance and lack of enthusiasm signal an underlying worry. Over time, recurring instances like this can strain social bonds. Some friends might roll their eyes at the "overly concerned" member, while that member might feel misunderstood or alone in their apprehensions. These rifts remain understated—no one may openly argue about climate anxiety, but the emotional gap is there, influencing group dynamics.

Community events tied to natural cycles or cultural traditions may also reflect subtle changes. Imagine a community known for its seasonal harvest festival, historically timed to coincide with a reliable moment in the agricultural cycle. If climate shifts alter that cycle—delaying or advancing harvest times—the community might adjust the festival schedule. This seems like a practical adaptation, but emotionally, it's a subtle sign that the world is not as it was. Attendees might feel a slight melancholy or a nagging sense that something is off, even if they can't pinpoint why. Over years, these small adjustments accumulate, creating a collective sense of unease. The community still celebrates, but the festival may feel less rooted, less certain, mirroring the collective anxiety that the stable future they imagined is now less guaranteed.

Subtle changes also emerge in personal choices. In consumer behavior, for example, someone might start feeling oddly uneasy selecting certain products at the store—excessively packaged goods or imported luxury items may now evoke a quiet discomfort rather than the straightforward pleasure they once did. The buyer might not stop purchasing these items immediately, but the seed of doubt is planted. Similarly, career decisions can shift. A recent graduate might feel drawn to environmental fields not out of a grand aspiration but because something nags at them that the world is changing and traditional careers tied to unsustainable industries might be risky bets. This sense of uncertainty may not explode into activism, but it influences the career path chosen, reflecting subtle anxieties internalized from broader environmental dialogues.

Another subtle sign might appear in what people avoid discussing. Silence can speak volumes. If a family once comfortably chatted about building a lakeside cabin for future vacations, now they might skirt around that topic, sensing that increasing drought or wildfires make that dream less secure. Without openly stating why, they shift conversations to safer ground. This avoidance points to unspoken fears. In workplaces, colleagues may avoid diving into long-term strategic plans if they involve relying on stable resource supplies. The lack of explicit mention doesn't mean no one is worried; it might mean everyone feels a low-level anxiety that makes them hesitant to commit to long-term visions. Over time, these avoidances hollow out certain narrative spaces, leaving future planning less robust, more cautious, and tinged with nervous restraint.

Identifying these subtle signs becomes easier when one pays mindful attention to emotional states and behavioral patterns. Keeping a journal might help individuals note when they feel inexplicably tense or sad, and link those feelings to recent environmental stories encountered. Observing changes in appetite, sleep quality, or exercise motivation can also yield clues. Maybe someone used to run outdoors for enjoyment, and while they still do, they now feel a slight heaviness in their chest when noticing signs of drought-stressed plants along the trail. These incremental shifts can be subtle keys, revealing that climate anxiety is whispering rather than shouting.

In relationships, small conflicts or misunderstandings might be traced back to this anxiety. If a couple disagrees on whether to buy a car or rely more on public transport, it might initially seem like a practical debate. But beneath practicalities, there could be emotional undercurrents—one partner feeling anxious about contributing to emissions, the other feeling that life's conveniences shouldn't vanish just because of abstract future concerns. Identifying these emotional layers helps prevent misattributing conflicts solely to personal differences. Instead, it acknowledges a shared stressor: environmental uncertainty. Doing so can foster greater empathy and open conversations that address the root cause—unspoken fears about the future—rather than letting frustration fester unexamined.

Social gatherings, too, can offer subtle cues. If discussions of climate news spark quiet discomfort or if certain friends always change the subject, it may indicate collective tension. While no one may declare "I am anxious about climate change," the group's dynamic reveals it—perhaps fewer people want to debate travel plans that involve carbon-heavy flights, or gift choices gravitate slowly toward more sustainable items without anyone making a big announcement about it. These shifts reflect the infiltration of climate anxiety into cultural norms, as people unconsciously adapt their preferences and behaviors to align with a world that feels less stable.

Even media consumption patterns can serve as subtle indicators. If a person once loved reading global travel magazines but now finds such content unsettling—imagining these far-flung destinations threatened by rising seas or extreme weather—that discomfort might show up as reluctance to subscribe or disinterest in travel documentaries. Similarly, someone might skip over certain environmental headlines to avoid that faint pang of dread. This selective attention signals an emotional coping mechanism, revealing anxiety at work even if externally everything appears calm.

In workplaces, changes might be reflected in how employees respond to corporate sustainability initiatives. If workers quietly welcome moves like reducing single-use plastics in the office kitchen or adopting greener procurement policies, their subtle relief at these measures hints that climate anxiety lurks in the background. Conversely, if employees resist such changes, it might indicate discomfort facing the moral implications of their company's environmental footprint. In either case, these workplace attitudes—mild approval, mild resistance, subtle hesitance—are emotional signposts pointing toward underlying environmental concerns.

Identifying subtle signs is not about pathologizing every nervous tick or attributing all emotional discomfort to climate anxiety. Humans experience stress from countless sources: personal health issues, financial struggles, relationship problems. Climate anxiety weaves into this tapestry rather than standing alone. But being aware of these influences helps differentiate between generic stress and stress arising from environmental instability. Someone noticing that their tension spikes specifically after reading climate reports or that their mood sours whenever extreme weather is discussed can start to understand the triggers and address them more directly.

This awareness can empower individuals and communities to take supportive steps. It might motivate someone to talk openly with a partner about fears for the future, rather than keeping them bottled up. It might prompt a workplace to host a brown-bag discussion on eco-anxiety, giving employees permission to voice their worries and brainstorm collective responses. It might inspire an individual to limit doomscrolling through alarming headlines and instead seek balanced sources of information that also highlight solutions and adaptation efforts. When subtle signs are recognized, proactive measures become possible, transforming latent anxiety into a driver for emotional management and constructive engagement.

On a cultural scale, as more people realize that these subtle emotional reactions are common, society can better integrate climate issues into mental health frameworks. Counselors and therapists may become trained to identify eco-anxiety not only when patients explicitly name it, but also when patients exhibit certain patterns—unexplained irritability after reading the news, minor sleep disturbances correlated with environmental triggers, or social withdrawal around topics linked to the future. Just as stress management techniques evolved in response to new workplace demands or technological pressures, eco-anxiety management might emerge as a recognized skill set, drawing upon mindfulness, narrative therapy, community support, and activism as routes to restoring emotional balance.

It's also worth noting that subtle signs can precede more overt expressions of fear or sorrow. They can act as early warning signals, giving individuals a chance to intervene before anxiety escalates. By tuning in to slight muscle tension, mild changes in mood, or small relational frictions, people can address their feelings sooner. Maybe they seek out an environmental book club for better understanding and camaraderie, or start volunteering at a local conservation project to channel anxiety into meaningful contribution. Early interventions can prevent more severe stress responses down the line, ensuring that climate anxiety remains manageable rather than overwhelming.

This process of recognition can also have a moral dimension. Realizing that even small emotional ripples stem from the planet's predicament may inspire more compassionate views toward oneself and others. Instead of feeling confused by one's own unease, a person can acknowledge that these sensations reflect empathy and concern for a world in flux. Instead of growing impatient with a friend who sidesteps climate discussions, one might understand that the friend could also be struggling internally. This understanding paves the way for kinder, more supportive interactions and reduces the stigma around discussing environmental fears.

In sum, identifying subtle signs of climate anxiety involves paying closer attention to the body's stress signals, the emotional colorations of everyday moments, and the nuanced shifts in how people communicate, socialize, and decide. From minor physical tension to faint mood dips, from small relationship frictions to quiet changes in consumption habits, all these hints collectively indicate that environmental uncertainty has seeped into the emotional fabric of daily life. Recognizing this infiltration allows for early, more gentle interventions—like talking openly with trusted friends, seeking supportive communities, adopting balance in media consumption, and taking small climate actions that restore a sense of agency and hope.

This attunement to subtlety also encourages flexibility and resilience. Instead of waiting for a crisis or a personal breakdown triggered by some catastrophic event, individuals and communities can begin adapting to the psychological challenges of climate instability now. They can learn from each other's experiences, share coping strategies, and gradually shift cultural norms to acknowledge and address these delicate emotional undercurrents. Over time, this greater self-awareness contributes to collective emotional health. People become less mystified by their reactions, less prone to denial, and more willing to channel worry into productive endeavors.

Identifying subtle signs is thus not a trivial exercise; it is a form of emotional literacy adapted to a changing world. Just as reading subtle nonverbal cues improves interpersonal communication, reading subtle environmental anxiety cues improves our relationship with the planetary context. Through this awareness, societies can cultivate understanding and compassion, encouraging proactive responses before anxieties escalate or harden into despair. It also contributes to a more holistic approach to climate engagement, where rational understanding of the issues merges with emotional intelligence, enabling a healthier, more integrated response to environmental change.

These subtle signs, in the end, remind us that environmental conditions are not abstract data points distant from personal life. They are felt physically, emotionally, and socially, even if faintly. By identifying these signs, we learn that climate instability is not a remote challenge to be filed under "future problems," but a current influence shaping hearts and minds. This realization can spur individuals to seek balance—neither plunging into panic nor numbing themselves to reality, but acknowledging that even small emotional signals warrant attention. With this awareness, we can step more lightly, speak more kindly, and plan more thoughtfully, honoring both the complexity of the climate crisis and the subtle ways it already shapes who we are and what we do.

Understanding That Even "Small" Anxieties Stem from Larger Changes

It's easy to think of climate anxiety as something that either hits hard—full-blown panic, despair, or rage—or barely registers at all. Yet this emotional spectrum is far more nuanced. Long before fear becomes overwhelming, or sorrow deepens into genuine grief, subtle hints of unease and moral discomfort often sprinkle day-to-day life. These small emotional ripples reflect not isolated quirks of mood, but rather the early, gentle signals that our internal worlds are responding to external shifts on a planetary scale. In other words, even the quietest flickers of anxiety about the environment can be understood as meaningful indicators of broader transformations quietly reshaping human expectations, behavior, and culture.

Consider a person who never explicitly identifies as anxious about climate change, yet finds themselves pausing for a split second before purchasing a cheap plastic trinket. Last year, they might not have hesitated; this year, after reading sporadic headlines about plastic pollution suffocating marine life, they feel a small but perceptible tug at their conscience. The moment is brief: a slight catch in the breath, a fleeting sense of discomfort. They might still buy the item, rationalizing that it's just one small thing. But the emotional trace remains. This subtle unease, although minor, is not random. It emerges from the underlying knowledge that something larger is happening in the world—something that questions the acceptability of previously unexamined habits.

Such experiences can manifest in countless tiny ways: a dimming of enthusiasm when planning a distant vacation involving multiple flights, a slight tension in the shoulders upon reading a social media post about yet another extreme weather event, a vague guilt when ordering out-of-season produce shipped from far away. Initially, these feelings may be so faint that individuals barely register them as climate-related. Perhaps they attribute a restless night's sleep to stress at work, not connecting the dots that earlier in the day they'd read an unsettling report about tropical forests reaching tipping points. Over time, as these small anxieties accumulate and coincide with environmental triggers, a pattern emerges. They reveal themselves as part of a subtle emotional response to living in an era of environmental instability.

This subtlety often results from the abstract, long-term nature of climate threats. Unlike immediate dangers that provoke visceral alarms, climate change unfolds slowly and globally. Because it's not an immediate predator or a sudden disaster—instead presenting as long-term shifts in weather patterns, biodiversity loss, and resource scarcity—the body and mind do not explode into clear panic. Instead, they respond with gentle nudges, low-level tensions, and mild worries that slip into daily consciousness. Over time, these mild responses can mold attitudes, shape choices, and influence how people view the future without ever crescendoing into obvious distress.

Small anxieties also reflect the difficulty of translating massive global issues into personal emotional language. It's challenging to maintain urgent fear day after day for something that might fully manifest decades hence, or mostly affect distant regions. Instead, one experiences small pangs of sadness reading about polar bears struggling to find ice platforms, or a twinge of frustration overhearing friends mock the seriousness of climate action. These micro-emotional responses act as emotional "tells," exposing that the mind, at some level, has registered that stability is under question. Each minor pulse of anxiety can be seen as a data point, a piece of evidence that we're adapting psychologically to a new ecological context even if we haven't consciously decided to feel concerned.

Such minor anxieties often arise at the intersection of personal behavior and moral knowledge. Many individuals carry mental notes of what they've learned: burning fossil fuels contributes to warming, deforestation accelerates biodiversity loss, plastic pollution chokes marine life. Consciously or not, this knowledge can cast a subtle shadow on decisions. The next time one fills the car's gas tank, a flicker of guilt may surface—slight, not crippling, but undeniably there. Buying clothing from fast fashion outlets might now come with a glimmer of discomfort, recalling that garment production often carries environmental costs. Although these feelings might not stop the transaction, their presence signals that moral cognition is blending with environmental awareness, shaping an internal narrative that even seemingly benign actions have broader implications.

Over time, this slow accretion of small anxieties can lead to incremental changes in behavior. The person who once dismissed the idea of bringing their own cup to a coffee shop might now do so, not because they had a major epiphany or breakdown, but because a series of small anxious nudges made it feel increasingly awkward not to. This shift might appear minimal from an outside perspective—just a single reusable cup—but it results from the internalization of concerns too subtle to be named as outright anxiety yet powerful enough to steer habits. Each slight worry about plastic waste, each minor cringe at the thought of overflowing landfills or ocean gyres, contributed to a gentle moral realignment.

The collective impact of these small anxieties can ripple through communities and markets. If many customers feel a mild discomfort purchasing products with excessive packaging, they may eventually gravitate toward brands offering minimal waste. If enough people make such choices, even marginally, retail and production practices adapt. Thus, subtle anxieties, while individually small, can aggregate into cultural shifts. Over time, what was once normal (like rampant single-use plastic) may become socially frowned upon, not because of a grand moral revolution announced by everyone simultaneously, but through the slow build-up of countless minor uneasy moments experienced by many individuals.

These small anxieties also shape conversations, even when not explicitly mentioned. Suppose a family once dreamed of a big retirement home near a vulnerable coastline. Now, they might start deferring that plan, joking lightly about sea-level rise, not pressing the point. This hesitation or avoidance might never become a serious debate. Yet the subtle emotional friction—a spouse's faint reluctance to commit, the children's gentle teasing about future floods—alters the family narrative. Over years, the dream is quietly retired or adapted to a safer location, reflecting how small anxieties influenced major life decisions without ever exploding into dramatic conflict.

In workplaces, such minor discomforts can guide professional choices. An employee might feel slightly uneasy whenever their company announces a partnership with a known polluter. Initially, the employee says nothing, but a residue of dissatisfaction collects. Over months or years, this slight moral dissonance might inspire them to seek jobs at greener companies, or to suggest sustainability initiatives at their current workplace. While at first this anxiety never reached a level of intense distress, it nudged the person along a path of incremental ethical alignment. When multiplied across many employees, small anxieties contribute to corporate cultures gradually leaning towards sustainable practices or facing internal pressure to reduce their environmental footprint.

There is also a generational dimension to these subtle signals. Younger people, who grew up amid constant climate discourse, may internalize low-level anxieties from an early age. Instead of experiencing a stark moment of realization, they absorb mild worries as part of their normal emotional baseline. They may find it odd that older generations fail to show similar concern, while older adults who lived in times of presumed environmental stability might register small anxieties as a new phenomenon, adding a mild generational tension. These differences in emotional starting points shape how societies evolve culturally, with younger cohorts more readily adapting their lifestyles in response to climate cues, even if their motivations never crystallize as fear but remain at the level of quiet discomfort.

Geography can amplify or modulate these minor anxieties. Someone in a coastal region may feel subtle worry each time a new property development goes up perilously close to the shoreline, knowing that rising seas will eventually pose a challenge. Those in drought-prone areas might grow used to feeling a slight twinge of concern with each short shower or casual mention of water use. No one may call it panic, but the steady hum of low-grade worry underscores the changed environmental context. Over time, regions with more visible climate impacts may breed more socially shared small anxieties that guide local norms—from discouraging lush green lawns to quietly accepting that certain foods are now harder to grow or procure locally.

Another subtle sign involves emotional labor related to climate information intake. Some people find themselves skimming fewer environmental articles or tuning out discussions on new climate reports, not because they are indifferent, but because each exposure provokes a faint sadness or worry. Rather than confront these emotions directly, they choose to limit their exposure to environmental news. The result is an avoidance behavior that might look like apathy but actually reflects a low-level anxious feeling that reading too much about climate change is emotionally taxing. This avoidance can prevent people from staying fully informed, creating a cycle where low-level anxiety encourages partial disengagement, thus complicating rational decision-making and civic participation.

Interpersonal relationships often serve as mirrors, reflecting small anxieties back to those who experience them. A friend might notice that someone bristles slightly whenever local politicians dismiss environmental measures. Although the bristle is subtle—maybe a tightened expression or a brief cessation of eye contact—it signifies emotional investment. If the friend gently inquires, "Why does that bother you?" the person may discover that indeed, these small anxieties carry deeper significance. Such conversations help name and validate what previously felt amorphous, enabling both parties to better understand how environmental issues permeate their emotional lives. Over time, acknowledging these slight reactions can strengthen bonds, as honesty and empathy replace silent discomfort.

Cultural artifacts—films, novels, art exhibits—often capture and magnify these subtle emotional states, providing language or imagery for what individuals feel but cannot articulate. A novel featuring characters who experience quiet unease about eroding landscapes resonates with readers who share those feelings, even if they never labeled them as climate anxiety. This cultural reflection reinforces that small anxieties are neither rare nor meaningless; they are part of a shared human response to instability. Engaging with such art can legitimize these feelings, reducing any sense of personal strangeness or inadequacy.

One might ask: If these anxieties are so subtle, why not simply ignore them? Why pay attention to minor discomforts that don't seriously disrupt daily functioning? The answer lies in what these feelings represent. They are like a soft alarm bell ringing in the background, alerting people that the world is no longer as it was. Ignoring them might preserve short-term emotional comfort but at the cost of missing opportunities for growth and adaptation. Just as noticing slight changes in one's health can prompt preventative care, noticing slight changes in emotional reactions to environmental cues can prompt proactive engagement—seeking information, joining community initiatives, adjusting personal habits, or supporting policy measures that address underlying ecological problems.

Small anxieties can guide moral evolution. Imagine a person who previously never thought twice about their carbon-intensive vacations. Now they feel a slight, recurring pang each time they book a flight. This repetitive mild discomfort can accumulate and eventually motivate seeking lower-impact travel options. Thus, small emotional signals, repeated over time, can catalyze moral shifts that might never have occurred if the person waited for a dramatic emotional breakdown. Incremental moral progress often grows from the fertile ground of subtle emotional tensions.

The interplay between personal life and global systems is also illuminated by these small anxieties. They show how even mundane acts—choosing a product, planning a trip, tossing out leftovers—intersect with global environmental conditions. Without overtly thinking "I am worried about climate change," a person's emotions and micro-choices hint that deep down, they recognize their actions contribute to or mitigate a larger problem. Thus, micro-anxieties can serve as emotional signposts, pointing to the interconnectedness of personal behavior and planetary health. This interconnectedness, though complex, becomes more accessible to the psyche through these incremental feelings rather than through abstract intellectual understanding alone.

In social movements, these slight anxieties may lay the groundwork for collective mobilization. Before people flood the streets in climate marches, they often experience a build-up of concern that begins with small worries. When many individuals share these minor anxieties, they create a reservoir of emotional energy that can be activated by a trigger—such as a particularly stark climate report or a visible environmental disaster. Without these underlying murmurs of disquiet, mass action might be harder to spark. Small anxieties thus function as a preliminary emotional infrastructure, ready to amplify into broader action when conditions demand it.

For some individuals, small anxieties reflect a mismatch between personal values and current lifestyle or societal conditions. These moments of tension—like feeling uncomfortable buying something known to have a high carbon footprint—signal that their values (caring for the planet, ensuring a stable future) conflict with habitual practices. Recognizing this conflict is the first step in resolving it. Instead of dismissing discomfort, using it as a prompt for reflection can lead to conscious choices that bring one's actions more in line with personal ethics. Such ethical refinement wouldn't occur if the person ignored these small hints.

In dialogues about the climate crisis, small anxieties can also function as a bridge for empathizing with those who may not outwardly declare themselves concerned. Just because a friend or coworker does not join environmental protests or speak passionately about climate policy doesn't mean they feel nothing. They might be experiencing subtle anxieties they don't fully understand or know how to express. Being sensitive to such possibilities fosters gentler conversations. Instead of accusing someone of indifference, one might explore whether they've ever felt a bit uneasy about certain environmental issues. This approach can open pathways to shared concern and collaborative thinking, rather than confrontation.

Recognizing small anxieties also helps counter arguments that climate concerns are exaggerated or alarmist. Some may claim that if the problem were real, everyone would be in panic mode. However, understanding that emotional responses to climate change often emerge as minor discomforts and moral twinges refutes this notion. People's emotional alarm systems simply do not always blare at full volume. By detailing how climate anxiety typically begins as subtle signals, one can argue that widespread mild unease and incremental shifts in behavior are, in fact, early indicators of a collective reckoning with environmental reality. Far from showing that fears are baseless, these subtle feelings confirm that something is nudging human consciousness.

Certain coping strategies specifically address these mild anxieties. Mindfulness can help individuals tune in to their subtle feelings, acknowledge them without judgment, and understand their origins. Talking with friends about small discomforts—like feeling odd about a certain purchase or noticing tension when discussing future vacations—can normalize these experiences and transform them into starting points for learning. Small anxieties may also be soothed by proactive environmental gestures—planting a garden that supports pollinators, attending a local sustainability workshop, or writing a letter to a representative about environmental policies. Each positive action can relieve a bit of the tension, reaffirming that even though threats are large and complex, personal contributions matter.

Over time, as these low-level anxieties accumulate and find expression in personal actions, cultural narratives, and policy pressures, society may move through phases of emotional adaptation. Initially, subtle anxieties feel like outliers, barely noticeable. Later, as more people recognize them, these emotions become common knowledge—no one is shocked to hear their neighbor voice quiet worries about future water availability or their sibling's hesitation about long-haul flights. Eventually, when such feelings become ubiquitous, the collective response may evolve into clearer political demands, stronger community initiatives, and long-overdue structural changes. The trajectory from small anxieties to large-scale adaptation reminds us that human emotional ecosystems evolve gradually, shaping collective futures as they do.

In understanding that small anxieties stem from larger changes, we appreciate that climate change's psychological footprint is not confined to dramatic episodes of fear or despair. Instead, it infiltrates everyday life gently, steering preferences, nudging moral reflections, and quietly altering conversational dynamics. People do not wake up one day declaring that their worldview is shattered by climate instability; rather, they gather fragments of discomfort, piece by piece, until a new picture emerges of a world no longer as certain or stable as once believed.

This recognition can reduce the sense of isolation or confusion some may feel. If one experiences faint worry at choosing imported strawberries, they are not being overly sensitive or irrational. They are reacting to a deeply rational understanding that global supply chains and agricultural systems face pressures that may impact availability, quality, or ethical sourcing in the future. Similarly, if someone feels a pang of guilt when driving a short distance, that pang arises from the knowledge that small emissions, multiplied billions of times, matter. These subtle moral tugs, while individually slight, weave together into a moral and emotional tapestry reflecting humanity's growing awareness of its ecological footprints.

In realms such as education and mental health support, acknowledging these small anxieties is vital. Teachers might understand that students' quiet questions about future careers or stable environments stem not just from personal ambition but from the looming presence of environmental uncertainties. Mental health counselors could learn to ask about subtle anxieties related to climate, gently guiding clients to connect those minor emotional strains to a larger narrative. Doing so demystifies the experience and empowers individuals to navigate their emotions more confidently and with greater self-compassion.

Economists, policymakers, and strategists might also glean insights from these subtle emotions. Understanding that low-level discomfort is widespread can inform more nuanced climate communication campaigns, emphasize incremental policy changes with visible short-term benefits, and create incentives that align with emerging values. Rather than waiting for public opinion to become vocally alarmist, decision-makers can act on these emotional undercurrents, recognizing them as early signals that citizens are prepared for change, even if they haven't yet staged massive protests or demanded radical shifts.

In personal relationships, couples or friends who sense minor anxieties creeping into discussions about future plans can address them openly. A partner who notices their spouse's hesitance about booking a certain trip can inquire compassionately, "Do you feel uncertain because of environmental issues?" Instead of brushing off tension, naming it builds trust and understanding. Over time, couples who acknowledge these small anxieties can collaboratively adjust their expectations—maybe choosing closer vacation spots or investing time in learning about eco-friendly travel options. These small adjustments not only ease emotional strain but also contribute to incremental cultural shifts where sustainable choices become more normal and less fraught.

Some may worry that by highlighting subtle anxieties, we risk amplifying them, turning faint worries into larger fears. Yet acknowledging them can have the opposite effect. When understood and discussed, mild anxieties lose some of their mystique. They no longer lurk as unarticulated unease. Instead, they become understandable responses to a complex reality, guiding thoughtful action or reflection rather than festering in silence. This process can mitigate the risk of small worries escalating into deeper distress. It also helps people accept that living in a changing world naturally introduces emotional complexity, and that grappling with it is part of adapting psychologically and culturally to new ecological conditions.

Furthermore, recognizing these minor signals can enhance resilience. Just as small changes in ecosystems—like the early arrival of spring or unusual bird migration patterns—can inform researchers that larger environmental shifts are underway, small anxieties can inform individuals that they need to develop coping strategies, join supportive networks, or seek reliable information. Early detection of subtle emotional shifts prompts earlier interventions. Whether it's setting limits on doomscrolling, attending a local climate workshop, or talking to loved ones about shared concerns, people who respond to these early cues can build emotional strength before adversity escalates.

While these subtle anxieties stem from larger changes, their impact is not confined to reinforcing helplessness. On the contrary, once identified, they can spur a sense of agency. The recognition that everyday choices connect to global challenges can inspire commitment to small but meaningful acts—like adopting more sustainable products, supporting local environmental initiatives, or writing letters to politicians. Such steps can restore a measure of moral coherence and emotional equilibrium, as individuals learn that even modest efforts matter in a collective journey towards sustainability.

In addition, acknowledging these small anxieties expands the definition of what it means to be environmentally engaged. Not everyone responds with overt activism. Some show their engagement through micro-emotional cues that shape their values and decisions. By appreciating that subtle apprehensions are a form of climate consciousness, we value quiet forms of eco-awareness alongside more outspoken activism. This inclusivity fosters a richer tapestry of public engagement—where extroverted marchers and more reserved individuals, both guided by small and large anxieties, work towards a common goal. Each person's emotional landscape, no matter how understated, contributes to the broader momentum for change.

There are also intersections with other societal shifts. As green technologies improve and local communities experiment with regenerative agriculture or renewable energy, people's small anxieties can gradually diminish. Seeing concrete solutions and adaptations that align with low-level environmental concerns can reassure individuals that their subtle fears are being addressed. Over time, when sustainable choices become cheaper, more convenient, and more culturally embraced, these small anxieties might evolve into a sense of cautious optimism or calm determination, reflecting that the world is indeed responding.

In fields like marketing and product design, acknowledging subtle anxieties can prompt companies to innovate responsibly. If many customers feel a tiny pang of discomfort buying single-use plastics, designers might create more appealing, low-impact alternatives. This quiet feedback loop—consumers feeling small anxieties, companies responding with better products—helps reshape entire supply chains without needing everyone to shout alarm. The moral and emotional markets respond to whispered worries as effectively as to shouted demands, given enough time and sensitivity.

Finally, these small anxieties serve as a reminder that human psychology, shaped by evolutionary constraints and cultural traditions, is flexible. We are capable of perceiving global changes even when they unfold gradually. If we weren't, these subtle signals wouldn't emerge. Our ability to feel mild unease at a distant environmental threat testifies to a latent capacity for long-term empathy and foresight. With practice and support, we can amplify the constructive sides of these emotions—using them as gentle prompts to seek knowledge, foster community, advocate for policy reforms, and reorganize social norms around caring for the planet.

In sum, while it might be easier to focus on glaring examples of climate anxiety—full-blown panic or despair—the reality is that countless individuals experience milder, almost invisible forms of environmental unease daily. Rather than dismissing these feelings, recognizing them illuminates the subtle ways environmental upheaval has seeped into ordinary life. Each minor worry, brief hesitation, or quiet twinge of guilt is a clue that we're all engaged in a gradual learning process. Through acknowledging these small anxieties, we can better understand the psychological shifts underway, respond thoughtfully, and ultimately navigate our changing world with sensitivity and grace.

This chapter has traced the subtle, pervasive influence of environmental instability on minds, bodies, and social structures. From faint physical tensions to mild but persistent emotional unease, and from small relational frictions to discrete shifts in personal and communal choices, these varied responses reveal that climate change is not a distant abstraction. Instead, it flows through every layer of daily life, registering as stress in the muscles, heaviness in the heart, and quiet worry in the marketplace of ideas. Individuals find themselves altering habits almost without realizing it, communities adjust traditions and festivals, workplaces face new moral considerations, and relationships incorporate new layers of uncertainty. By recognizing that even minor anxieties and tensions stem from larger global changes, we see that environmental challenges have become intertwined with personal well-being, moral reasoning, and cultural identity. This awareness invites a more empathetic understanding of ourselves and each other, encouraging supportive dialogues, incremental adaptations, and renewed efforts to align human actions with planetary needs.

Chapter 4: Context from the Past and Around the World

Historical Examples: How Past Societies Coped with Environmental Upheavals

Human beings have long inhabited dynamic environments, and while modern society wrestles with the emerging complexities of climate change, past civilizations also faced daunting ecological shifts. Their responses—ranging from adaptation and innovation to migration, cultural evolution, and in some cases, collapse—offer a trove of lessons. Understanding how earlier peoples confronted environmental upheavals can provide perspective, shedding light on humanity's capacity for resilience, foresight, cooperation, and, at times, tragic miscalculation. Although historical conditions differed from today's globalized, industrialized era, these earlier narratives remind us that environmental stressors have always demanded tough choices and moral reckonings, prompting individuals and groups to reconsider values, resource management, governance, and community ties.

One early example lies along the banks of the Nile, where ancient Egyptian civilization flourished for millennia thanks in large part to the predictability of annual floods. These floods replenished farmland with nutrient-rich silt, enabling stable agriculture. Yet even in this seemingly stable environment, periodic episodes of lower floods, prolonged droughts, or sudden climate anomalies tested Egyptian resilience. In response, the state and religious institutions developed robust granaries, planned irrigation systems, and stored surplus grain to buffer against lean years. Priests and leaders interpreted environmental changes through religious narratives, forging cultural frameworks that reassured the populace that environmental variances were part of a cosmic order. Though not immune to hardship, Egyptians mitigated environmental crises through centralized coordination and social cohesion, demonstrating that understanding riverine rhythms, bolstering food security, and maintaining cultural unity could help a society endure centuries of fluctuating conditions.

Turning to the Fertile Crescent, we find ancient Mesopotamian city-states that rose and fell in tandem with shifting rainfall patterns and river courses. Early irrigation systems turned deserts into productive farmland, supporting dense urban centers. However, these irrigation works required constant maintenance and careful water management. Salinization of soils, driven by over-irrigation and insufficient drainage, gradually undermined crop yields. Rulers and communities attempted to cope by rotating crops, seeking new canals, or relocating fields. Some adjustments worked temporarily, but in the long run, ecological degradation contributed to political instability and migrations of populations seeking more hospitable lands. These complexities show that while engineering marvels can temporarily offset environmental challenges, ignoring underlying ecological limits invites cascading problems that simple technical fixes cannot always resolve. The Mesopotamian experience underscores that even highly organized societies must respect ecological boundaries, or else face a slow unraveling of their agrarian base.

Across the Atlantic centuries later, the Maya civilization thrived in a region of what is now southern Mexico, Guatemala, and Belize, mastering the art of agriculture in tropical forests. For many centuries, the Maya built complex city-states with monumental architecture, sophisticated calendars, writing systems, and trade networks. Their society was not static; it experienced various environmental pressures, including droughts that intensify periodically. Archaeological and paleoenvironmental studies suggest that episodes of prolonged drought played a significant role in the so-called "Classic Maya collapse," a centuries-long process involving the abandonment of many major urban centers. Faced with dwindling water sources, Maya communities tried to adapt by constructing reservoirs, shifting agricultural practices, and in some areas, resettling. But when climate stressors overlapped with social inequalities, warfare, and overexploitation of resources, the resilience of these once-flourishing city-states waned. The Maya did not simply vanish; many descendants survived, adapting through smaller-scale social organizations. Still, their story illustrates how intricate societies reliant on delicate environmental balances can fracture when faced with persistent climatic adversity, especially if power struggles and uneven resource distribution prevent coherent responses.

In contrast, the Norse settlements in Greenland during the medieval period highlight another dimension of adaptation—or the lack thereof. Around the late 10th century, Norse colonists established farmsteads in Greenland's marginal environment, thriving initially on livestock grazing and trade in valuable walrus tusks. For centuries, they managed to maintain their European-style lifestyle in a subarctic setting, but as the climate shifted during the Little Ice Age—bringing colder temperatures, shorter growing seasons, and pack ice that interfered with shipping routes—the Norse failed to significantly adjust their cultural and dietary habits. Unlike the Indigenous Inuit populations who relied heavily on marine resources, the Norse clung to pastoralism and exported luxury goods. Over time, dwindling trade connections with Europe, reduced forage for cattle, and harsher winters eroded their food security and social cohesion. Eventually, the settlements disappeared. Their fate suggests that cultural rigidity and reluctance to learn from local ecological knowledge can spell disaster when environmental conditions no longer match ancestral norms. Adaptation requires openness to change, not only technologically but culturally and morally.

Half a world away, societies in East Asia navigated environmental fluctuations in their own ways. In medieval and early modern China, for instance, droughts and floods frequently posed existential threats to agrarian communities. Dynasties rose and fell amid cycles of climatic instability. Yet Chinese states invested heavily in large-scale infrastructure—canals, dikes, and granaries—plus bureaucracies dedicated to disaster relief. Confucian philosophical frameworks encouraged rulers to care for their subjects, reinforcing moral duties to store grain, regulate markets, and distribute relief in tough times. While these measures did not eliminate the hardships of environmental upheavals, they limited their worst impacts. The moral imperative for just governance in the face of natural challenges helped legitimize state interventions and fostered a social contract attuned to ecological stresses. This interplay between political responsibility, moral values, and environmental risk management demonstrates that cultural ethics can drive institutional preparedness, easing the emotional and social toll of ecological uncertainties.

Elsewhere, on the islands of the Pacific, Indigenous communities developed intricate traditions to cope with variable climates and limited resources long before outside contact. Polynesian and Micronesian navigators honed their understanding of winds, currents, and seasonal cycles, enabling them to voyage across vast ocean distances and find new islands when local conditions grew unsustainable. Many island societies employed sophisticated techniques like agroforestry, fishpond aquaculture, and taro terraces to maintain stable food supplies despite shifting rainfall and tropical cyclones. These communities embedded environmental knowledge into oral traditions, rituals, and taboos that governed resource use, ensuring sustainability was reinforced by cultural norms. Although European colonization and globalization disrupted these systems, the resilience encoded in their traditions underscores the power of cultural adaptation and communal memory in facing environmental uncertainty.

Another case worth considering is the Ancestral Puebloans of the American Southwest, who inhabited regions in what is now Colorado, New Mexico, Arizona, and Utah. Over centuries, these communities built cliff dwellings and managed fragile environments prone to drought. They developed techniques for water conservation, soil retention, and mixed agriculture—corn, beans, squash—adapted to unpredictable rainfall patterns. Cultural practices likely included sharing and redistributing surpluses, reinforcing social ties that provided mutual aid in lean years. Yet repeated droughts and the longer cycle of aridity around the 13th century tested these systems to their limits. Many Puebloan peoples migrated, relocating to more reliable water sources and reorganizing their communities. While some settlements were abandoned, descendants survived by adopting more dispersed and mobile strategies. Their story shows that resilience can involve flexibility: when confronted with insufficient rainfall and environmental stress, communities can opt to move or redefine their patterns of living, rather than clinging to past arrangements.

These historical examples highlight that environmental upheavals are not new challenges. Yet we must acknowledge differences between then and now. Past societies confronted mostly regional or localized climate shifts, not a globally interconnected crisis driven by industrial emissions. Their technologies, population densities, and energy systems were incomparable to modern complexities. Still, their experiences hold insights. For one, they prove that societies often recognized environmental instabilities—through failed harvests, migrations of prey, salinized fields, or shifting rainfall—and attempted varied coping mechanisms: diversifying economies, building infrastructure, migrating, or forging cultural narratives that encouraged cooperation, sacrifice, and ethical stewardship.

Moral dimensions surface repeatedly. In numerous cultures, resource allocation and responses to scarcity triggered reflections on justice, fairness, and communal duties. In situations where elites monopolized resources or neglected vulnerable populations, climate stresses accelerated social breakdown. Conversely, where moral frameworks and governance structures fostered equitable distribution of food or invested in shared infrastructure, societies stood better chances of weathering environmental disruptions. Thus, moral and ethical considerations often lay at the heart of environmental resilience. While ancient societies lacked the scientific data or global treaties of today, they still grappled with questions of responsibility: who should store grain, who should decide when to allocate water, how to enforce rules that prevent overhunting or deforestation?

Another repeated theme is the importance of cultural flexibility. Civilizations that adapted their lifestyles, diets, settlement patterns, and even belief systems to changing ecological conditions fared better than those that insisted on maintaining familiar ways. Cultural rigidity—like the Norse in Greenland unwilling to emulate Inuit survival techniques—proved perilous.

Conversely, Polynesian navigators who could sail to distant atolls, or Indigenous peoples who employed rituals to moderate resource use, benefited from cultural repertoires that embraced environmental variability. Such versatility suggests that current societies, facing global climate change, might find strength in cultural creativity and openness to adopting knowledge from other traditions or from local ecosystems.

Memory and knowledge transmission also stand out as crucial. In many traditional communities, elders passed down ecological wisdom accumulated over generations. These memories guided people through cyclical changes, ensuring that responses to droughts, floods, or plagues were not improvised but informed by precedent. While historical lessons were sometimes insufficient for unprecedented conditions, they often offered a baseline for comparison. Modern civilization, by contrast, might rely on scientific models rather than folklore, but the principle remains the same: learning from past experiences—both successes and failures—enhances our ability to anticipate and mitigate future challenges. The difference now is that we have access to global archives, archaeological data, and interdisciplinary research that can illuminate how and why certain responses worked.

Another dimension is the emotional and psychological resilience displayed by past peoples. Although we lack transcripts of their inner thoughts, the endurance of communities through droughts, resource collapses, or climatic cooling implies a capacity to handle stress and uncertainty. Their cultural narratives, religious beliefs, and communal rites likely helped process fear and sadness. Whether the Egyptians framed poor harvests as tests from the gods, or the Maya carved reliefs depicting cosmic cycles, these symbolic frameworks allowed individuals to place environmental upheavals in a broader moral or cosmological context. This helped transform anxiety into motivation—rebuilding irrigation canals, redistributing surplus grains, or seeking alliances to secure resources. Though we cannot directly compare ancient mental health strategies to modern therapy, we can admire their ingenuity in making sense of and coping with a precarious world.

The variety of historical examples also highlights that no single strategy guaranteed success. Sometimes grand engineering works like Mesopotamian irrigation systems enabled centuries of prosperity before collapsing under ecological strain. Other times, minimalist strategies—such as the mobility of nomadic groups—outlasted more complex, sedentary civilizations when conditions changed radically. Political structures mattered too. States or chiefdoms that facilitated collective action, stored resources, and enforced fair rules had a leg up over fragmented societies where environmental stress led to conflict or disintegration. Thus, resilience emerged not from any single factor but from an interplay of technology, culture, ethics, and governance.

For contemporary readers, the diversity of past experiences can foster humility and a willingness to learn from different traditions. While our global crisis differs in scale and nature—fueled by industrial emissions and affecting every corner of the planet—the underlying human challenge of responding to environmental instability remains. Acknowledging that past societies also struggled, adapted, or collapsed under environmental pressures sets a realistic context. We are not the first to face uncertainty; we stand in a lineage of trial and error. This perspective may calm some of the existential dread by reminding us that humans have long been problem-solvers who modify their practices, sometimes creatively, to meet new realities.

At the same time, these historical precedents underscore the moral gravity of today's choices. Past societies often had smaller populations and less capacity to damage global systems. Their environmental problems were more localized, and even their failures affected fewer people. Today, billions depend on stable climate conditions for food security, infrastructure reliability, and geopolitical harmony. Knowing how high the stakes are—given our large populations, sprawling economic networks, and advanced technologies—amplifies the importance of learning from history. If limited resource management and cultural wisdom helped ancient communities endure droughts or storms, what might we achieve with the benefit of global knowledge, scientific forecasting, and unprecedented resource exchange, provided we muster the political and moral will?

Another valuable insight is that environmental upheavals often accelerated social and cultural evolution. Past crises sometimes led to new religious movements, political reforms, or economic innovations that ultimately stabilized societies in altered conditions. For instance, after periods of instability, some civilizations reorganized their resource distribution or adopted new crops better suited to changing climates. The idea that adversity spurs adaptation could inspire modern policy frameworks that embrace agile governance, diverse energy portfolios, and flexible social safety nets. The past suggests that rigidity leads to vulnerability. Accepting change and responding creatively to ecosystem signals is a time-tested survival strategy.

Yet, not all historical lessons are comforting. Some societies failed to adapt, leaving ruins for archaeologists to ponder. Easter Island is a frequently cited example, where extensive deforestation and resource depletion—exacerbated by isolation and limited trade—contributed to social strife and population decline. While debates continue about the exact nature and timing of these changes, the general narrative warns of overshoot and collapse when ecological limits are ignored. Similarly, certain drought-related collapses in the American Southwest or the Southeast Asian Angkor state illustrate that no complex society is immune to environmental constraints. These cautionary tales underline the urgency of our global predicament: ignoring signals or delaying action can carry dire consequences.

From a psychological standpoint, knowing that others before us confronted environmental uncertainties can provide some emotional anchoring. The anxiety we experience—whether mild or severe—has echoes in the worries of ancient farmers fretting over erratic rains, or coastal fishermen fearing shifting currents. Humans have always expressed concern over unpredictability and sought ways to fortify their well-being. Our current crisis may dwarf earlier ones in complexity, but it also offers us richer tools for understanding and responding. We can integrate lessons from multiple past societies, blending their strategic responses with modern science and human rights frameworks.

In practical terms, studying historical examples might encourage a return to practices that proved resilient in the past. Agroforestry, diversified crop rotations, or community-based water management have deep roots in pre-industrial civilizations that understood how to maintain fertility and buffer against drought. Re-learning these techniques can complement cutting-edge innovations, producing a hybrid approach where traditional ecological knowledge merges with advanced technologies. This synergy can yield robust adaptation strategies less vulnerable to the pitfalls of purely modern, industrial solutions that often externalize costs or degrade ecosystems.

Crucially, historical precedents reveal that responses to environmental stress weren't solely technical. They included storytelling, religious rituals, and moral principles that influenced how people perceived their responsibilities. By reintroducing moral and cultural dimensions into climate discussions, modern societies can move beyond technocratic solutions. Recognizing that past societies often handled environmental threats through community cooperation, ethical norms, and shared myths, we might remember that rallying collective willpower and forging common values is as essential as developing wind turbines or carbon capture systems.

We must, however, avoid romanticizing the past. Many ancient societies were unequal, patriarchal, or engaged in wars that environmental pressures could intensify. Their coping strategies sometimes involved harsh measures, forced migrations, or ritual sacrifices. Understanding historical complexity helps us avoid simplistic solutions. Just because a certain approach worked for the Maya does not mean replicating it directly today. Instead, we refine our moral compass by examining both the strengths and limitations of past responses.

By illuminating these earlier struggles, successes, and failures, history expands our moral imagination. If ancient peoples managed to survive on limited technologies and knowledge, can we—armed with global data, advanced engineering, and a moral understanding that embraces all humanity—forge a path that avoids the worst outcomes? If past societies sometimes failed due to lack of cooperation, might our interconnected world overcome that hurdle by fostering international treaties, sustainability frameworks, and cultural exchanges that share best practices?

A final insight is that historical examples demonstrate environmental upheavals need not signify inevitable decline. They can be turning points. For communities that adapted wisely, crises led to new balances between humans and nature. For those that ignored warnings and entrenched destructive habits, decline ensued. Today's global society stands at a similar crossroads, confronted not by a localized drought but by a worldwide phenomenon affecting every region differently. Past examples show that the future is not predetermined: it lies in how we choose to respond, morally and practically, to the signals we receive. By reflecting on the breadth of human environmental histories, we equip ourselves with humility, courage, and an appreciation that resilience is possible, though never guaranteed.

Cultural and Spiritual Frameworks for Navigating Uncertainty

All human societies have wrestled with uncertainty. Even before the term "climate change" existed, shifting weather patterns, resource fluctuations, and environmental surprises challenged communities to make sense of the world's unpredictability. Long ago, people confronted sudden droughts, plagues of locusts, and unseasonal storms without the modern tools of climate science. Instead, they turned to what they had: cultural narratives, religious teachings, spiritual rites, myths, rituals, and moral philosophies. These frameworks did more than provide comfort; they guided decisions about resource management, cooperation, sacrifice, and adaptation. By examining how cultures and spiritual traditions have historically navigated environmental uncertainty, we gain insight into how moral imagination, communal identity, and faith can help individuals and groups cope with the anxiety and disorientation that arise when natural baselines shift.

In many ancient communities, environmental phenomena were not viewed as separate from human life. Instead, they were integrated into cosmologies where gods, spirits, or ancestral presences governed ecological cycles. In parts of Indigenous North America, for instance, peoples understood weather, animal migrations, and seasonal changes as reflections of a living, interconnected cosmos. Ceremonies aimed to maintain balance, appealing to spiritual entities for guidance, rainfall, or bountiful harvests. When conditions became difficult—extended droughts or dwindling game—shamans, elders, and spiritual leaders performed rituals seeking harmony. Although these rituals did not alter atmospheric chemistry in a scientific sense, they could alter community morale and ethics. By framing hardship as part of a moral or spiritual test, they encouraged patience, solidarity, and a sense of shared destiny. Such attitudes mitigated panic and despair, replacing them with determination and a willingness to endure.

In some African cultures, ancestral veneration and moral codes linked to the land helped communities interpret environmental changes. If rainfall declined, people might consult diviners or hold communal supplications to ancestors, believing these ancestors could influence natural forces. This belief system fostered a sense that resource scarcity or erratic weather patterns had moral dimensions—perhaps the community had strayed from righteous conduct or failed to respect certain taboos. While this may sound simplistic from a modern scientific perspective, the effect was to embed environmental stewardship in moral conduct. Behaving ethically, maintaining social harmony, and treating the land with reverence were seen as ways to restore equilibrium. Such frameworks offered emotional stability, for they told individuals that uncertainty could be managed through collective moral effort, rather than leaving them feeling helpless.

In the Pacific Islands, complex navigational traditions and ecological knowledge were often underpinned by spiritual beliefs that guided voyaging and resource use. When islanders faced changes in currents, shifting fish stocks, or unpredictable storms, their cultural frameworks—myths of ocean deities, chants passed down across generations, and ceremonially recognized elders—helped them adapt. By interpreting the ocean as a spiritually animated entity and treating navigation as a sacred skill, communities bolstered their emotional resilience. Pilgrimages, offerings, and songs evoked a sense of connection with forces larger than human endeavors. In uncertainty, people anchored their identities in cultural wisdom that affirmed they had navigated tough seas before and could do so again.

Major world religions have also provided spiritual scaffolding for coping with environmental unpredictability. Consider the Abrahamic faiths—Judaism, Christianity, and Islam. Each tradition contains narratives in which humans learn to trust in divine providence, adapt to hardships, or morally purify themselves in response to environmental trials. For instance, stories of famines and plagues in the Hebrew Bible are not just historical anecdotes; they are moral and spiritual lessons. People interpret these trials as tests of faith, calling for repentance, renewed covenant with the divine, and compassion toward the vulnerable. Such interpretations create social cohesion: when drought strikes, believers might fast, pray, give alms, and strengthen communal bonds. While modern believers may blend spiritual comfort with scientific strategies, the underlying principle remains that uncertainty can be met with ethical introspection and mutual support grounded in spiritual convictions.

In parts of South and Southeast Asia influenced by Hinduism, Buddhism, or various local spiritual traditions, the concept of impermanence often features prominently. Environmental changes, droughts, or unexpected monsoons might reinforce the idea that all things are transient. This belief does not trivialize suffering; rather, it encourages emotional flexibility. If life is inherently dynamic, then adapting to new conditions is part of being human. Moral principles such as compassion, non-harm, and detachment from material excess can guide responses. Instead of panicking about unstable future conditions, practitioners might lean on meditative practices, cultivating equanimity and the understanding that clinging too tightly to any single environmental state leads to distress. Such spiritual frameworks reduce anxiety by teaching that uncertainty is not an aberration but a fundamental aspect of existence, and that moral and mental discipline can lessen suffering.

Traditional Chinese philosophies, including Confucianism and Daoism, also offer wisdom. Confucian thought emphasizes the moral duty of rulers and elites to provide for the populace and maintain harmony under Heaven's mandate. If environmental signs—droughts, floods, crop failures—indicated cosmic displeasure, the remedy lay in ethical governance, just laws, and benevolence. This correlation of moral governance with environmental stability urged leaders to prevent corruption and care for the poor. Daoism, on the other hand, highlighted natural cycles and the virtue of following the Dao (the Way), which meant aligning human life with nature's flow. Accepting uncertainty and change as part of the cosmic order could help individuals remain calm and adapt gracefully. Through festivals, spiritual gatherings, and shared texts, these traditions reassured people that environmental shifts were not random punishments but signals to adjust behavior, restore virtue, or realign human affairs with natural principles.

Indigenous peoples worldwide often consider land and environment as kin. Rather than viewing nature as a background resource or a stage for human drama, they see it as a community of beings with whom humans hold reciprocal responsibilities. Environmental changes thus have moral significance: if animals migrate differently or plants fail to bloom at usual times, these signs might indicate a disturbance in the human-land relationship. Ceremonies to apologize to animals for overhunting, or rituals to thank the earth for its fertility, are not quaint traditions—they are moral dialogues restoring equilibrium and trust. This relational perspective means uncertainty can be approached as a breakdown in communication that must be repaired through moral and spiritual actions, not merely a technical puzzle. Emotions like anxiety transform into motivation to mend relationships with non-human entities, and in doing so, restore emotional balance.

Such frameworks offer comfort and ethical direction, but they are not without complications. Sometimes spiritual interpretations can lead to scapegoating or reinforcing hierarchies. If a group believes that drought is punishment for immoral behavior, some might blame marginalized members or rival factions, fostering conflict instead of unity. Yet in many cases, cultural and spiritual traditions encourage inclusivity, shared responsibility, and mutual care. The moral narratives often emphasize that all must cooperate, share resources, and act compassionately to restore environmental balance. This collective ethic can mitigate fear and despair by dispersing emotional burdens across the community—no one shoulders uncertainty alone.

Modern environmental ethics movements sometimes borrow from these older traditions, blending scientific understanding with spiritual or cultural wisdom. The rise of "ecospirituality," which fuses environmental stewardship with spiritual practice, or the growth of religious networks calling for climate justice, shows that contemporary societies seek moral anchors similar to those of the past. People turn to rituals—tree plantings, interfaith prayers, community feasts that celebrate local harvests—reinvigorating cultural expressions that place environmental care at the heart of communal identity. By doing so, they reduce feelings of helplessness in the face of global threats, reaffirming that uncertainty calls not for paralysis but for collective moral effort.

Artistic expressions also play a role. Myths, poetry, dances, and songs historically guided communities through environmental unpredictability. For instance, epic narratives in various cultures might recount how heroic ancestors overcame environmental challenges through wisdom, patience, and moral courage. These stories serve as touchstones, reminding people that resilience is possible, that former generations survived despite hardships. Even now, contemporary artists and writers craft new myths, speculative fictions, and spiritual meditations on climate change, encouraging audiences to approach uncertainty with creativity rather than dread.

Another dimension is how cultural and spiritual frameworks often integrate environmental signs into calendars, festivals, and rites of passage. By embedding environmental cycles into these events—celebrating the return of certain animal species, marking the start of monsoon season with special ceremonies—communities learn that change is not always catastrophic but can be managed rhythmically. When anomalies occur—late rains or fewer migrating birds—the community notices and responds collectively. Anxiety may arise, but it meets pre-existing moral and ritual channels that help process this emotional energy constructively. Thus, cultural rhythms anticipate and absorb some of the emotional shock of uncertainty.

These traditions also foster moral restraint. For instance, taboos limiting when or how certain animals can be hunted or when forests can be harvested ensure that societies do not exacerbate environmental volatility. Although these rules sometimes originate from spiritual stories or sacred mandates, their ecological benefits are clear. By institutionalizing respectful and moderate use of resources, communities reduce the risk of sudden scarcity that would fuel anxiety. The moral framework thus converts potential despair into disciplined behavior. When everyone agrees that overfishing is not just imprudent but morally wrong, people find emotional reassurance in the knowledge that their collective actions safeguard stability.

Ritualized communication—such as asking permission from nature spirits before harvesting a tree or performing blessings before sowing seeds—can transform anxiety into a moment of spiritual negotiation. If individuals believe they have respectfully requested nature's cooperation, they gain a sense of psychological relief. Rather than feeling powerless amid climatic variables, they feel engaged in a moral dialogue. Even if scientifically this does not alter weather patterns, psychologically it imbues humans with agency and purpose. Agency, even symbolic, combats despair by making adaptation a sacred duty rather than a burdensome chore.

In modern pluralistic societies, the challenge is to integrate these diverse cultural and spiritual lessons into a global context where multiple faiths, philosophies, and narratives coexist. Different communities may draw on distinct traditions—some praying to saints for rains, others holding meditation retreats to cultivate inner calm amid environmental stress, still others reviving Indigenous ceremonies lost under colonial suppression. While this diversity can be confusing, it also offers a rich menu of coping strategies. No single tradition holds a monopoly on resilience. Just as environmental problems are global, the moral and spiritual toolbox for navigating uncertainty can be global too. Cultural exchange, interfaith dialogues, and comparative studies of environmental ethics can help societies learn from each other's experiences, enriching the emotional and moral repertoire needed to handle climate anxiety.

Contemporary spiritual movements also adapt old teachings to new realities. Some religious leaders emphasize stewardship of creation, framing climate action as a moral obligation central to faith. Others highlight interdependence and compassion, arguing that caring for distant ecosystems and future generations is a spiritual calling. Secular moral philosophies similarly evolve, using humanistic ethics to assert that love, justice, and responsibility should guide our responses to environmental instability. The result is a growing moral consensus that transcends cultural boundaries: uncertainty calls for empathy, cooperation, humility, and courage.

By centering on moral and spiritual values, societies can counter the purely technical framing of environmental issues. While technology and policy matter immensely, they are not emotionally self-sufficient. People need reasons to care that reach into their hearts and cultural identities, not just practical calculations of cost and benefit. Cultural and spiritual frameworks provide these reasons. They say: we cherish this land because it is our sacred home; we protect these species because they belong to a larger moral order; we share resources because that's what goodness demands. Such narratives transform anxiety into resolve, confusion into moral clarity, and isolation into community.

Another insight is that cultural and spiritual frameworks need not be static. Just as past civilizations adapted or reinterpreted their myths when encountering new environmental conditions, today's communities can revise spiritual teachings to address novel climate challenges. New rituals might emerge that bless solar panels as gifts of the sun's energy, or prayers might be offered for policymakers to find just solutions. Cultural creativity and spiritual imagination are not relics of a bygone era; they remain living tools that can evolve as our circumstances demand.

Some might worry that relying on cultural and spiritual frameworks distracts from the rational, science-based actions needed to reduce emissions or adapt infrastructure. Yet these spheres need not compete. On the contrary, moral narratives and spiritual practices can complement science by encouraging greater emotional resilience and public support for informed policies. If a community believes morally that protecting wetlands is a sacred duty to maintain the balance of creation, that belief can bolster political will to implement restoration projects that scientists deem ecologically beneficial. Thus, the moral legitimacy conferred by cultural and spiritual frameworks can enhance, not hinder, rational decision-making.

There is also the question of how individuals who do not adhere to any particular faith or cultural tradition might benefit from these frameworks. Even secular individuals can draw inspiration from humanistic philosophies that echo spiritual principles: reverence for life, commitment to fairness, or the idea that uncertainty challenges us to become more compassionate. Simply understanding that past peoples found moral and spiritual anchors under environmental stress can reassure modern skeptics that it's normal and healthy to seek meaning beyond mere data. Whether one calls it spirituality, ethical humanism, or cultural narrative, the idea is to connect personal emotions to a transcendent moral horizon that fosters resilience.

Cultural and spiritual frameworks also promote patience. Rapid climate shifts test short attention spans and impatience. Faith traditions that emphasize long-term cycles, ancestral wisdom, or divine timelines encourage endurance. People taught to think in generations rather than election cycles may handle uncertainty better because they do not expect instant gratification. Instead, they view environmental care as a long pilgrimage, with small steps accumulating over time. This patience reduces anxiety by tempering unrealistic expectations of quick solutions, replacing them with faith in steady moral effort.

The communal nature of many spiritual practices provides another buffer against despair. Group prayers, shared ceremonies, community feasts, and collective rites bring people together physically and emotionally. During uncertain times, these communal bonds reduce isolation, distribute emotional burdens, and affirm that everyone navigates uncertainty together. Such solidarity is essential when scientific warnings alone might leave individuals feeling overwhelmed. Knowing that one's moral community stands united in confronting environmental challenges helps convert private fear into collective resolve.

In some traditions, humor and storytelling also play roles. Folktales about trickster figures who adapt to changing landscapes or gods who test humans with challenging seasons can lighten the emotional load. Humor allows people to acknowledge uncertainty without succumbing to terror. Storytelling weaves meaning into chaos, showing that while change is constant, humans can respond creatively and with moral integrity. These narrative forms channel anxiety through culturally familiar patterns, making it more manageable.

As societies accelerate their responses to climate change, cultural and spiritual frameworks may guide how we structure climate education. Instead of presenting climate facts in purely technical terms, educators can explore moral parables, spiritual metaphors, and historical examples that illustrate why caring about distant ecosystems and unborn generations aligns with long-standing virtues. Students can learn that just as ancient communities revered the sanctity of rivers or forests, we too can cultivate a respectful stance toward the Earth. By embedding environmental lessons in moral and spiritual contexts, we speak to both intellect and heart, nurturing a holistic form of engagement less prone to apathy or cynicism.

In global forums, interfaith dialogues on climate issues have already begun. Religious leaders unite to declare that environmental stewardship is a common moral ground shared by diverse beliefs. Such dialogues highlight that while theological details differ, the moral imperative to prevent suffering, uphold justice, and preserve life resonates across traditions. The moral weight of uncertainty, fear, and adaptation finds transcendence in these gatherings as participants draw on sacred texts, ancient wisdom, and collective conscience to assert that climate action is not just a technical necessity but a spiritual calling. By doing so, they harness deep emotional energies that rational arguments alone may not fully activate.

Still, these frameworks do not guarantee success. It's possible to have strong cultural or spiritual convictions yet fail to implement necessary ecological measures if social inequalities, political corruption, or entrenched interests stand in the way. Cultural and spiritual values must be coupled with institutional reforms, policy decisions, and technological advances. Yet the emotional and moral backing provided by these frameworks can boost public support for bold measures. They provide an internal compass guiding leaders and citizens alike through moral dilemmas posed by, say, relocating communities from flood-prone areas or rationing resources during severe droughts.

From an emotional standpoint, understanding that many cultures have historically turned to faith, narrative, and ritual to process uncertainty normalizes our current quest for meaning amid climate instability. We are not alone in feeling unsettled. Our ancestors also grappled with unpredictability and found solace and direction in moral stories, sacred rites, and communal ethics. This lineage can reassure us that emotional turmoil under environmental stress is a longstanding human condition, and that culturally mediated responses—whether ancient or newly conceived—can ease fear, restore hope, and guide moral action.

In observing contemporary movements that fuse environmental activism with spiritual retreats, eco-theology discussions, green pilgrimages, or ceremonies honoring the Earth, we witness cultural and spiritual creativity flourishing anew. People gather to plant trees in memory of lost species, light candles for environmental healing, or sing hymns updated with verses about stewardship and humility. These evolving traditions build bridges between past wisdom and future needs, making moral courage feel attainable. They remind individuals that uncertainty, while challenging, can be metabolized emotionally through shared symbols and moral commitments. The result is a kind of emotional ballast that keeps despair at bay, not by denying difficulty, but by situating it in a larger moral and spiritual landscape.

As knowledge of climate change's complexities grows, so does the recognition that technical solutions alone cannot resolve the psychological toll of uncertainty. Cultural and spiritual frameworks, whether ancient, modern, or syncretic, offer a profound means of integrating moral understanding, communal solidarity, and emotional resilience. They transform fear into reverence, sorrow into empathy, and confusion into purpose. By drawing on these reservoirs of meaning, humanity can face environmental upheavals not as isolated, panic-stricken individuals but as moral communities grounded in principles of care, justice, and respect for life's mysteries.

Drawing Strength and Perspective from Diverse Traditions

As climate instability affects every corner of the globe, individuals and communities seek ways to ground themselves emotionally and morally, looking beyond their immediate experiences for guidance and inspiration. In a world that sometimes feels unmoored by environmental change, drawing strength and perspective from diverse traditions can help anchor people in something larger than themselves—lineages of thought, practice, and wisdom that stretch across continents and centuries. Each cultural or spiritual inheritance, whether ancient or newly evolving, can offer tools for facing uncertainty, forging solidarity, and cultivating resilience. By exploring how different societies have long responded to environmental stresses, how they've interpreted moral duties to the Earth, and how they've sustained hope amid hardship, we expand our own repertoire of emotional and ethical strategies for the climate era.

Diverse traditions do not present a single monolithic template. Rather, they form a mosaic of narratives, rituals, philosophies, and ethical codes. One might turn to Indigenous teachings that emphasize reciprocity with non-human beings, learning to see forests, rivers, and animals not as resources but as relatives. Another might study monastic traditions in Asia, where spiritual discipline and contemplative practices transform anxiety into equanimity through acceptance and compassion. Others may find insight in pilgrimage routes that historically connected communities across ecological gradients, reminding travelers of the immense variety and vulnerability of life on Earth. By comparing these threads, we discover that multiple pathways exist for engaging with environmental uncertainty—pathways that do not rely solely on scientific projections or policy debates, but on moral imagination, empathy, and reverence for creation.

Take, for instance, Indigenous knowledge systems deeply rooted in local ecologies. Many Indigenous peoples conceive of land as alive, imbued with spiritual significance and woven into their identities and histories. In these communities, environmental stewardship emerges naturally from the understanding that human well-being depends on maintaining respectful relationships with other species and natural forces. When faced with changing rainfall patterns or migrating animal populations, Indigenous traditions often respond through adaptive practices informed by generations of experience. If climate shifts upend predictable patterns, spiritual leaders, elders, and storytellers re-interpret ancestral knowledge to guide new responses. The emotional reassurance lies in knowing that their cultures have survived past fluctuations by weaving moral obligations, community bonds, and ecological awareness into daily life. Such an integrated view suggests that rather than feeling helpless, people can draw on inherited moral compasses that treat adaptation as a collective spiritual duty, not just a technical challenge.

Comparatively, consider the wisdom embedded in religious pilgrimage routes like the Camino de Santiago in Europe. Historically, pilgrims walked these paths partly in search of spiritual growth, partly to connect with cultural heritage, and partly to witness the changing landscapes of their region. Today, as shifting rainfall or hotter summers alter the pilgrimage experience, travelers can reinterpret these journeys as opportunities to witness the Earth's fragility firsthand. The tradition of pilgrimage—steeped in humility, perseverance, and openness to challenge—can help modern sojourners process climate anxiety. By framing environmental changes as spiritual tests or moral calls to awaken compassion, pilgrims tap into a tradition that transforms hardship into an impetus for moral reflection and greater solidarity with all life encountered along the way.

In East Asia, traditions influenced by Buddhism, Daoism, and Confucian thought emphasize harmony, balance, and moral cultivation. Instead of seeing climate instability as an alien intrusion, these philosophies may encourage reframing it as an invitation to restore equilibrium.

Daoist ideals, for instance, suggest following the natural flow (Dao) rather than attempting to dominate it. When confronted with erratic weather or resource scarcities, a Daoist perspective might guide individuals to align themselves more closely with nature's patterns, simplifying desires, reducing excess, and cultivating humility. Emotional stress arising from uncertainty can be eased by contemplating impermanence and relinquishing the illusion of total control. Similarly, Confucian ethics, which highlight moral responsibilities of leaders and citizens alike, can inspire policies and community initiatives that treat environmental care as an extension of caring for one's family and society. By blending personal virtue with communal welfare, these traditions transform anxiety into a moral impetus, pushing leaders and ordinary citizens to act ethically for collective benefit.

In African communities with strong ancestral traditions, moral lessons are often passed through oral stories that link human behavior to environmental outcomes. If a tale recounts how disrespecting certain animals led to ecological imbalance, listeners learn that their actions carry moral weight. As climate patterns shift, retelling and reinterpretation of old stories can guide people to find renewed relevance. A narrative about ancestors who conserved seeds during a past famine might inspire modern communities to protect biodiversity amid unpredictable seasons. The emotional comfort here lies in connecting present dilemmas to historical narratives where moral courage, cooperation, and reverence for life helped societies endure. Rather than succumbing to fear, people can think: "Our ancestors faced hardship and found a moral path through it, and so can we."

For those within Abrahamic faith traditions—Judaism, Christianity, Islam—the concept of stewardship under a Creator's watchful eye can transform environmental uncertainty into a moral call. Even if these texts were written in times less anthropogenically altered, contemporary interpreters find guidance in passages urging care for the poor, justice in resource distribution, and humility before divine creation. Religious leaders might stress that harming the environment constitutes harm to the vulnerable, future generations, and other beings sharing the Earth. Instead of leaving believers adrift in despair over melting ice caps, these traditions offer a moral framework that equates caring for creation with fidelity to divine mandates. Prayer, fasting, charity, and community gatherings become tools not just for spiritual growth but for forging emotional resilience and solidarity. Believers can say, "If we strive toward righteousness and compassion, we answer our uncertainties with faith in a moral order that rewards good deeds and mindful stewardship."

Hindu and Buddhist teachings, informed by karmic principles and concepts of interdependence, can also lend emotional support. If environmental disruptions highlight the consequences of exploitative behaviors, karmic thinking suggests these outcomes arise from past actions' moral weight. Individuals and societies reap what they sow. Such an understanding promotes a sense that addressing climate challenges is not merely practical but ethically required to restore balance and generate beneficial karma for the future. Meanwhile, Buddhist meditation practices help individuals embrace impermanence and reduce suffering by cultivating inner peace, even amid external instability. The emotions sparked by climate anxiety—fear, sadness, anger—become objects of observation rather than sources of torment. Seeing that suffering arises from attachment to expected outcomes, practitioners can navigate uncertainty with more calmness and moral clarity. Instead of feeling hopeless, they might feel urged to act compassionately, transform greed into generosity, and embrace lifestyles of moderation.

Across Polynesia, Micronesia, and Melanesia, seafaring traditions and kinship-based resource management systems taught communities that survival depended on reading subtle environmental cues and acting collectively. If winds changed or reefs declined, communities adjusted their fishing methods, sometimes guided by spiritual stories that framed the ocean as a living entity deserving respect. Emotional resilience grew from a cultural narrative that valorized adaptation, resilience, and flexible responses to challenges. Modern attempts to revitalize Indigenous marine management practices and integrate them into national policies can draw on these old frameworks, showing that moral understandings of reciprocity and gratitude foster psychological well-being. Individuals feel less alone when they know their ancestors faced similar uncertainties and overcame them through ethical norms and shared responsibilities.

Modern secular philosophies can also provide moral grounding. Humanism, for example, emphasizes human responsibility, dignity, and empathy without relying on supernatural sources. Humanistic approaches might draw from historical examples—like communities that overcame resource bottlenecks through cooperation—to assert that humanity can collectively find moral strength in rational understanding and mutual care. Environmental humanists argue that by recognizing our interdependence and moral obligations to all humans and non-humans, we can transform fear into purposeful action. Small anxieties become stepping stones to solidarity: if people acknowledge their worries and work together guided by principles of fairness, equity, and compassion, they create emotional resilience.

As climate concerns accelerate, interfaith and intercultural dialogues arise more frequently, bringing together leaders from various traditions to affirm common ground. They might identify shared virtues: stewardship, humility, reverence, compassion, and justice. By acknowledging these commonalities, societies gain moral momentum. People see that their small acts of environmental kindness—reducing waste, aiding neighbors in floods—are not isolated gestures but part of a global moral chorus. Such cross-cultural learning expands emotional capacity by affirming that multiple paths converge on the same moral destination: care for life and readiness to adapt ethically to new conditions. This knowledge soothes existential dread by presenting a united human front, enriched by countless traditions rather than divided by differences.

Artistic expressions and ceremonies can also draw on multiple traditions for strength. Hybrid rituals might emerge that blend Indigenous blessings with music from world faiths, poetry from secular humanists, or meditative silence borrowed from monastic practices. These syncretic forms create communal experiences that honor environmental realities and moral imperatives simultaneously. Standing under trees once revered in animist traditions, praying with language inspired by Abrahamic prophets, reflecting quietly on impermanence as taught by Buddhism, participants forge emotional connections that transcend cultural borders. In doing so, they weave a moral and emotional safety net that can catch individuals who stumble under the weight of climate anxiety.

Historical lessons reinforce the value of looking outward to diverse traditions. Civilizations that recognized moral wisdom in neighboring cultures occasionally borrowed strategies or symbolic practices that aided adaptation. Today, technology allows people to access countless spiritual teachings, cultural stories, and ethical philosophies with ease. Instead of relying on a single narrative, one can explore various sources for emotional comfort and guidance. A person overwhelmed by data on glacier retreat might find solace in Shinto traditions revering mountains and water, or in the quiet resolve of Quaker meetings, or in the lyrical reflections of Sufi poets praising creation's unity. Each tradition adds a layer to the tapestry of resilience, ensuring that no single thread bears the entire emotional load.

Critically, embracing diverse traditions does not require abandoning rational approaches or scientific understanding. On the contrary, integrating moral and spiritual dimensions with empirical facts creates a more comprehensive response. Science teaches what is happening and suggests technical options; moral and spiritual frameworks teach why it matters and how to remain emotionally balanced. They help transform sterile statistics into moral catalysts, moving from intellectual comprehension to heartfelt conviction. By internalizing these moral narratives, individuals grow more persistent in their pursuit of sustainability, less susceptible to despair, and better able to inspire others to join collective efforts.

Another advantage of learning from diverse traditions is that it cultivates empathy and respect. Recognizing that different communities have found meaningful ways to cope with environmental changes fosters humility. One realizes no single cultural or spiritual system holds all answers. Instead, humans have experimented for millennia with moral stories and spiritual practices that help interpret, respond to, and sometimes mitigate environmental stresses. This realization encourages listening, humility, and intellectual openness. When faced with new climate challenges—say novel pests, unpredictable monsoons, melting permafrost—people can recall that their ancestors or those of others faced their own trials, and their responses can inform today's moral choices.

Even in a secular age, many non-religious individuals can still appreciate the aesthetic and symbolic power of rituals drawn from various traditions. A community reeling from a severe flood might hold a candlelight vigil, inspired partly by a local religious congregation's practice of prayerful reflection and partly by Indigenous rites thanking the land's spirits. Everyone contributes something: the believer finds comfort in divine connection, the skeptic finds meaning in human solidarity, the child finds hope in seeing adults act with kindness. This synergy shows that moral and emotional nourishment can flow through channels that transcend doctrinal boundaries, uniting people around shared values of love, care, and responsibility.

As environmental instability persists, the emotional strain may intensify. Without moral and spiritual frameworks, people risk becoming jaded, nihilistic, or numb. Drawing on diverse traditions provides long-term stamina. When a particular tradition's narrative grows less resonant for some individuals, they can supplement it with insights from another source. A farmer who relied on personal faith for comfort might also adopt certain Indigenous land stewardship principles. A city dweller who drew strength from historical epics might incorporate meditative breathing exercises from Eastern philosophies to cope with daily reports of extreme weather. Adaptability in emotional and moral toolkits mirrors the adaptability needed ecologically, reinforcing that resilience is about variety, diversity, and openness, both environmentally and psychologically.

Cultural and spiritual frameworks remind people that they are more than passive spectators of a changing world. They have roles—as moral agents, caretakers, and participants in sacred or meaningful relationships with the Earth. Instead of a narrative where humans face environmental disruption as helpless victims, these traditions suggest that uncertainty is a call to action, a test of character, or a stage in a cosmic story where humans can learn humility and forge deeper compassion. This narrative inversion empowers individuals to face anxiety with a sense of purpose rather than defeat.

Moreover, diverse traditions often celebrate cycles—of seasons, birth and rebirth, death and renewal. By embracing cyclical thinking, people learn that hardship and uncertainty are not endpoints but moments in a larger process. Even if climate trends appear grim, these cyclical metaphors encourage hope that human creativity and moral courage can intervene. Just as droughts in ancient cultures sometimes preceded periods of innovation or migration that rejuvenated societies, today's climate challenges might precede moral and cultural renaissances guided by lessons from the past. Small communities experimenting with regenerative agriculture or restoring local ecosystems can view their efforts as part of an evolving moral tradition that updates ancient wisdom for modern times.

The presence of intergenerational narratives in many traditions also helps maintain perspective. By seeing themselves as stewards for unborn descendants, people anchor their anxiety in a timeline that transcends personal lifespans. Worries about future coastal inundations or disease patterns lose some sting when framed as moral imperatives to protect future children and grandchildren. Many Indigenous stories emphasize seven generations ahead, encouraging long-term thinking. Eastern philosophies highlight karma and cycles of rebirth, extending moral responsibility beyond immediate relations. Abrahamic faiths stress covenantal relationships spanning countless generations. Drawing from these intergenerational visions helps tame anxiety by showing that the moral community is not just those alive today but also those who came before and those who will follow. Thus, environmental care becomes a sacred trust passed across time, diminishing feelings of isolation.

By connecting local efforts to global cultural tapestries, individuals and communities can find courage in the knowledge that they contribute to a human legacy of resilience. If Polynesian navigators crossed vast oceans guided by stars and spiritual beliefs, if pastoral tribes balanced grazing with reverence for the land's regenerative powers, if medieval monks preserved learning and ethical principles through centuries of turmoil—then today's inhabitants facing climate-induced uncertainty join a lineage that overcame adversity through moral fiber and collective action. Such awareness infuses present actions with historical significance, suggesting that current struggles against warming temperatures or vanishing wetlands continue humanity's millennia-old conversation with nature and the moral law.

In a world where technology and information proliferate, moral and emotional clarity can still feel scarce. Drawing strength and perspective from diverse traditions cuts through this noise. It provides narrative scaffolding that explains why fear must not yield to apathy, why sadness can motivate empathy rather than resignation, and why anger can channel into advocacy rather than nihilism. Whether one resonates more with spiritual ceremonies, ethical philosophies, indigenous ecological knowledges, or interfaith dialogues, each path leads toward the understanding that humans have always shaped and been shaped by their environments. Accepting uncertainty with moral grace and emotional steadiness is not new; it is an ancient art continually rediscovered and renewed.

No single tradition or narrative will eliminate climate anxiety. Challenges remain daunting, and no magic incantation can halt the rising seas. Yet recognizing that multiple moral and spiritual legacies exist offers choices. People can experiment, borrow, and blend traditions as needed. A secular activist might find unexpected comfort in a Sufi poem celebrating creation's unity. A Catholic environmentalist might enrich their moral framework by learning from Zen Buddhist meditative practices that calm worry. A tribal community revitalizing old ceremonies might inspire an urban congregation to adopt rituals that honor seasonal changes. This cross-pollination fosters a global moral ecosystem, robustly diverse, in which emotional resources flow freely across cultural boundaries.

Such fluid exchange respects differences without forcing uniformity. Just as biodiversity strengthens ecosystems, cultural and spiritual diversity strengthens humanity's collective emotional resilience. Each tradition contributes unique moral nutrients that help communities thrive under climate pressure. If one tradition emphasizes humility before cosmic cycles, another highlights justice for the vulnerable, another cherishes gratitude for daily gifts of nature, another teaches forgiveness and compassion. Embracing this moral biodiversity provides the psychological scaffolding needed to face an uncertain future with dignity.

At times, the complexity of the climate crisis might tempt people to retreat into cynicism or despair. But knowing that countless human generations have turned to moral wisdom, spiritual narratives, and cultural ethics to overcome adversity reminds us that moral creativity is a renewable resource. The process may feel slow, as small communities rediscover old rites or craft new ones suited to rising seas and shifting biomes. Yet each small initiative—an interfaith climate vigil, a cultural exchange program on sustainable farming, a spiritual retreat focused on environmental healing—helps build a collective emotional treasury that reduces isolation and fear.

In this process, the fear of standing alone against an impersonal climatic force diminishes. People realize they stand among billions, each holding fragments of insight from their own cultural or spiritual wellsprings. The moral imagination can assemble these fragments into a map guiding action and emotional steadiness. By acknowledging that uncertainty has always tested humanity and that cultural and spiritual frameworks have historically provided moral compasses, we gain the confidence to navigate today's storms. Rather than floundering in emotional turmoil, we find reassurance in the fact that we are continuing a long human tradition of responding morally, creatively, and compassionately to environmental challenges.

By honoring these diverse traditions, we expand our understanding of what it means to engage ethically with the planet. This engagement transcends short-term political debates or narrow economic calculations. It becomes a matter of inherited wisdom, inherited dreams, and inherited responsibilities. Emotional burdens lighten when individuals know they need not invent moral codes from scratch. Instead, they can apply lessons tested by time and complexity, blending them with modern insights into ecology and technology. The result is a robust emotional framework that helps people remain hopeful, committed, and compassionate even when headlines grow grim.

Throughout history, spiritual leaders, storytellers, philosophers, and community elders have acted as moral beacons in times of ecological upheaval. Now, as climate changes intensify, modern societies can tap into similar roles, whether filled by interfaith coalitions, cultural preservationists, or educators bridging old and new teachings. By doing so, people transform anxiety into a resource—an impetus to deepen moral understanding, cultivate empathy, and strengthen communal bonds. Instead of succumbing to despair, they become stewards of a rich moral legacy, well-equipped to face the unknown.

These historical glimpses reveal that environmental upheavals are neither new nor insurmountable. From the Egyptians who carefully stored grain to cushion against droughts, to the Mesopotamians who struggled with salinized soils and social disarray, to the Maya and Norse whose cultural patterns influenced their survival or collapse, each case testifies that environmental challenges and moral choices have always been entwined. Societies that recognized ecological limits, adapted their practices, and aligned resource use with ethical considerations tended to fare better than those that clung rigidly to outdated norms or ignored the warning signs nature offered. Past peoples found meaning, community cohesion, and resilience through shared narratives, religious beliefs, technological innovation, and willingness to learn from neighbors or adopt new ways of life. Their experiences suggest that climate-related anxieties and uncertainties are part of the human condition, yet they need not end in despair. Instead, they can prompt creativity, moral courage, and cooperation, enabling communities to craft responses that balance cultural identity with ecological wisdom. By learning from those who preceded us, we broaden our sense of what is possible—understanding that resilience emerges from a union of technology, ethics, governance, and shared narratives that give shape to collective efforts in uncertain times.

Part II: Finding Emotional Ground

Chapter 5: Immediate Relief When Fear Spikes

Simple, Science-Backed Grounding Techniques (Breathwork, Mindfulness)

When anxiety suddenly surges—whether triggered by a distressing headline about record-breaking temperatures, a moment of realization about disappearing habitats, or just an ambient sense that the world's ecological balance is off—many people find themselves searching for immediate ways to regain calm. These episodes can feel overwhelming, as though the mind and body have been hijacked by fear. Yet, amidst this storm of emotion, simple, science-backed grounding techniques offer a lifeline, a set of practical tools to steady the nerves, restore focus, and gently guide attention back to the present moment.

The key word here is "present." Anxiety often thrives on projections into the future, worries about what might happen decades from now or on distant continents. In those intense moments, the mind rushes ahead, conjuring catastrophic scenarios. Grounding techniques counter this tendency by anchoring awareness in what is real and tangible right now—your breath, your senses, your immediate surroundings. They aim to pause the runaway train of anxious thought, offering a respite that allows for more measured responses. Instead of drowning in panic, one learns to ride the waves of emotion and emerge with greater clarity and resilience.

Among the simplest and most powerful grounding methods is breathwork. The human nervous system responds to the pace and depth of breathing; rapid, shallow breaths can signal distress, while slow, steady inhalations and exhalations encourage calm. Focusing on the breath can break the feedback loop of escalating anxiety. When fear spikes, it's common to notice breath becoming rapid and shallow, perhaps centered high in the chest. Simply bringing attention to this pattern—observing the inhale and exhale without judgment—starts to shift physiology toward relaxation.

A tried-and-true breath technique involves lengthening the exhale. For instance, breathing in for a count of four and exhaling for a count of six or eight can engage the parasympathetic nervous system, which governs the "rest and digest" response. By making the out-breath longer than the in-breath, you send a signal to the brain that there is no immediate danger. Over time, this simple practice reduces heart rate and lowers levels of stress hormones like cortisol, gently guiding the body from a state of alarm toward one of safety.

Some find it helpful to visualize each breath. Inhaling, they might imagine drawing in a sense of stability, exhaling, releasing tension or fear. Another approach is to pair breathing with simple words or phrases. Inhaling with the thought "I am here," exhaling with "I am safe." These mental anchors help prevent the mind from drifting into catastrophic thinking. They restore a sense of agency—indeed, the breath is something you can always influence. Even amid chaos, you can choose how to breathe, and thus regain some measure of control.

While breathwork directly influences physiology, mindfulness complements this approach by refining where attention goes. Mindfulness encourages observing thoughts, sensations, and emotions without attaching judgment or panic to them. It's not about banishing anxiety altogether—after all, feelings triggered by climate uncertainty are legitimate. Instead, mindfulness teaches acceptance: acknowledging the presence of fear, naming it, but not letting it dictate behavior. Rather than wrestling with anxious thoughts, you watch them come and go, as if they were clouds passing through an open sky.

This perspective shift eases the intensity of fear spikes. When a frightening image of a flooded city or scorched forest pops into the mind, mindfulness counsels stepping back, noticing the thought rather than becoming entangled in it. "This is worry," you might say internally, "this is fear arising." By labeling it, you create a small gap between yourself and the emotion. You stop identifying wholly with the anxiety. This gap can be enough to prevent escalation. It frees mental space to consider, "What can I do right now?" or "How can I care for myself in this moment?"

Research supports the effectiveness of mindfulness for anxiety management. Neuroscientists have found that regular mindfulness practice—be it meditation, body scans, or mindful walking—can reduce activity in the amygdala, the brain's fear center, over time. It also strengthens connections in parts of the brain involved in emotional regulation. So while a single mindful breath won't solve systemic climate issues, it can retrain the nervous system to remain steadier in the face of unsettling information. This emotional steadiness, in turn, enhances the capacity for clear thinking and constructive action.

Grounding techniques aren't limited to breathwork and mindfulness. They can also involve using the five senses to root oneself in the present. For example, if fear spikes after reading a dire environmental forecast, you might put down your device and focus on something tangible and immediate: the feel of the chair beneath you, the scent of fresh air through an open window, the color of the floor, the sounds of distant traffic or chirping birds. Naming these sensory details—"I see the pattern in the wooden table," "I feel the texture of my shirt," "I smell tea brewing in the kitchen"—helps displace anxious ruminations with concrete realities. By engaging the senses, you remind yourself that, in this moment, you're safe and physically intact.

Another grounding exercise known as the "5-4-3-2-1" technique can be especially effective. It involves identifying five things you can see, four you can touch, three you can hear, two you can smell, and one you can taste. This sequential focus on the senses directs attention outward, breaking the cycle of inwardly spiraling thoughts. It's a tactile, immediate way to confirm that no matter what future scenarios swirl through the mind, right now you exist in a specific place and time, connected to the material world.

Some people incorporate gentle movement into their grounding practices. A short walk, focusing on the sensation of each footstep, can calm racing thoughts. Stretching the arms overhead, rolling the shoulders, feeling the body's weight shift—these physical acts center attention in the body, not in anxious speculation. Yoga poses, even a simple forward bend or a seated twist, synchronize breath with motion, reinforcing the message that you have a say in how your body responds. Movement-based grounding can dissolve some of the nervous energy that accumulates when fear flares.

Journaling can also serve as a grounding practice. When fear spikes, writing down what you're feeling provides a tangible outlet. You might note the trigger: "Just saw a news article on coral bleaching," then name the emotion: "Feeling sadness and fear," and describe physical sensations: "Heart racing, shoulders tense." By articulating these observations on paper, you slow down mental spiraling. The act of writing externalizes thoughts, making them more manageable. You might also scribble down a few reassuring truths: "Right now, I am sitting in my living room," "I have survived many challenges before," or "I can look up local climate groups tomorrow." These small affirmations can help redirect energy toward potential steps rather than getting stuck in panic.

Some may prefer audio-guided grounding techniques. Listening to a short, calming meditation recording that guides attention through the breath, or soothing nature sounds that help envision stable ecosystems, can restore equilibrium. Hearing a gentle voice reminding you to breathe slowly and focus on bodily sensations can break the isolation that often accompanies fear. It feels as though someone is guiding you back to solid ground, reminding you that anxious states come and go, that no emotion is permanent.

Science supports these methods by explaining how the nervous system toggles between states of arousal and rest. The sympathetic nervous system (SNS) produces the fight-or-flight response during fear, while the parasympathetic nervous system (PNS) restores calm. Techniques like breath regulation stimulate the vagus nerve, a key conduit to the PNS, encouraging a shift from SNS dominance to a more balanced state. Understanding this biological mechanism demystifies why focusing on slow breathing or mindful attention works. It's not mere wishful thinking—there's a physiological foundation behind it.

Regular practice enhances effectiveness. Just as muscles strengthen with exercise, the ability to ground oneself quickly during anxiety spikes improves with repetition. Initially, trying a mindful breath when fear surges might feel forced or awkward. Over time, as the brain learns that these interventions reliably lead to calmer states, the response becomes more natural. Eventually, you need not struggle so much; a few steady breaths or a quick sensory scan might suffice to prevent panic from snowballing. This skill becomes a protective resource, a mental toolkit that can be carried wherever you go.

Grounding techniques, though powerful on their own, can be woven into daily routines. Even on days when fear isn't spiking, practicing a short mindfulness meditation in the morning or taking three slow breaths before starting work can build a baseline of emotional steadiness. Then, when an anxiety trigger appears—a difficult news segment, a conversation about future climate policies—you're not starting from scratch. You've already taught your nervous system to return to a state of calm more easily. This baseline resilience is crucial when dealing with ongoing environmental uncertainties that may generate frequent emotional perturbations.

Also, these methods complement rather than replace other coping strategies. If climate-related anxiety persists, seeking professional help—such as a therapist well-versed in eco-anxiety—may be wise. Advocating for structural changes and joining community resilience projects can address underlying causes of dread. Grounding techniques do not solve systemic climate problems, but they empower individuals to face them from a calmer vantage point. Emotional steadiness can make civic engagement and advocacy more sustainable. Instead of burning out in frantic cycles of panic and despair, you can approach actions methodically, fostering sustained involvement in long-term solutions.

For some, pairing grounding techniques with personal values intensifies their impact. When breathing slowly, you might think, "I breathe in strength, I breathe out fear," connecting each breath to a moral aspiration. Or while performing a mindful body scan, you might recall your motivation—caring for future generations, protecting beloved species, or honoring ancestral traditions of stewardship. Integrating moral purpose into these practices can transform them from mere stress management tactics into expressions of hope and commitment, further mitigating fear's intensity. The idea is to link immediate emotional relief with broader ethical intentions, ensuring that calmness serves not as an escape but as a springboard for compassionate action.

Imagery can also enhance grounding. Visualizing a stable, balanced ecosystem—a lush forest, a healthy coral reef, a resilient wetland—while practicing slow breathing can replace catastrophic mental images with more positive, life-affirming ones. This mental shift harnesses the mind's capacity to influence emotional states. If reading dire headlines conjures doom-laden images, counter them with images of restoration efforts, of communities planting mangroves, or solar panels gleaming under the sun. The point is not to deny reality but to remember that recovery and renewal are also part of nature's story. Integrating such imagery into grounding practices fosters a sense of balance, reminding you that not all trends move downward, that resilience and adaptation are also natural phenomena.

Another subtle approach involves affirmations. When fear spikes, repeating gentle affirmations—short phrases that reflect truths or aspirations—can stabilize the mind. Something like, "I am doing my best in difficult circumstances," or "We can find strength in community," or "Change is possible, and I can contribute." These affirmations anchor thought patterns in constructive territory, preventing anxious rumination from dominating. While affirmations alone won't solve ecological crises, they influence emotional tone, making fear more manageable and less debilitating.

In group settings, practicing these techniques together enhances their effect. A family facing anxiety over what the future holds might agree to pause occasionally, taking three shared breaths before discussing difficult news. A community workshop might teach mindfulness sessions geared toward eco-anxiety, showing participants that they are not alone. When people see others using the same grounding tools, a supportive atmosphere emerges, normalizing emotional vulnerability and empowering collective resilience. Mutual encouragement amplifies the benefits—knowing a friend or coworker also turns to slow breathing during tense moments can inspire you to do the same, fostering emotional solidarity.

Adapting these techniques to personal taste is also beneficial. Some may prefer focusing on breath counts, while others like to hum quietly to feel vibrations that soothe the nervous system. Some might close their eyes to block out visual distractions, while others might find keeping eyes open and focusing on a comforting object more grounding. Experimentation leads to a personalized toolbox of methods that feel most natural and effective. The variety ensures that if one technique proves less effective in a certain context—perhaps it's hard to concentrate on breath in a noisy environment—you have alternatives, like a quick five-senses scan or a short mantra.

Remember that progress occurs gradually. Early attempts may yield modest relief. Over time, as you repeatedly interrupt fear spikes with grounding techniques, your brain learns new patterns. Instead of going from a slight worry to full-blown panic, you might notice that after a few mindful breaths, the surge stabilizes at mild concern. Gradually, what once induced severe anxiety may evoke a more manageable emotional response. This incremental improvement fosters confidence. As your confidence grows, you feel less powerless. Knowing you possess tools that reliably steady your emotions transforms anxiety from an all-consuming threat into a challenge that can be met with skill.

Though these techniques focus on the individual level—on what you can do when fear assaults your senses— their implications reach further. A world filled with individuals who can keep their emotional balance amid troubling headlines is more likely to think clearly and act collectively. If fear paralyzes everyone, coherent action falters. If, however, people learn to calm themselves, they can then engage thoughtfully in dialogues, policy debates, and collective problem-solving. Emotional resilience is therefore not just a personal benefit; it's a cornerstone for fostering rational, compassionate societal responses to climate challenges.

For some, integrating these techniques into daily rituals makes them more accessible in moments of acute need. Practicing slow breathing for a few minutes each morning, or concluding the day with a brief mindful check-in, can wire the habit into daily life. Then, when anxiety flares, you won't struggle to recall instructions; you'll naturally revert to the familiar pattern. Routine practice lays neural groundwork, ensuring that calming responses become ingrained. Over time, less conscious effort is required, and one can deploy these strategies swiftly, without scrambling for instructions in the midst of fear.

As science continues to study anxiety, more nuanced techniques might emerge. Biofeedback devices, for instance, could guide users to slow their heart rate and deepen their breathing more effectively. Virtual reality experiences might simulate calming landscapes that synchronize with the user's breath, reinforcing a sense of safety. While these technologies are not strictly necessary, they illustrate that as understanding grows, so does the potential to refine and customize grounding techniques. The underlying principle remains consistent: focusing attention on the present and controlling physiological responses to restore balance.

Ultimately, simple, science-backed grounding techniques remind people that even in an era of global instability and daunting environmental projections, individuals retain some agency. No one can fully predict the future or singlehandedly prevent sea-level rise, but each person can decide how to respond emotionally at a given moment. This agency reduces helplessness. The knowledge that you can slow your breath, observe your thoughts, and choose where to place your attention builds a foundation of inner strength.

Such strength matters because the climate crisis is not a short drama concluding in a few months. It is a decades-long, multifaceted challenge that requires sustained mental and emotional endurance. The ability to cope with fear spikes ensures that emotional exhaustion does not derail long-term engagement. When anxiety arises, these grounding methods give you a break—time to process, regain perspective, and return to the work of mitigating harm and cultivating solutions.

In an era saturated with alarming information, grounding techniques serve as a filter. They prevent emotional overwhelm from spiraling into paralysis or despair. By steadying the mind and body at critical junctures, they help maintain a healthier emotional baseline from which informed choices and meaningful actions arise. While not solving the root causes of climate anxiety, these tools empower individuals to navigate environmental uncertainty with greater resilience, clarity, and compassion.

Sensory Exercises, Journaling, and Nature-Based Calming Practices

When anxiety intensifies, especially the kind related to climate change and environmental uncertainty, it can feel as though one's mind is trapped in a loop of distressing thoughts. While breathwork and mindfulness offer powerful avenues for regaining composure, sometimes expanding the toolkit to include sensory exercises, journaling, and nature-based practices can further enhance emotional stability. These approaches help anchor attention in tangible experiences, engage the body's innate calming mechanisms, and harness the restorative power of the natural world. By integrating these methods, individuals can cultivate a richer tapestry of responses to fear, moving beyond momentary relief toward sustained emotional resilience.

Sensory exercises leverage the fact that anxiety often takes the mind on a journey far from the present moment—into imagined future catastrophes or replays of disturbing headlines. By directing attention to immediate sensory input, these exercises ground people in what is real right now. They can be as simple as holding a small object—a smooth stone, a piece of fabric—and focusing on its texture, temperature, and weight. When the mind tries to catastrophize, returning awareness to this object reminds the nervous system that, at this moment, one is safe. Touch is a direct and primal sense; it bypasses linguistic complexity and directly reassures the body.

Another sensory exercise involves tasting something with full attention. Instead of mindlessly sipping tea or coffee, dedicate a moment to really taste it—notice the aroma before sipping, feel the warmth on the tongue, identify subtle flavors. This anchored observation slows mental chatter. Similarly, smelling a calming scent—lavender oil, a piece of fresh citrus peel—can interrupt anxious cycles. Scents have a potent link to emotional centers in the brain, making them ideal for invoking feelings of comfort or recall of a safe place. Each whiff anchors the mind, preventing it from drifting too far into fear's domain.

Sound can also serve as a sensory anchor. Listening mindfully to ambient noises, whether the hum of an appliance, distant traffic, birdsong, or wind rustling leaves, pulls attention outward. By naming each sound—"I hear a car passing," "I hear leaves shaking"—the mind breaks from anxiety's narrative and returns to the here and now. Some people use soundscapes—recordings of rain, ocean waves, or forest ambience—to foster calm. Others might hum softly or sing a quiet tune, feeling the vibrations in their chest, reestablishing a sense of physical presence that counters abstract dread.

Visual exercises can complement these approaches. Instead of letting the mind run rampant after reading upsetting environmental forecasts, pause and look around carefully. Identify colors, shapes, patterns, or interesting details in the immediate environment. Notice how light falls on objects, how shadows form. This visual grounding interrupts anxious rumination. If possible, look outside a window or step onto a balcony, observing trees, clouds, or even architectural lines. Seeing that the world still contains stable patterns, beauty, and continuity can counterbalance fearsome scenarios. This visual reassurance doesn't deny problems but reminds that complexity includes resilience and continuity alongside change.

While sensory exercises ground attention externally, journaling directs attention inward, translating complex feelings into written words. When anxiety peaks, the mind can feel chaotic and tangled. Writing provides a structured outlet. Begin by describing the trigger: "Just read an article about Arctic ice thinning faster than expected." Then name feelings: "I feel uneasy, sad, and powerless." Next, explore thoughts without censoring: "I'm worried about future generations; I fear that nothing we do will matter." This step externalizes mental clutter. Once these emotions and ideas sit on paper, they become more manageable, more clearly defined. The act of naming fears breaks their silent hold. Journaling transforms intangible worry into concrete text, allowing reflection and gradual release.

Another journaling technique is to write a dialogue with the anxious part of oneself. Imagine anxiety as a character. Ask it: "Why are you here today? What do you want to tell me?" Let the pen record its answers. Then respond with compassion. This exercise encourages self-understanding and reduces internal conflict. Instead of treating anxiety as an enemy, approach it as a messenger. Maybe anxiety says, "I'm here because the headlines on melting glaciers scare me," and you can reply: "I hear you. It's scary. But we're going to do what we can—support policies, join a community group, take care of ourselves—and we don't have to solve everything alone." Such dialogues foster a sense of agency and comfort.

Journaling can also include gratitude lists or acknowledgments of what remains stable and good. Though environmental worries loom large, there are still moments of kindness, technological progress, restoration projects, and communities acting responsibly. Writing about these positives balances the emotional scales. Highlighting small wins or personal strengths—like your capacity to learn, adapt, or find meaning in helping others—infuses the narrative with hope. Over time, re-reading journal entries can reveal patterns: certain triggers recur, certain coping techniques prove effective. This feedback loop refines emotional strategies.

Another journaling approach involves setting intentions. After acknowledging fears, write down a simple intention: "Today, I will spend 10 minutes researching local climate initiatives," or "This week, I will make a small donation to a reforestation project." Concrete intentions counter helplessness by turning anxiety into motivation. Each completed intention reinforces the sense that while uncertainty exists, constructive responses do too. The journal thus becomes a place not only to process emotions but to chart small steps forward.

Nature-based calming practices link sensory grounding and journaling with the healing potential of the natural environment. Countless studies show that spending time in green spaces lowers stress levels, improves mood, and promotes psychological well-being. In a time of environmental worry, reconnecting physically with the natural world offers a potent antidote to despair. A short walk in a park, feeling grass underfoot and observing plant life, reminds that nature continues to persist and adapt. This direct contact can restore perspective: yes, ecosystems face threats, but nature is not solely defined by collapse. Resilience and regeneration persist, visible in a single thriving oak, a blooming wildflower, or a colony of busy pollinators.

If going outside is challenging, even a potted plant by a window, a small herb garden on a balcony, or photographs of natural landscapes can help. Tending to a living plant can ground attention in growth processes that unfold at a slower, steadier pace. Observing leaf buds appear, flowers bloom, and seeds germinate counters the frantic, crisis-driven timeline that anxiety imposes. Nurturing something alive confirms one's capacity to contribute positively, affirming a sense of purpose amidst uncertainties.

For those able to access natural settings, activities like forest bathing—spending quiet, unhurried time in wooded areas—can deeply soothe the nervous system. Forest bathing, originated in Japan as "shinrin-yoku," encourages individuals to stroll slowly, breathe deeply, and engage senses fully. Feel the texture of tree bark, listen to birds, inhale the scent of pine needles. Such immersion lowers cortisol, enhances immune function, and calms racing thoughts. The presence of other life forms, each following their own rhythms, places human anxieties in a broader ecological context. This perspective can diminish the sense of isolation and helplessness. Instead of feeling alone against climate change, one realizes they are part of a rich tapestry of life, each species adapting in its own way.

Similarly, practicing mindfulness outdoors—be it sitting beside a river or gazing at clouds passing overhead—merges sensory grounding with natural stimuli. The environment itself offers a gentle soundtrack and dynamic visuals. Watching a stream flow, noticing sunbeams shifting through leaves, or feeling a gentle breeze caress the skin encourages relaxation without effort. These natural cues communicate stability: water flows, seasons turn, plants regenerate. Though large-scale patterns are changing, local encounters with resilient pockets of nature offer emotional respite, reminding people that not everything is unraveling at once, and that nature's capacity for renewal can inspire human resilience as well.

Journaling after or during nature visits can solidify this experience. Describing the scene—the sound of insects, the pattern of moss on a stone—integrates sensory input with reflective thought. If anxious thoughts resurface, writing them down amid descriptions of nature's beauty can contextualize them. A worry about rising seas might appear alongside notes of a resilient coastal marsh pioneering new growth after a storm. This juxtaposition does not deny the problem but places it in a landscape where adaptation and persistence are visible truths. As a result, journaling bridges internal emotions with external evidence of survival and evolution.

Some might incorporate nature-based rituals, drawing on cultural traditions or personal creativity. For instance, leaving an offering of gratitude at the base of a tree, quietly acknowledging the life it supports, can transform anxiety into reverence. Or listening to natural sounds—rainfall, wind, rustling leaves—while performing simple stretches or yoga poses can center the mind and calm the body. Movement synchronized with natural rhythms resituates humans as part of ecological communities rather than isolated agents facing abstract threats.

Even something as humble as watching ants carry crumbs provides a metaphor for cooperative effort and persistence in challenging conditions.

Over time, these practices—sensory exercises, journaling, and nature-based calm—become habits that can be summoned at will. In a sudden spike of anxiety after reading grim climate projections, one might close their eyes, inhale slowly, exhale longer, then open their eyes to notice a beam of sunlight on the floor. Feeling a hint of calm, they jot a few lines in a journal: "Right now, I'm scared, but I'm breathing steadily. Outside, I can hear sparrows. I know small actions matter." This simple sequence transforms raw fear into a manageable emotional moment, preventing panic from escalating and maintaining a thread of hope.

In urban environments with limited green space, even small nature encounters—potted herbs, window boxes with flowers, a patch of sky visible between buildings—can help. Tending a tiny garden or placing a succulent on a windowsill is not trivial. Such micro-nature experiences engage the senses, requiring observers to slow down and appreciate subtle life processes. These quiet acts of care and observation counter the frantic pace of alarming news cycles. Similarly, visiting parks or green rooftops, exploring community gardens, or volunteering in local ecological projects can fortify emotional resilience. Direct involvement in nurturing life around us reduces helplessness by demonstrating that individuals can contribute to environmental well-being, even if modestly.

Combining journaling with sensory details during these nature-based experiences deepens their effect. Writing about how the soil smells after rain, or how leaves dance in the wind, captures moments of sensory clarity. If anxious thoughts creep in—"What if these species vanish in future climates?"—acknowledge them, but also note the present resilience: "Even if future challenges loom, this garden thrives now, and my hands can support its growth." Through such interplay, anxiety loses some of its sting, as it's balanced by lived experiences of stability and positive engagement.

Cultural traditions can amplify these benefits. Some communities have longstanding ceremonies that honor seasonal transitions—harvest festivals, spring rituals, or solstice celebrations. Participating in such events, or crafting personal mini-ceremonies, can anchor emotional well-being in collective practice. For instance, lighting a candle at dusk each day and spending a moment reflecting on one's feelings, journaling a few lines, then listening to nighttime sounds can become a calming ritual. Over weeks, this ritual trains the nervous system to associate uncertainty with gentle, nurturing routines rather than panic. The body learns that fear does not have to dominate; it can be met with consistent, reassuring practices.

Integrating these methods into daily life builds resilience cumulatively. Sensory exercises might only take a few minutes—touching a cool pebble, inhaling the scent of a favorite essential oil, listening closely to a morning bird's song. Journaling might happen briefly in the evening, just a paragraph summarizing feelings encountered and small gratitudes found. Nature-based practices might occur weekly: a weekend walk in a park, or a monthly trip to a nearby forest trail. None of these habits need to be grand or time-consuming. Their power lies in regularity and sincerity. Over time, these small interventions shape emotional landscapes, transforming reactivity into responsiveness.

When climate anxiety spikes, one can use a quick mental checklist: "Am I breathing fast? Let's slow it down. Are my shoulders tense? Let me stretch. Am I caught up in catastrophic thoughts? Let me name one thing I see, one texture I feel, one sound I hear. Can I write down what I'm feeling? Can I recall a recent walk outside that felt calming?" By stepping through these simple actions, fear loses momentum. Instead of sinking deeper, the individual gently rises toward balance.

These approaches also complement other emotional support systems. If one attends a climate support group or therapy sessions, the skills honed through sensory grounding, journaling, and nature exposure enhance what is learned there. They become portable strategies that work in real time—between therapy appointments, after a tough conversation, or upon waking from a troubling dream. The synergy between personal coping methods and external support networks strengthens overall resilience. External validation—hearing from others that these methods help them too—reinforces their effectiveness and encourages sharing of best practices.

In times of crisis—intense drought, an extreme weather event, a heartbreaking news story about ecosystem collapse—these grounding techniques become lifelines. Instead of feeling drowned by collective despair, individuals can breathe, center their senses, write down their fears and hopes, and perhaps step outside to touch living foliage or feel the earth under their feet. Such gestures remind them that although the future is uncertain, they inhabit a present moment still full of tangible life, potential actions, and moral choices. Anxiety does not vanish instantly, but it transforms into a more bearable companion, one that can be understood, managed, and ultimately channeled into constructive engagement rather than paralysis.

Some might integrate creativity into journaling—drawing sketches of leaves or clouds, penning a short poem that merges fear with hope. This creative expression extends the benefits of writing. Instead of treating words as a bare record, creativity turns journaling into an art form that reclaims emotional agency. Even when describing climate concerns, one can highlight nature's colors, textures, and rhythms, blending anxiety with appreciation. This fusion fosters emotional complexity rather than one-dimensional fear. People learn that multiple truths coexist: danger and beauty, loss and possibility, sorrow and determination. Recognizing this complexity reduces the starkness of fear, making it just one element in a larger emotional palette.

Over time, as these methods become ingrained, individuals notice their emotional thresholds changing. What once triggered intense panic might now provoke milder concern, quickly soothed by a few grounding breaths or a short journal entry. The nervous system grows more adept at returning to equilibrium, like a well-trained muscle that recovers swiftly after strain. This adaptability is crucial for long-term emotional health in the face of ongoing environmental uncertainties that may persist for decades. Instead of long, draining bouts of anxiety, people experience shorter, more manageable episodes of worry that are quickly addressed with the learned techniques.

These practices also highlight that individual emotional management contributes to collective capacity. If widespread fear paralyzes societies, meaningful action stalls. But if many individuals learn to navigate their emotions skillfully, their combined stability supports rational debate, sustained activism, policy innovation, and community resilience projects. Just as resilient ecosystems contain diverse species that contribute to overall stability, resilient communities contain individuals who know how to steady themselves and each other under stress. Sensory exercises, journaling, and nature-based grounding thus form part of a moral ecology that nurtures both private well-being and public engagement.

In multicultural environments, these techniques can be shared and adapted across traditions. A friend from another cultural background might introduce a particular nature-based ritual or a sensory meditation method. Communities might organize journaling workshops or sensory exploration sessions in public parks, encouraging neighbors to link environmental awareness with emotional self-care. By exchanging approaches, people learn that while climate anxiety is common, so too are solutions that tap into human ingenuity and compassion. Such cultural cross-pollination can yield inventive hybrids—like journaling while sitting in a community garden, listening to multilingual poems celebrating ecological resilience, or blending aromatic herbs used in one culture's healing traditions with mindfulness practices from another.

Even digital tools can support these methods. Apps offering guided journaling prompts tailored to eco-anxiety, or soundscapes recorded in various ecosystems, can be at one's fingertips when fear spikes. Virtual communities share tips: "When I feel overwhelmed, I hold my pet's paw and focus on its softness," "I look at a leaf's veins and imagine nature's complexity," "I write down one positive ecological story I learned today." The abundance of such shared wisdom fosters hope—if so many are finding ways to cope, none are truly alone in navigating these emotional storms.

At its core, the integration of sensory grounding, journaling, and nature-based practices reaffirms human capacity for adaptation on multiple levels. Evolution endowed us with senses attuned to the environment, cultures developed storytelling and art to process emotions, and individuals discovered personal rituals that bring solace amid uncertainty. By drawing on these deep human inheritances, people facing climate anxiety acknowledge that while technology and policy matter enormously, emotional resilience springs from timeless sources: the feel of earth underfoot, the pen capturing inner voices, the calm exhalation that softens tension, and the willingness to engage all senses in remembering that life persists and can inspire respectful care.

Strategies to Return to the Present Moment When Overwhelmed

In a time when unsettling information about environmental changes and unpredictable futures floods the senses, many find themselves spinning into worry, fear, or sadness that seem to have no end. In such moments, emotions can feel like a heavy current, pulling attention into realms of hypothetical disaster and lost stability. Yet it's possible to stem this tide by developing strategies that harness the power of the present moment. When the mind drifts into catastrophic visions, techniques for returning to the here and now can restore balance and sanity. These approaches do not deny real challenges; rather, they anchor one's mental footing so that acknowledging problems doesn't mean being consumed by them.

Returning to the present moment may sound simple, but in the grip of anxiety, it can be remarkably challenging. Modern life trains people to multitask, anticipate, and plan ahead. Climate-related stress amplifies that forward-looking tendency, often dragging thoughts into dire "what ifs." Overcoming this habit means cultivating practices that gently guide awareness back to immediate experience. The goal is not to escape reality but to reclaim agency over attention. When fear and uncertainty run rampant, focusing on what can be done right now—observing a breath, feeling the ground beneath the feet, listening to a passing sound—empowers individuals to pause panic and move forward more thoughtfully.

One foundational strategy is learning to identify early signs of mental drift into anxious territory. Often, a subtle inner tension, rapid breathing, or scattered thoughts indicate that the mind is slipping away from the present. By noticing these early cues without judgment, people can intervene sooner rather than later. For example, if reading a distressing climate article triggers racing thoughts, pause immediately. Acknowledge: "I'm feeling overwhelmed," or "I notice my thoughts racing." Naming the experience breaks the unconscious loop and positions oneself as an observer rather than a helpless participant.

After naming the state, choose a grounding technique that resonates. Many find it helpful to use the five-senses approach. Look around and identify something visually appealing or neutral—a pattern on the wall, a tree outside the window. Describe it silently: "I see the texture of the bark, the way sunlight filters through leaves." Then shift to hearing: "I hear distant voices, a car passing, a bird chirping." Touch something: feel the temperature of an object, note its hardness or softness. Inhale a scent if available, even if subtle, and consider taste if appropriate. This sensory inventory roots the mind in concrete reality. By carefully naming these stimuli, the mind's frantic energy disperses, replaced by calm observation. These few moments anchor awareness in the body's immediate environment, halting the spiraling forward into frightening futures.

Breathing techniques complement sensory focus. When tension escalates, breath often becomes shallow and quick, fueling anxiety. Consciously slowing and deepening the breath counteracts this. Inhale through the nose for a count of four, hold briefly, exhale for a count of six or eight. This extended exhalation engages the body's relaxation response. While exhaling, silently repeat a calming phrase, like "I am here, I am safe." Coordinating breath with a reassuring mantra steadies both mind and body. If thoughts wander back to panic, acknowledge their presence, then refocus on the counting and the comforting words. Over time, this trained response makes returning to the present more automatic and less laborious.

Another strategy involves using physical movement to reinforce presence. Anxiety can freeze the body or send it into restless fidgeting. Instead of letting that energy go uncontrolled, channel it into purposeful motions. Stand up if seated, feel feet on the ground, roll the shoulders, do a brief stretch. Notice how the body feels when doing these actions—muscles elongating, tension releasing. Movement breaks the cycle of mental rumination by demanding attention to balance, coordination, and sensation. Even a short walk, focusing on each step, can bring awareness away from catastrophic thoughts and into the cadence of footsteps and the air against the skin. Physicality acts as a magnet for attention, pulling it out of abstract fears and into lived reality.

Journaling can also help refocus on the present. When the mind races into long-term projections—melting ice caps decades from now, food insecurity in distant regions—pause and write down what is being felt. Describe the emotion: "Right now, I feel a knot in my stomach thinking about future generations." Then list immediate truths: "In this room, it's quiet. I'm breathing evenly. I had breakfast this morning. I have people who care about me." This contrast between fear-driven what-ifs and tangible here-and-now facts reminds that not all is lost or dire at this moment. The act of writing slows the mind, requiring each word to form on paper. This deceleration creates space for calm reason to reemerge.

Journaling can also include action-oriented questions: "What small step can I take today?" Maybe it's researching a local environmental group or turning off unnecessary lights. Identifying a tiny, achievable action breaks the loop of helplessness. By focusing on what can be done immediately, the mind shifts from despair to mild empowerment. The present moment, once a scene of distress, transforms into a platform for meaningful steps. Anxiety thrives on powerless futures; returning to the present highlights personal agency.

Nature exposure is another potent catalyst for presence. When overwhelmed, a brief encounter with the natural world—be it a stroll in a park, pausing by a window with a view of a tree, or holding a living plant—reconnects individuals to ongoing life processes that are immediate and tangible. Observing a leaf's veins, feeling the rough bark of a tree trunk, or noticing how sunlight shifts colors can all anchor attention. Instead of being swept away by environmental headlines, one acknowledges nature as it is right now: alive, intricate, still functioning in many places despite challenges. This direct, sensory relationship with nature reminds that change unfolds in real time, and that humans remain part of ecosystems rather than isolated victims of catastrophe.

Combining nature with movement intensifies the grounding effect. A mindful walk, noticing each footstep on soil or pavement, feeling the breeze, listening to rustling leaves—these sensations tether anxious thoughts to lived experience. Instead of drifting into abstract fear about future droughts, the mind becomes fascinated by present details: the pattern of a leaf, the texture of a path, the interplay of light and shadow. Such simple beauty counters the grimness of anxious narratives. While it doesn't erase climate issues, it prevents them from eclipsing all other experiences. The mind learns that even while problems loom, life's immediate richness persists.

Some individuals find rituals help them return to the present. Lighting a candle, placing a hand over the heart, or reciting a short verse can signal the nervous system to shift gears. Rituals create a pattern the mind recognizes as safe and intentional. If every time anxiety spikes, a person gently closes their eyes, takes three slow breaths, and says a phrase like, "I return to this moment," over weeks and months, this sequence becomes a reliable anchor. The brain forms a link: fear arises, the ritual follows, calm returns. Rituals tap into ancient human tendencies to respond to symbolic actions with emotional realignment.

Similarly, using props or tools can reinforce presence. A small grounding object—like a worry stone or a seashell—carried in a pocket provides a tactile reminder. When anxiety surges, touching this object recalls the intention to remain present. The object can store an emotional memory of calm, so feeling its texture recalls moments of previous steadiness. This external cue simplifies the mental work required, allowing for faster recovery. Such tools help make returning to the present more accessible, turning an abstract concept into a tangible gesture.

Visualizing safe, stable scenarios can also break the cycle of anxious projections. Instead of letting the mind race into unknown futures, deliberately picture a serene place personally meaningful—a childhood garden, a calm lakeside, a quiet library. Engage all senses in this visualization: feel the grass, smell the flowers, hear gentle water lapping, see soft light. This mental scene functions like a sensory exercise inside the mind, a refuge from agitation. Once calm, attention can return to the external present with renewed clarity. The image is not escapism but a strategic pause, allowing for emotional reset.

In social contexts, seeking brief connection can help maintain presence. Speaking a reassuring word to a friend, calling a loved one, or joining a short online support session can anchor emotions. Others often help reflect reality more accurately than an anxiety-inflamed mind. Hearing a calm voice say, "I know it's scary, but you're not alone," anchors the present with communal strength. Another person's stability can help break fear's momentum, preventing isolation from fueling panic. A swift check-in with a community group focused on environmental well-being can remind that meaningful collective actions are underway, stabilizing hope and grounding emotions.

For some, humor can also provide a pathway back to the present. Gently laughing at the absurdity of certain doomsday scenarios, or sharing a lighthearted anecdote, breaks the spell of fear. Humor doesn't trivialize serious concerns but can lighten emotional weight, making it easier to think practically. When people find something to smile about—a clever cartoon, a witty remark—the mind relaxes slightly, allowing presence to reassert itself. Even laughter's physical effects—muscles relaxing, endorphins releasing—can shift one's emotional state back to a calmer baseline.

At times, balancing present awareness with acknowledging uncertainty is crucial. Denying future threats is not helpful. Instead, techniques for returning to the present moment provide emotional room to assess challenges more rationally. Once calm, individuals can distinguish between immediate fears and long-term strategies, deciding which steps matter now. With the mind grounded, listing possible responses—personal habit changes, supporting policy initiatives, joining local adaptation efforts—feels less overwhelming. The present becomes a stable platform from which to build small, incremental changes, rather than a chaotic whirlpool of panic.

Practicing these strategies regularly enhances their effectiveness. If someone only tries to anchor themselves after hours of doomscrolling, the habit may not stick. But incorporating periodic "presence breaks" throughout the day—pausing every few hours to breathe slowly, notice sensations, or record a brief journal entry—trains the nervous system to return to calm states more readily. Over time, emotional set points shift, and it takes more to trigger overwhelming fear. The result is greater emotional endurance when confronted with tough news or difficult environmental discussions.

To maintain variety, people can rotate techniques. If on one day focusing on breath feels stale, try a sensory walk the next. If journaling becomes routine, introduce a new prompt: writing about one thing in the immediate environment that symbolizes resilience. If nature exposure is limited, close your eyes and imagine a natural setting you know well. Flexibility ensures that these strategies remain fresh and effective. Different triggers might call for different responses—

some might find breathing is best when feeling physically tense, while journaling suits moments of mental overload, and nature visits help when visual media triggers despair.

Cultural or personal preferences influence which methods resonate. Some might adapt these techniques to religious practices, integrating a short prayer or a meditation on impermanence before sensory grounding. Others may prefer purely secular approaches. Some might blend techniques: first taking three calming breaths, then naming sensory details, then writing a short note in a journal. The synergy of multiple strategies can deepen their impact. Experimentation helps find the right combination for each individual's temperament and life circumstances.

Over time, one may observe subtle but profound shifts in perspective. Initially, anxiety might feel like an unstoppable force, pushing the mind into catastrophic timelines. After practicing present-moment strategies, fear episodes might become shorter or less intense. One learns that while fear arises, it can also subside with intentional effort. This fosters confidence in one's emotional resilience. Confidence does not solve the environmental crisis, but it prevents emotional collapse, enabling sustained engagement with solutions.

As societies confront long-term challenges, widespread adoption of these techniques could influence collective emotional tone. If more people can remain present rather than succumbing to panic, decision-making might improve. Leaders who can stay grounded under stress set an example, demonstrating that while gravity of environmental issues is real, losing one's emotional center is not inevitable. Calmness under pressure can inspire trust, fostering cooperation and more constructive public discourse. The present moment strategies learned individually ripple out into a more stable communal emotional climate.

In educational settings, teaching children and adolescents these coping methods can shape future generations' emotional literacy. Instead of growing up helpless before frightening predictions, young people could learn that fear, while normal, is manageable. A short breathing exercise before discussing heavy topics in a classroom could help students process difficult information without melting into despair. Sensory breaks and journaling assignments encourage them to articulate emotions and explore moral responses. Over years, these skills accumulate, producing adults who navigate uncertainty with poise rather than paralysis.

Workplaces can also benefit. Employees involved in climate research, environmental NGOs, or green tech companies face persistent emotional strain. Integrating short mindful breaks, nature walks during lunch, or journaling prompts can maintain morale. When staff feel overwhelmed by the enormity of their tasks, recalling that they can recenter in the present moment restores mental bandwidth. This emotional agility supports creativity, problem-solving, and sustained commitment to projects that may span decades.

It is important to acknowledge that these techniques do not erase pain or render climate threats trivial. Returning to the present moment does not dismiss the future's significance or the injustice of environmental damage. Instead, it ensures that acknowledging tough truths does not spiral into incapacitating anxiety. By standing firmly in the present, individuals affirm their capacity to respond meaningfully. They grant themselves permission to rest, to find comfort, and to act from a place of balanced awareness. This balance allows continued engagement: activism rooted in calm determination rather than frantic desperation, planning informed by reason and compassion, not panic.

The gradual mastery of present-moment strategies resembles learning an art. Early attempts may feel clumsy or ineffective. Doubts arise—"How can naming the color of the wall help me cope with vanishing coral reefs?" Yet over time, the nervous system learns these cues, and the results become tangible. Moments of near-panic might reduce to a mild flutter that's quickly soothed. The wall's color is no magic cure, but the act of focusing on it displaces catastrophic visions long enough to restore composure and, with composure restored, rational thought and moral courage follow.

Combining presence techniques with supportive communities multiplies their impact. Online forums or local groups can share experiences: "When I start feeling overwhelmed, I practice the 5-4-3-2-1 sensory method, and it really helps," says one member. Another suggests, "I go outside and pick up a fallen leaf, examining its veins. Instantly, I feel calmer." Hearing others' successes validates that these strategies are not mere placebos; they are tested by diverse people facing similar anxieties. By building a culture that normalizes emotional self-care, societies become better at weathering emotional storms together.

Language evolves as people internalize these methods. Instead of saying, "I'm spiraling into panic," someone might say, "I'm going to pause and ground myself now," or "I'll take a sensory inventory to reconnect." Over time, such phrases become part of everyday vocabulary, reducing stigma and encouraging preemptive emotional regulation. Colleagues might remind each other to "take a grounding moment" before delving into stressful climate reports, making emotional steadiness a shared priority rather than a private struggle.

Artworks, songs, or brief guided meditations can be created around these principles, making them accessible to those who find written instructions less compelling. A short animation showing a character applying present-moment techniques, or a playlist of soothing sounds curated for moments of crisis, can reinforce their adoption. Cultural creativity—short poems that remind one to breathe, images that visualize calm, local artists hosting sensory workshops—enriches the landscape of solutions and ensures they appeal to diverse tastes and learning styles.

Over months and years, these strategies alter one's relationship with fear. Fear no longer seems an enemy that must be conquered, but a signal that the mind needs a rest and a recalibration. Each time a person successfully returns to the present, they accumulate positive emotional memory. The brain builds a library of experiences: "I survived that anxious episode by focusing on the here and now, so I can do it again." This library reduces dread next time anxiety looms. Confidence grows, and with confidence, the moral courage to face challenges without collapsing inward emerges stronger.

In a landscape of global uncertainty, such personal emotional resilience is invaluable. It prevents burnout among activists, despair among concerned citizens, and passivity among those who feel powerless. By learning to return to the present moment when overwhelmed, individuals maintain their psychological stamina. This stamina translates into long-term engagement with climate solutions, whether personal lifestyle changes, community adaptation measures, political advocacy, or scientific research. Emotional steadiness supports the sustained effort required, ensuring that people remain devoted to meaningful action rather than succumbing to nihilism or apathy.

Just as ecosystems depend on diverse species and their interactions for resilience, emotional resilience may arise from a diversity of strategies, all focused on presence. By mixing breathwork, sensory grounding, journaling, and nature encounters, each individual can find a blend that resonates. Over time, this internal diversity ensures that if one approach falters, another picks up the slack. If breath control feels forced one day, perhaps journaling flows naturally that day. If writing seems burdensome, maybe a walk outdoors quickly restores balance. This adaptability mirrors the ecological principle that variety and flexibility enhance survival under changing conditions.

As these methods become second nature, one can witness how even highly alarming information no longer leads straight to panic. Instead, it might trigger a pause—a few deep breaths, a mindful observation of a tangible object, jotting down a brief reflection—before deciding how to respond. Instead of reacting instantly with despair, individuals respond thoughtfully, considering their options, reaffirming their values, and maintaining a sense of agency. In this way, returning to the present moment is not a retreat from reality but a strategic pause that allows moral clarity and rational planning.

The overarching message is that while climate uncertainty is real and challenging, it need not dominate one's emotional life. Simple, accessible techniques exist to recenter attention in the now, reducing emotional turbulence and making room for steady, constructive action. This emotional skill enriches the human capacity to cope with long-term, complex issues. By grounding ourselves repeatedly, we forge habits that fortify mental health and build collective resilience, ultimately enabling more thoughtful, persistent engagement with the world's ecological challenges.

These approaches serve as a reminder that even in the face of alarming forecasts and unsettling predictions, individuals retain the ability to influence their own emotional landscapes. By slowing the breath, focusing on tangible sensations, acknowledging anxious thoughts without allowing them to define one's identity, and redirecting attention to the here and now, people reclaim a measure of calm and control. This steadiness in the present moment is not an escape from pressing environmental realities, but rather a foundation that supports thoughtful, compassionate engagement with them. In strengthening personal resilience, these simple, scientifically grounded techniques empower individuals to navigate uncertainty with greater clarity, dignity, and resolve.

Chapter 6: Reframing Your Thoughts

Identifying Common Cognitive Distortions (Catastrophizing, All-or-Nothing Thinking)

In moments of distress, especially when grappling with fears about environmental futures, the mind's thought processes can become skewed by patterns known as cognitive distortions. These mental filters warp perceptions, transforming uncertainties into monstrous visions or simplifying complex realities into stark binaries. Two common distortions frequently encountered in the context of eco-anxiety are catastrophizing and all-or-nothing thinking. Recognizing these distortions is an essential first step toward regaining emotional balance, as it allows individuals to question their initial reactions, correct flawed assumptions, and guide their thoughts toward more nuanced and constructive perspectives.

Cognitive distortions do not arise from ignorance or weakness; they are part of the human mind's attempt to cope with uncertainty and risk. The climate crisis, with its global scale, long timelines, and far-reaching consequences, presents unparalleled challenges for processing information. Under such conditions, the brain seeks shortcuts, albeit imperfect ones, to manage complex emotional responses. Distortions often emerge when fear and worry run ahead of rational analysis, provoking patterns that intensify anxiety rather than clarify reality. Learning to identify and name these distortions can counteract their influence, opening the door to more realistic, compassionate, and solution-oriented thinking.

Catastrophizing is one of the most common distortions linked to environmental fears. It involves leaping to the worst possible outcome at the slightest indication of trouble. Reading a headline about a species decline might trigger thoughts such as "This proves everything is collapsing; the entire ecosystem is doomed." A small setback in policy negotiations might lead to internal monologues like "We will never fix this; humanity is headed for extinction." Instead of acknowledging incremental progress, complexity, and adaptive efforts underway, catastrophizing highlights only the direst scenarios. This distortion magnifies each negative piece of news, reducing the ability to see positive developments or incremental improvements. Over time, catastrophizing can drain motivation and foster hopelessness, leaving individuals feeling paralyzed by dread.

Why does catastrophizing take hold so easily? From an evolutionary standpoint, being vigilant for threats and preparing for the worst once aided survival. Yet, in an era where threats are complex and long-term, this ancient bias backfires. The mind's tendency to focus on negative possibilities can overshadow balanced assessments. When confronting climate change—an immense problem without straightforward, immediate fixes—focusing excessively on worst-case scenarios can become a habit. Media coverage of disasters, policy failures, or melting ice caps provides endless fodder for catastrophe-oriented thinking. Without consciously interrupting this pattern, fear-driven narratives become self-reinforcing, eroding confidence and emotional resilience.

All-or-nothing thinking, another pervasive distortion, also features prominently in discussions of environmental issues. This distortion frames situations in absolute terms: success or total failure, hero or villain, salvation or doom. An environmental policy that doesn't achieve its entire set of goals might be dismissed as utterly worthless. If an individual can't live a perfectly sustainable lifestyle—eliminating all plastic, never driving a car—then they might feel like a complete hypocrite or a moral failure. Such rigid polarization ignores the complexity of incremental progress, partial solutions, and the moral ambiguity inherent in large-scale challenges. It also imposes unrealistic standards, guaranteeing that most efforts appear insufficient, fueling frustration and discouragement.

Why does all-or-nothing thinking flourish in this context? Climate change is a grand moral and existential crisis, leading people to yearn for clear-cut heroes and perfect solutions. This desire for moral clarity tempts the mind to categorize actions and outcomes into neat boxes: perfect or pointless, ethical or evil, saved or doomed. But reality seldom fits these neat labels. Progress in environmental matters often comes in increments—partial emissions reductions, gradual shifts in consumer habits, policies that mitigate some damage while leaving other issues unresolved. Embracing complexity requires patience and humility, while all-or-nothing thinking offers a brittle kind of certainty that breaks under real-world nuance.

These distortions often reinforce each other. Catastrophizing can feed all-or-nothing thinking, and vice versa. For instance, if someone catastrophizes about the future—envisioning total ecological collapse—they might then adopt an all-or-nothing stance, believing that unless perfect solutions appear immediately, nothing matters. Conversely, rigid binary thinking can intensify catastrophizing: if one perceives a small policy setback as a total defeat, this might confirm the catastrophic narrative that all is lost. Breaking free from these interwoven patterns demands awareness, compassion for oneself, and the willingness to reframe one's thoughts.

The first step in identifying catastrophizing is to notice when language in thoughts becomes absolute or overly dramatic. Words like "never," "always," "totally destroyed," "no hope," or "doomed" indicate the mind is pushing toward extreme conclusions. In these moments, it can help to pause and question: "Is this really the only possible outcome?" or "Am I certain that no solutions exist?" Asking such questions challenges the assumption that the worst-case scenario is inevitable. Sometimes writing the thought down helps expose its extremity. For example, turning "We will never solve climate change, and everyone is doomed" into a written sentence often reveals its sweeping nature and invites more balanced reflection.

Similarly, identifying all-or-nothing thinking involves catching words and phrases that split reality into strict dichotomies—"perfect or worthless," "complete success or utter failure," "fully green or total hypocrite." Recognize that sustainability and climate action frequently progress along a spectrum. Is it really true that using a car occasionally invalidates all other eco-friendly habits, or that a policy reducing emissions by 30% is meaningless if it doesn't hit 100% targets immediately? Acknowledging that partial improvements still count, that moral effort lies on a continuum, and that mixed results can still represent forward movement, disrupts the binary trap.

Having recognized these distortions, what next? Developing strategies to counter them can restore perspective. One method involves seeking counterexamples or alternate scenarios. If catastrophizing insists that everything is getting worse, look for evidence of positive change: communities restoring wetlands, new technologies lowering carbon footprints, youth-led movements influencing policy. Even if these successes are partial, they undercut the notion of absolute doom. For all-or-nothing thinking, consider that moral and ecological growth often emerges incrementally. Celebrate small victories. If an individual reduces meat consumption significantly, that does not become irrelevant because they are not fully vegan. If a country invests heavily in renewables, that step is not worthless just because fossil fuels remain part of its energy mix. Acknowledge complexity and gradual improvement, not as excuses but as realistic assessments of how progress unfolds.

Reframing is a powerful tool. Instead of "If we don't fix everything now, we fail," one could think, "We are taking important steps, and while we still have far to go, each effort lays the groundwork for further improvements." Instead of "Climate changes mean total collapse," try "The future poses severe challenges, but humans have adapted and can develop creative solutions, though it won't be easy." These reframes don't deny seriousness; they moderate extreme conclusions so that problem-solving mindsets can flourish. In this more balanced mental space, fear coexists with possibility, and anxiety doesn't drown out hope.

Mindfulness practices aid in catching distortions early. By regularly meditating or performing body scans, individuals become more attuned to subtle shifts in mood and thought patterns. When catastrophizing thoughts appear, a mindful stance observes them: "This is catastrophizing, a distortion, not an accurate reflection of all realities." Labeling the distortion reduces its power. Similarly, when all-or-nothing judgments arise—"I must be either perfectly green or I'm a fraud"—mindfulness can note: "This is that rigid thinking again, I can choose a more moderate interpretation." Over time, the brain learns not to automatically accept these extreme narratives.

Cognitive Behavioral Therapy (CBT) techniques offer structured ways to challenge distortions. For example, one might use a thought record to capture a catastrophic or all-or-nothing thought, then list evidence for and against it. If the thought is "No one cares about climate action," gather evidence: People have installed solar panels, engaged in protests, passed some environmental laws. This contradicts the absolutism. Recognizing that distortions ignore contrary evidence weakens their grip. Likewise, for binary thinking, noting nuances—maybe a policy isn't perfect but does reduce emissions—breaks the false dichotomy and reveals progress as a spectrum.

Practicing self-compassion is crucial. When noticing a distortion, it's easy to judge oneself harshly: "I'm thinking catastrophically again, I'm failing at coping." Instead, acknowledge that these patterns are common, especially amid frightening information. Gently remind yourself that the mind is trying to protect you by overestimating danger or seeking simple moral narratives. Thank your mind for its concern, then steer thoughts toward more balanced ground. Self-compassion creates a supportive internal environment where changing cognitive habits feels less like a battle and more like a gradual reorientation.

Peer support can also help dismantle distortions. Talking with friends, family, or community members about environmental anxieties can reveal that others share similar fears but also retain hope and perspective. Listening to someone else describe how they find meaning in incremental successes challenges the assumption that partial steps are useless. Hearing about a new local initiative can counter the notion that everything is doomed. These social correctives help recalibrate one's cognitive filters by exposing them to alternative viewpoints not dominated by extreme narratives.

Media literacy and information management complement these strategies. Catastrophizing often intensifies when constantly exposed to alarming news without balance. Setting boundaries around media consumption—limiting time spent doomscrolling, seeking sources that also report on solutions or positive trends—can reduce the raw material that feeds distortion. Similarly, curating a more nuanced information diet helps undermine all-or-nothing perspectives. If every story encountered is about failure, consider reading about adaptation measures, indigenous land stewardship successes, green innovations, or diplomatic breakthroughs. Diversifying inputs makes it harder for the mind to fixate on extreme endpoints.

Embracing complexity is another antidote. Realizing that environmental scenarios span a range of outcomes—some worse, some better—shifts the mental frame away from absolute doom. Acknowledge that uncertainty cuts both ways: negative feedback loops exist, but so do innovations and ethical awakenings. Maybe not all coral reefs will vanish; some may adapt or benefit from restoration efforts. Perhaps some regions will mitigate floods through new infrastructure. Admitting uncertainty does not mean ignoring the severity of risks, but it does prevent concluding that the worst case is the only case. Complexity demands humility, which dissolves all-or-nothing stances and reduces catastrophic leaps.

Revisiting past environmental progress can further challenge distortions. Historically, societies have curbed pollution, saved species from extinction, phased out harmful chemicals, and improved air quality. While current challenges exceed past triumphs in scale, these success stories prove that positive change is possible. They inject nuance into black-and-white thinking, reminding that not all efforts fail, not all trajectories lead to collapse. Reflecting on past gains encourages identifying partial solutions now worth pursuing, acknowledging that even modest improvements save lives, habitats, and chances for larger reforms later on.

Art and storytelling can also help counteract distortions. Fiction, films, and documentaries that depict both environmental struggles and efforts to restore ecosystems or build sustainable cities provide a richer narrative. Instead of reinforcing catastrophic outcomes or moral absolutes, these stories show characters grappling with complexity, evolving their strategies, and finding hope in small alliances. Consuming narratives that model complexity and resilience trains the mind to accept that the world is neither wholly doomed nor purely saved—it's always in flux, shaped by human choices that can shift outcomes.

Personal symbolic actions can reinforce more balanced thinking. Planting a tree, reducing waste, supporting green businesses—these actions, though small, ground the idea that not everything is pointless. If all-or-nothing thinking says, "If I can't solve it all, why bother?" a simple tree-planting retorts, "This helps sequester carbon, provides habitat, and symbolizes faith in the future." Tangible deeds combat the notion of helplessness and challenge catastrophizing by demonstrating that incremental improvements exist and matter. Similarly, participating in community projects—clean-ups, restoration efforts—visually contradicts the narrative of total collapse. Seeing neighbors working together solidifies the fact that progress doesn't require perfection.

Another technique involves reframing time horizons. Catastrophizing often fixates on distant, unverified scenarios. Bringing attention back to immediate steps—what can be done this week, this month—shrinks the temporal scale and reduces the chance of spiraling into distant doom. Similarly, rejecting all-or-nothing judgments about a policy's final outcome and focusing on the next review stage, or the next incremental target, reintroduces a sense of process. Change is often iterative, unfolding step by step, and acknowledging this reduces pressure to attain instant perfection.

It's important to remember that challenging cognitive distortions is not about denying legitimate concerns. The climate crisis is serious, and acknowledging its severity is rational. The goal is to prevent the mind from exaggerating negative outcomes to the point of paralysis or from imposing impossible moral standards that discourage effort. Balanced thinking invites sober acknowledgment of problems while maintaining emotional room for hope, problem-solving, and moral growth. Without catastrophizing, one can say, "The situation is dire, but not yet fixed forever in a negative outcome; our actions still matter." Without all-or-nothing thinking, one can say, "This effort may not solve everything, but it's a meaningful contribution and a step in the right direction."

Over time, repeated efforts to identify and correct distortions shape mental habits. What once came instinctively—leaping from a piece of bad news to total despair—becomes less automatic. As neural pathways strengthen around more balanced thinking, catastrophizing and rigid binaries lose their grip. This progression increases emotional resilience, allowing individuals to stay engaged with environmental issues without collapsing under their weight. Instead of giving up after reading a tragic climate report, one might acknowledge sadness, consider partial solutions, and take a break to breathe and reassess.

Mindfulness practices integrated with cognitive techniques further reinforce progress. Observing thoughts like "This will never get better" as transient mental events rather than truths disempowers the distortion. The mind learns to watch the thought float by without merging with it. Similarly, observing emotions without judgment helps recognize when fear inflates probabilities or oversimplifies outcomes. Gradually, the mind differentiates between facts and distortions, creating space for more reasoned responses.

In addition, relating these cognitive shifts to personal values grounds the process. If caring for the Earth and future generations matters deeply, then abandoning hope is not an option. Recognizing that catastrophizing undermines moral engagement and that all-or-nothing thinking discourages incremental changes reframes cognitive corrections as acts of moral fidelity. Tuning up one's mental filters to allow complexity, partial wins, and uncertain but possible better futures aligns thought patterns with chosen ethical commitments. With this alignment, the effort to avoid distortions is not just self-protection—it's part of living out cherished ideals under trying circumstances.

Social support remains vital. Sharing realizations about cognitive distortions with friends or community members encourages mutual reinforcement. Hearing someone say, "I've been catastrophizing too, and it helped when I reminded myself that not all outcomes are settled," breaks isolation. Watching others manage their fears more gracefully proves these efforts are realistic. As more people adopt these mental shifts, collective narratives can become more balanced, neither sugarcoating problems nor surrendering to fatalism. This collective cultural resilience can inspire policies and initiatives that reflect nuanced thinking—pursuing large-scale mitigation even when progress is incremental, celebrating partial victories while pressing for further improvements.

Creative expressions can record personal progress in challenging distortions. Some might write poems contrasting the mind's dark fantasies with the subtle hope found in collaborative action. Others might paint images representing complexity rather than collapse—mixing colors rather than painting in stark black or white. Artists who depict both the fragility and adaptability of nature can help viewers escape all-or-nothing frames and see that life is never wholly defined by a single narrative. Such artistic engagement cements the understanding that mental flexibility and emotional balance are part of cultural evolution as well as personal growth.

Over time, noticing fewer catastrophic thoughts or less all-or-nothing rhetoric in one's internal dialogue offers tangible evidence of progress. Anxiety episodes may become less frequent or less intense. More space appears for constructive planning—like deciding to volunteer for a local environmental group, adopting greener personal habits without feeling they must be perfect, or supporting incremental policy reforms. As thought patterns shift, hope emerges not as naive optimism but as a balanced perspective acknowledging difficulties while embracing potential improvements.

In professional contexts, these lessons can inform communication strategies. Advocates, educators, and policymakers who understand cognitive distortions can present information in ways that minimize panic and rigid thinking. They can emphasize incremental achievements, acknowledge uncertainties honestly, and highlight diverse scenarios. By doing so, they empower audiences to remain engaged rather than feeling crushed by fear. Emotionally intelligent communication respects psychological realities, helping individuals navigate complexity without resorting to extreme mental shortcuts.

No single technique cures all distortions instantly. Rather, it's a process of gentle correction—each time fear tries to shout "Doom!" or "All efforts are useless!," responding with "Wait, let's look at evidence, complexity, and incremental gains." Each time a policy setback tempts one to declare total failure, reframing the event as a learning step that may lead to future improvements. Each time catastrophic images loom large, recalling that the future is not fully written, that efforts can still alter trajectories, and that humanity has a track record of unexpected creativity. Gradually, the mind evolves from a host of distortions into a more even-handed evaluator of reality.

This journey is inherently moral. By resisting catastrophizing, individuals honor the truth that not all hope is lost. By challenging all-or-nothing thinking, they respect the moral worth of partial solutions and small positive actions. Embracing complexity and nuance is an ethical stance, rejecting easy despair and rigid judgments in favor of open-mindedness and persistence. This stance strengthens the emotional backbone needed to sustain involvement in long-term environmental endeavors.

In living through uncertain times, one must remember that how one thinks about challenges matters as much as the challenges themselves. Recognizing cognitive distortions and working to correct them keeps emotional responses proportionate, energizes moral courage, and supports realistic problem-solving. It's a way of meeting the climate crisis not with despair or denial, but with steady, compassionate clarity. Each thought reframed, each distortion softened, improves not only personal emotional health but the collective capacity to face a complex future with integrity and purpose.

Tools from Cognitive Behavioral Therapy (CBT) and Acceptance and Commitment Therapy (ACT)

Navigating the emotional turbulence provoked by environmental uncertainty often requires deliberate psychological tools. While simple grounding techniques and reframing can provide immediate relief, more structured therapeutic approaches, such as those derived from Cognitive Behavioral Therapy (CBT) and Acceptance and Commitment Therapy (ACT), offer deeper frameworks for understanding and reshaping one's relationship with fear. These therapeutic models were originally developed for various forms of anxiety, depression, and stress, but their principles can adapt readily to the anxieties triggered by the climate crisis. By learning to identify, challenge, and transform distressing thoughts, or to accept uncomfortable feelings without letting them dictate behavior, individuals can foster sustained emotional resilience that supports ongoing engagement with environmental issues.

CBT rests on a foundational idea: how we think influences how we feel and act. In situations involving climate-related fears, the mind may become trapped in patterns that amplify anxiety—catastrophizing, all-or-nothing thinking, or selective focus on negatives. CBT tools help recognize these distortions, evaluate their accuracy, and replace them with more balanced, grounded interpretations. The goal is not to adopt rosy optimism detached from reality, but to refine perception so that emotional responses match actual circumstances more closely. By adjusting thought patterns, CBT enables more constructive actions and reduces emotional overwhelm.

One core CBT technique involves thought records or worksheets. When a distressing thought arises—such as "We are doomed, everything is collapsing"—one writes it down. Then, systematically assess its evidence, consider alternative explanations, and note how thinking differently might change emotional intensity. For example, noting that while some systems face severe strain, many communities worldwide are implementing adaptation measures, restoring habitats, developing clean energy, and collaborating across borders to mitigate damage. Acknowledging these counterpoints does not deny severity but shrinks the mental space in which catastrophic assumptions run unchecked. Over time, this process trains the brain to spot cognitive distortions early and respond with balanced reasoning.

CBT also encourages behavioral experiments. If anxiety insists that no effort matters, test that assumption. Perhaps join a local environmental group or make a lifestyle change—like reducing single-use plastics—and observe outcomes. While a single action won't solve global crises, it may yield positive experiences: meeting supportive people, seeing measurable local improvements, or feeling more empowered. These experiences contradict fatalistic beliefs. By gathering real-world evidence through experimentation, CBT reduces the mind's reliance on unchallenged assumptions and reinforces more nuanced perspectives.

In addition, CBT often suggests identifying core beliefs that underlie anxiety. A core belief might be "Humans are incapable of doing the right thing," or "I must control everything to feel safe." In the climate context, such beliefs foster despair or unrealistic expectations. Challenging a core belief means asking: is it universally true that humans cannot cooperate for the greater good? History and current examples show otherwise. Perhaps the belief formed from selective attention to failures and neglect of successes. Modifying core beliefs from rigid absolutes to flexible viewpoints—"Humans have done both terrible and wonderful things; we can learn"—enables more stable emotional foundations and sustained motivation.

While CBT focuses on altering thought patterns, ACT takes a different but complementary approach. ACT posits that suffering often arises not merely from the content of thoughts, but from struggling against them or insisting that unpleasant feelings must vanish before taking action. Instead of trying to eliminate anxiety, ACT encourages acceptance: acknowledging fear, sadness, or uncertainty as natural responses in challenging times. Acceptance does not mean resignation; rather, it means making space for discomfort without letting it dictate one's choices. In the climate context, this could mean recognizing that worry about the future is understandable and will likely recur, but refusing to let that worry halt personal or collective efforts toward solutions.

One ACT technique is called cognitive defusion. When anxious thoughts appear—such as "This effort is pointless"—instead of wrestling with them, you learn to observe them as mental events passing through consciousness. Instead of "I am doomed," you might say internally, "I'm having the thought that I am doomed." By inserting this small phrase—"having the thought"—you create a subtle distance between the self and the thought. The thought ceases to be an absolute truth and becomes a product of the mind's activity. This distance reduces the thought's power. If "I am doomed" is just a thought, not a fact, one can still act in ways that align with values and hopes.

ACT also emphasizes identifying personal values and using them as guides. Values in the environmental context might be stewardship, compassion for future generations, reverence for biodiversity, or commitment to justice. ACT invites individuals to clarify what truly matters to them, then choose actions consistent with those values, regardless of whether anxiety subsides. Instead of waiting until fear disappears, ACT says: act anyway, guided by what you hold dear. Over time, this approach transforms anxiety from a barrier into a background noise that does not prevent living meaningfully. The distress is acknowledged, but it does not run the show.

Another ACT concept is willingness. Willingness means opening up to the full range of experiences—pleasant and unpleasant—associated with confronting climate issues. Rather than trying to avoid distressing information, one can allow sadness or worry to arise while still participating in community projects, reading about solutions, or speaking up in discussions. This stance resembles surfers riding waves rather than fighting the ocean. They know waves will come and go. Similarly, emotions ebb and flow. Being willing to feel discomfort for the sake of living ethically and purposefully aligns with moral courage. Climate uncertainty may provoke fear, but willingness means acknowledging that fear as part of the journey, not a sign to retreat.

Both CBT and ACT offer practical exercises that can be integrated into daily life. For example, a CBT-based exercise might involve listing catastrophic thoughts encountered after reading environmental news, then writing down less extreme alternatives. Over time, this exercise trains the mind to question initial reactions and consider balanced viewpoints. An ACT-based exercise might be a brief meditation where one visualizes anxious thoughts as leaves floating down a stream—observing them come and go without grabbing them. This visualization builds the skill of defusion, allowing thoughts to pass without entanglement.

Combining CBT and ACT principles can create a robust emotional toolkit. One might use CBT to identify and dispute distortions—ensuring thinking is not dominated by unfounded absolutes—while using ACT to accept that some anxiety will remain, yet it doesn't prevent constructive steps. For instance, a person might recognize the distortion in believing "If I can't fix everything, I've failed," challenge that thought by acknowledging the value of partial solutions, and simultaneously accept that feeling uneasy about the future is natural. Instead of waiting to feel no anxiety, they proceed to help with a local tree-planting initiative, embracing complexity rather than craving emotional perfection.

ACT's emphasis on values-based action is especially helpful in the face of long-term challenges like climate change. Anxiety might say, "It's too big, why bother?" but if your value is preserving life or ensuring justice for vulnerable communities, then action aligns with identity, not with fleeting emotional states. By acting according to values, individuals build meaning and resilience, even if anxiety persists. Values become the compass, anxiety a passing storm. Over time, consistent value-driven actions reduce helplessness and remind that individual and collective efforts can shape trajectories, even if incremental.

Moral clarity in ACT contrasts with the all-or-nothing mindset discussed earlier. ACT encourages embracing incremental changes without demanding they solve everything at once. By accepting partial progress and persisting anyway, people escape the trap of seeing small efforts as futile. Similarly, CBT's tools help recognize that catastrophizing ignores multiple outcomes. Together, they enable balanced reasoning and moral perseverance.

Both CBT and ACT acknowledge that thoughts are mental events, not absolute dictates. CBT does so by challenging distorted content, ACT by disentangling thoughts from identity. In eco-anxiety, this distinction matters. Many people feel morally burdened by their worries or ashamed of their inability to remain calm. Understanding thoughts as events or products of a stressed mind reduces self-criticism. One can say, "I'm having anxious thoughts," rather than "I'm weak for feeling anxious." This shift fosters compassion toward oneself, crucial for maintaining stamina amid ongoing environmental stressors.

As emotional patterns improve, individuals may find more room for creativity and innovation in their responses to climate problems. Freed from paralyzing fear or rigid thinking, they can imagine adaptive strategies, mobilize communities, or communicate effectively. The emotional resilience built through CBT and ACT enhances the capacity for rational debate, technological exploration, and policy advocacy. Instead of an anxiety loop that stifles action, one gains a stable mental platform from which to engage the complexities of climate solutions.

Group settings can integrate these therapeutic tools as well. Environmental organizations might host workshops introducing CBT or ACT principles to their members. This could involve role-playing scenarios where catastrophic thoughts are challenged collectively, or guided exercises in mindfulness and values clarification. By sharing these approaches, groups reduce burnout, support each other emotionally, and maintain a spirit of hope and resolve. Collective well-being improves when fear does not dominate internal narratives, allowing effective collaboration, sustained campaigns, and better negotiation of solutions that require patience and compromise.

Cultural adaptation of these tools ensures broader accessibility. While CBT and ACT developed in Western psychological traditions, their core insights resonate universally: thoughts affect emotions, acceptance aids resilience. Practitioners can translate exercises into different cultural contexts, incorporating local metaphors, spiritual references, or traditional rituals. For example, a community that reveres ancestors might adapt defusion exercises by imagining anxious thoughts as transient visitors who come and go, guided by ancestral wisdom that endures. Another community might blend mindfulness with longstanding meditation traditions, reinforcing acceptance through culturally familiar practices. Such adaptation ensures these techniques align with diverse moral and spiritual frameworks.

As individuals get comfortable with these approaches, they begin recognizing patterns in their mental lives. A person might notice that reading certain types of news triggers catastrophizing more than others, or that discussing policy failures with friends fuels all-or-nothing judgments. With awareness, they can prepare. Maybe before reading a heavy climate report, they commit to a short ACT-based defusion exercise. Or when entering a policy debate, they remind themselves that partial improvements are not zero. Over time, emotional regulation becomes a skill applied proactively, not only in crisis moments.

These therapies highlight that emotional resilience is learnable. Initially, practicing cognitive restructuring or accepting feelings without resistance might feel awkward. But as neural pathways strengthen, the new habits become more natural. This gradual transformation offers hope. Just as societies can learn to adapt infrastructures or revise policies in response to climate signals, individuals can learn to adapt their mental landscapes. Anxiety ceases to be an unalterable fate; it becomes a state that can be influenced, managed, and guided toward productive ends.

Some might worry that accepting anxiety or recognizing distortions means losing vigilance. On the contrary, balanced thinking and acceptance support rational vigilance. Freed from catastrophic and rigid thinking, people can assess risks more accurately. Fear remains an important signal, alerting to dangers, but it no longer controls the narrative. With tools from CBT and ACT, individuals maintain alertness without letting anxiety spiral into despair. This preserves clarity and moral focus, enabling decisive actions rather than futile hand-wringing.

Building a supportive environment enhances these techniques' efficacy. Friends, family, or community members aware of CBT and ACT principles can remind each other when distortions appear. A simple prompt—"Are you catastrophizing?" or "Is there a more balanced view?"—can restore equilibrium. Encouraging one another to accept difficult emotions rather than fighting them can break isolation. Over time, such communal reinforcement normalizes healthier cognitive patterns, spreading resilience beyond individuals to entire networks of people concerned about the planet.

Educators and communicators can also integrate these approaches. When explaining climate information, acknowledging common distortions helps audiences navigate emotional responses. Policy briefs, public talks, or workshops can mention that absolute doom narratives may exaggerate certain dynamics, and that while serious, not all data points to inevitable collapse. Presenters can gently challenge all-or-nothing expectations by celebrating incremental policy gains and partial transitions to renewable energy. Transparency about complexity can preempt distortions and invite a more stable emotional reception of facts.

In families, parents might help children handle eco-anxiety by sharing CBT concepts like identifying thoughts that predict worst-case scenarios and testing them. Or by introducing ACT's idea that feeling worried about the planet is normal, yet does not prevent acting kindly toward nature and making greener choices. Children raised with these mental tools may approach environmental challenges in adulthood with confidence that their emotional states, while intense, can be understood and guided. Such early education sets a foundation for lifelong emotional resilience in uncertain times.

Personal reflection is key. Integrating CBT and ACT techniques involves experimenting to find what resonates. Some may prefer structured thought-challenging exercises, writing down distorted thoughts and counterarguments. Others might lean toward ACT's acceptance and values-based approaches, placing emphasis on living meaningfully despite ongoing worry. The choice depends on temperament, culture, and personal style. Many find a blend helpful: starting with CBT to clarify thought patterns, then applying ACT principles to embrace emotional complexity and commit to actions aligned with values.

Over time, consistent use of these therapies rewires emotional habits. The immediate panic once triggered by certain headlines may diminish in intensity. Instead of ruminating on catastrophic outcomes, the mind might pause, note the fear, and respond with balanced reasoning. Instead of dismissing partial achievements as failures, the mind can appreciate them as steps forward. Instead of waiting for perfect emotional comfort, individuals take meaningful steps while carrying anxiety lightly. This new emotional repertoire supports a healthier, more engaged relationship with the climate crisis—one that acknowledges threats without succumbing to despair, and values incremental progress without demanding unattainable purity.

As these therapeutic tools become second nature, individuals gain confidence that they can face future environmental developments with greater stability. As more people adopt such skills, collective emotional resilience grows, creating fertile ground for long-term adaptation and policy innovation. Emotional well-being and effective climate action reinforce each other: clear thinking and moral courage emerge more readily from a mind not trapped by distortions or paralyzed by aversion to discomfort. By employing CBT and ACT principles, people can evolve from anxious spectators to active participants in shaping the future.

Tools from CBT and ACT, while stemming from clinical psychology, resonate with universal human experiences of uncertainty and moral striving. Climate anxiety is one among many challenges that test emotional stamina. The good news is that humans have developed robust psychological frameworks to handle inner turmoil. Adopting these methods does not trivialize the climate crisis but ensures that engagement arises from grounded, courageous reasoning rather than panic or rigid dogma. With practice, identifying distorted thoughts, embracing acceptance, clarifying values, and committing to meaningful actions becomes a fluid process, enabling individuals to move through fear and act with integrity and compassion.

Learning to Hold Complexity: Balancing Alarm with Openness to Solutions

As news about environmental change accumulates, many find themselves grappling with layers of conflicting emotions—fear, sadness, anger, frustration, and occasionally hope. The mind often craves simplicity, wanting to categorize the world as either on the brink of collapse or easily salvageable. Yet reality rarely abides such neat divisions. Environmental challenges unfold amid a tapestry of partial successes, incremental improvements, ongoing injustices, scientific breakthroughs, political inertia, inspired activism, and cultural shifts. Navigating this complexity demands the capacity to hold alarm side by side with openness to solutions. Instead of letting fear drive a narrative of inevitable doom, or clinging to naive optimism that erases real problems, learning to embrace complexity empowers individuals to respond with informed courage and measured resolve.

Complexity means acknowledging that multiple truths can coexist. Climate models project serious risks, yet human societies have addressed formidable problems before and made measurable progress in certain areas. Sea levels may rise, but that does not preclude adaptation measures like restoring wetlands or adopting floating infrastructure. Forest degradation is widespread, yet reforestation efforts, sustainable agriculture, and indigenous land stewardship provide counterpoints. Embracing complexity is not about minimizing threats. Rather, it involves recognizing that while alarm is justified, it need not eclipse a nuanced understanding that includes windows of opportunity, moral agency, and adaptive capacity.

The tendency to oversimplify arises partly from emotional overwhelm. When faced with dire projections, the mind may resort to catastrophic visions that dismiss any possibility of mitigation. Alternatively, some deny severity to maintain comfort, focusing solely on good news and ignoring harsh realities. Complexity lies between these extremes, admitting gravity without surrendering to fatalism, acknowledging steps forward without denying ongoing harm. This balanced stance requires intellectual humility and emotional flexibility. It means admitting uncertainty: no single narrative—doom or salvation—encompasses the whole truth. Instead of final verdicts, one must learn to work with probabilities, partial outcomes, and evolving scenarios.

Balancing alarm with openness to solutions also involves resisting the lure of instant conclusions. For instance, discovering that a particular ecosystem faces grave threats might spark despair. Yet pausing to consider scientific research on species adaptation, local conservation initiatives, policy reforms, and the resilience of natural systems can complicate that initial despair. Complexity encourages looking at multiple layers: yes, coral reefs face bleaching events, but researchers experiment with heat-resistant corals and improved marine policies. This does not guarantee success, but it refutes the notion that all paths lead to total ruin. Complexity honors the moral challenge of acting despite uncertainty. Rather than waiting for a guaranteed outcome, people invest effort in uncertain ventures because partial improvements matter and incremental changes can shift trajectories.

One key step is questioning all-or-nothing judgments. When reading unsettling data, notice if the mind leaps to "nothing can be done" or "we must achieve perfect solutions immediately." Complexity means recognizing value in partial solutions. For example, reducing emissions by 30% is not everything scientists recommend, but it's still meaningful progress that buys time for further improvements. A habitat restored in one region does not solve global biodiversity loss, but it preserves genetic reservoirs, stabilizes local ecosystems, and demonstrates that positive interventions are possible. Instead of dismissing these steps as inadequate, complexity acknowledges them as pieces in a larger puzzle. Each step forward can facilitate additional measures, alliances, and cultural shifts that build toward greater achievements.

Another aspect of complexity is accepting that environmental narratives unfold over decades, even centuries. Alarm often inflates the immediacy of threats to total irreversibility, while solution-oriented optimism may downplay long-term struggles. Complexity recognizes timescales: some impacts may be locked in, but how societies adapt and minimize further damage will shape long-term outcomes. Complex thinking acknowledges that trends can bend over time. Political climates shift, technologies advance, cultural norms evolve. Just as ecological systems sometimes surprise researchers with resilience or unexpected regeneration, human systems can pivot under pressure. Alarm urges vigilance; openness to solutions invites sustained, adaptive effort rather than declaring outcomes prematurely.

Embracing complexity also means tolerating discomfort. Holding multiple truths—severe problems alongside genuine responses—feels mentally demanding. Simplicity seduces because it reduces cognitive load, offering clear villains or neat endings. Complexity demands acknowledging moral ambiguity: some policies help in one area but cause trade-offs in another; certain technologies reduce emissions but raise ethical questions about resource extraction. Accepting complexity is not a passive shrug of resignation. It involves active engagement with moral dilemmas: if a wind farm displaces a community, how to weigh clean energy benefits against social costs? If reforestation sequesters carbon but affects traditional land uses, how to navigate these tensions?

The willingness to sit with complexity fosters nuanced thinking. Instead of discarding partial solutions as greenwashing or celebrating them uncritically, one can evaluate them on multiple axes—environmental impact, social justice, long-term viability. Complexity encourages asking: what progress is this measure achieving, what are its limitations, and what complementary actions are needed? This perspective avoids the trap of cynicism that lumps all efforts as futile and the trap of complacency that overlooks ongoing harm. It enables a flexible, iterative approach where strategies evolve and improve over time.

Learning to hold complexity also enhances communication and coalition-building. Extreme narratives—either total doom or assured salvation—alienate those who find them too simplistic. Embracing complexity resonates with people who acknowledge the seriousness of issues but also want space for constructive action. Policymakers, activists, scientists, and citizens can find common ground by discussing trade-offs, incremental steps, and evolving scenarios. Instead of polarizing debates into fatalistic or dismissive camps, complexity-based dialogue allows participants to refine goals, negotiate compromises, and celebrate progress without losing sight of remaining challenges.

From a psychological standpoint, holding complexity provides emotional steadiness. Catastrophic fears can paralyze, while naive optimism can shatter when confronted with hard facts. Complexity grounds expectations in a realistic assessment: yes, major difficulties lie ahead, but incremental measures can mitigate harm and create pathways forward. This balanced outlook reduces emotional whiplash—where each new piece of news yanks emotions from despair to euphoria or back again. A complex perspective acknowledges that ups and downs are natural. Some initiatives fail, others succeed partially, and over the long run, cumulative efforts shape the future.

Techniques from previous frameworks—like CBT and ACT—support complexity thinking. When catastrophic thoughts arise, complexity-oriented reframing might list ongoing mitigation projects, showing that outcomes are not binary. When encountering all-or-nothing judgments, complexity encourages seeing the gray zones: a policy might cut emissions in one sector while needing improvement in another. Mindfulness helps remain present with conflicting emotions—alarm and hope—without forcing a resolution. Acceptance of complexity also aligns with ACT's emphasis on values: acting ethically even in uncertain conditions, acknowledging that moral worth is found in persevering, adapting, and learning, rather than in guaranteeing perfect solutions.

Real-world examples illustrate complexity's power. Consider renewable energy transitions. Early on, skeptics argued that wind and solar could never displace fossil fuels. Alarmists might say that since renewables can't fix everything at once, it's hopeless. But complexity reveals that wind and solar costs have dropped dramatically, installations soared, and while fossil fuels persist, a cleaner energy mix is emerging. This does not solve climate change overnight, but it undermines claims of absolute futility. Complexity recognizes that transformation often happens unevenly, through partial successes that open doors for more ambitious steps later.

Agricultural shifts, too, show complexity at work. Industrial farming practices degrade soil, pollute waterways, and contribute to emissions. Alarmist thinking might conclude all food systems are doomed, while all-or-nothing mindsets might demand immediate perfection. Complexity acknowledges movements toward regenerative agriculture, soil health initiatives, local food networks, and innovative agroforestry. These approaches do not cure all ills at once, but they improve some conditions, preserve biodiversity, and demonstrate viable alternatives. Complexity embraces these efforts as sources of learning and incremental improvement rather than dismissing them as inadequate. Seeing complexity encourages sustained support, refinement, and scaling of these initiatives over time.

In the realm of adaptation, coastal communities exploring living shorelines, wetlands restoration, and managed retreat from flood zones exemplify complexity. Alarm might say rising seas doom all coasts. Denial might say technology will fix everything easily. Complexity admits coastal areas face tough choices, but layered solutions—like marshland restoration plus policy shifts—

can reduce risks incrementally. Not perfect or total, these measures mitigate harm and buy time for further innovations. Over decades, layered strategies can shift outcomes from catastrophic to challenging but manageable. Embracing complexity allows communities to proceed with measured hope, not blind faith or abject surrender.

Cultural examples highlight complexity's moral dimension. Indigenous wisdom often teaches that humans have long engaged with changing ecosystems, neither mastering them fully nor passively enduring them. They've developed cultural practices that balance alarm—recognition of vulnerability—with openness to adaptive strategies, knowledge exchange, and respectful stewardship. Complexity thus resonates with ancestral approaches that never expected a static world, instead valuing continuous learning, moral responsibility, and mutual care. Modern societies can learn from these traditions, seeing uncertainty not as an excuse for despair but as an invitation to remain morally alert and ethically flexible.

At a personal level, practicing complexity involves refining emotional literacy. When reading dire climate predictions, instead of concluding all hope is lost, acknowledge distress, then remember partial efforts, resilience stories, and incremental reforms. Let conflicting emotions coexist: sorrow for damage already done, appreciation for restoration projects, worry over political inertia, and recognition that some policies show promise. Holding these emotions side by side frees the mind from binary traps, allowing it to proceed thoughtfully. This emotional agility prevents burnout, keeps the door open for constructive thinking, and encourages seeking allies, learning new skills, or joining community initiatives.

Balancing alarm with openness to solutions also aids decision-making. If alarm dominates, one might withdraw from engagement, overwhelmed by despair. If naive optimism dominates, one might overlook ongoing harms and fail to push for necessary reforms. Complexity ensures decisions arise from a realistic assessment of challenges and possibilities. This might mean advocating strongly for emissions cuts while supporting adaptation measures, or simultaneously pressing for environmental justice alongside technological innovation. Complexity-driven decisions recognize that we must work on multiple fronts—mitigation, adaptation, cultural change—and that these processes unfold gradually, demanding patience and persistence.

This perspective also reshapes dialogue around targets and timelines. Setting ambitious emissions goals is crucial, but complexity acknowledges that meeting them involves setbacks, partial compliance, policy revisions, and unexpected breakthroughs. Instead of interpreting any shortfall as total failure, complexity encourages treating these milestones as part of a learning curve. Countries or organizations that miss a target can adjust strategies, strengthen commitments, or invest in new solutions. This approach aligns with moral perseverance, refusing to yield to cynicism when outcomes do not align perfectly with initial hopes. Complexity fuels sustained effort, interpreting difficulties as lessons rather than as final verdicts.

Embracing complexity also respects the diversity of cultural, political, and economic contexts. Climate solutions vary across regions, influenced by local conditions and histories. Alarm might push a one-size-fits-all narrative, demanding immediate uniform changes. Complexity understands that different communities find different entry points into sustainability. One region may start with renewable energy, another with reforestation, another with policy reforms. Each path has merits and limits. Recognizing this variety reduces friction and encourages sharing best practices. Complexity translates into ecological pluralism: acknowledging that multiple routes toward greener futures can coexist, each contributing partial solutions that cumulatively shift trajectories.

Narratives that embody complexity can inspire audiences. Instead of films, books, or documentaries that depict only apocalypse or utopia, storytellers can craft nuanced tales showing communities grappling with trade-offs, forging alliances, experiencing partial wins and losses, yet persisting over time. Such stories mirror reality more closely, helping people accept complexity as normal rather than confusing. When complexity becomes culturally familiar, individuals resist panic or oversimplification, improving emotional preparedness for future developments. Engaging with complex narratives trains emotional muscles to handle uncertainty without defaulting to extremes.

Certain spiritual or philosophical traditions echo complexity. Daoist thought, for instance, emphasizes balancing opposites, acknowledging change as constant and urging flexible responsiveness. Applying such insights to climate uncertainty encourages acceptance that conditions evolve, that moral agency resides in adapting gracefully. Similarly, traditions highlighting impermanence remind practitioners that even dire situations can shift, just as stable conditions may erode. Complexity resonates with these teachings, validating that alarm about threats and hope from incremental improvements can coexist within a grand cycle of change and adaptation.

From a psychological perspective, complexity offers a refuge from emotional exhaustion. Clinging to an all-doom narrative drains energy. Believing in a neat rescue scenario that never materializes fuels disillusionment. Complexity, by acknowledging multiple streams of progress and setbacks, creates a stable emotional baseline. Instead of oscillating between despair and denial, complexity-driven mindsets maintain a steady engagement, recognizing that each step forward counts. This stability reduces burnout, allowing long-term commitment to environmental action. Emotional health improves when people feel no need to force certainty onto inherently uncertain situations.

In educational contexts, teaching young people to handle complexity provides durable skills. Students who learn early that environmental issues involve messy realities—partial solutions, ongoing learning, moral trade-offs—are better equipped for adulthood. They won't be shocked or demoralized by incremental progress or partial failures. Instead, they see them as normal phases in social and ecological evolution. This mindset fosters resilient future leaders who can integrate science, ethics, politics, and economics without flinching at contradictions or partial outcomes. Complexity training becomes a form of emotional inoculation against despair.

In activism and policymaking, acknowledging complexity can refine strategies. Campaigns framed entirely around catastrophe may mobilize initial attention but risk leaving supporters feeling helpless if quick results don't emerge. Campaigns that highlight incremental victories, show learning curves, and admit ongoing challenges maintain morale over time. Similarly, policymaking that involves iterative targets—updating goals as technology improves or as understanding deepens—feels more authentic and achievable. Complexity informs adaptive governance, where policies evolve rather than fail entirely, and new data leads to adjustments rather than abandonment.

Media narratives that embrace complexity can also help. Instead of sensationalism or sugarcoating, journalists can present both the gravity of climate findings and stories of adaptation, technological gains, environmental justice movements, and community resilience. Complexity in reporting provides audiences a fuller picture, reducing knee-jerk despair or complacency. Over time, public discourse matures, guided by pluralistic understanding rather than simplistic headlines. Citizens become better informed and less susceptible to emotional extremes, more capable of supporting nuanced policies.

At the personal level, practicing complexity can mean regularly seeking balanced information sources. After reading a grim forecast, deliberately look up ongoing restoration projects or policies showing some success. This is not to deny problems but to integrate multiple perspectives into a broader mental map. Building a habit of cross-checking dire predictions with incremental improvements prevents one narrative from monopolizing the mind. Emotions become more stable when they reflect a realistic mixture of alarm and possibility.

Daily reflection exercises can cultivate complexity. Before concluding the day, think about a worrying environmental story encountered and pair it with a positive development or a partial solution learned. This mental pairing teaches the brain that worry and hope often co-occur. Over time, it becomes second nature to see not just the negative but also the kernel of resilience or ingenuity that can inform the next step. Complexity becomes the default lens, making emotional balance more accessible even in tough times.

Dialogues with friends, family, or colleagues also provide opportunities to model complexity. When someone says "We're doomed," gently remind them of counterexamples or incremental gains without dismissing their alarm. Conversely, if someone expresses blind optimism, acknowledge the progress they highlight but also point out ongoing challenges. By practicing this balanced communication, everyone grows more comfortable with complexity. This reduces polarization and supports respectful, informed discussions where multiple viewpoints integrate into richer understanding.

Complexity aligns with moral realism: acknowledging that while outcomes are uncertain, moral obligations persist. Despite not knowing if global warming can be kept below certain thresholds, people can still strive ethically, valuing incremental emissions cuts and harm reductions. Complexity says that even if not all species can be saved, preventing the extinction of some still matters deeply. This moral stance protects against cynicism. Instead of giving up because perfection is elusive, complexity respects moral worth in making conditions "less bad" or "less risky," understanding that better outcomes often arise from cumulative modest improvements.

Ultimately, complexity does not offer guaranteed comfort. Accepting it means admitting that no single solution or policy ensures an ideal future. But it does mean avoiding the emotional traps of absolute despair or naive cheerleading. Complexity legitimizes the emotional weight of alarm while also legitimizing the rational hope found in ongoing efforts. It encourages acting today—even amid uncertainty—investing in strategies that could yield better conditions tomorrow.

Complexity-based engagement might not feel as emotionally neat as a narrative of guaranteed doom or instant rescue, but it is more aligned with reality's texture.

By learning to hold these contradictions—recognizing grave dangers yet also acknowledging human ingenuity and partial successes—individuals cultivate a kind of emotional maturity. This maturity fosters resilience, enabling them to stick with climate action for the long haul. It also allows them to integrate new information without collapsing into despair or denial. Complexity-based thinking is dynamic, updating as new evidence and initiatives emerge. Over time, complexity transforms despair into determined pragmatism, anger into focused advocacy, and apathy into constructive participation.

In a world reeling from complexity's demands, discovering strength in it reframes uncertainty. Complexity becomes not a burden but a resource, teaching patience, moral tenacity, and creativity. Its embrace paves the way for nuanced policies, robust public dialogue, sustainable activism, and personal emotional stability. Rather than a crisis of meaning, environmental uncertainty becomes a field of moral endeavor where steady engagement, learning, and incremental wins matter. Complexity doesn't promise easy outcomes, but it ensures that alarm is not wasted and that openness to solutions remains kindled, guiding humanity through an evolving landscape with courage and humility.

These approaches teach that even when anxious thoughts distort perception, it is possible to regain clarity and moral steadiness. Recognizing catastrophic and all-or-nothing patterns prevents being trapped by fear or unrealistic demands, opening room for more balanced appraisals. Utilizing CBT and ACT principles encourages reframing thoughts, accepting emotional complexity, and acting according to deeply held values rather than transient distress. Embracing the nuances and uncertainties of environmental challenges allows alarm to coexist with pragmatic hope, reminding that progress often arises from incremental efforts, learning curves, and moral perseverance. By refining thinking habits and emotional responses, individuals can engage with the climate crisis from a place of resilience, creativity, and ethical commitment—building a sturdier foundation for long-term dedication to the work of shaping a livable future.

Chapter 7: Building Lasting Resilience

Long-Term Self-Care: Sleep, Nutrition, Exercise, Routine

As engagement with environmental issues deepens, many people encounter emotional turbulence that does not dissipate overnight. Ongoing climate uncertainties, distressing headlines, and moral dilemmas accumulate pressure on minds and bodies. Faced with long-term challenges that may span decades, maintaining resilience requires more than immediate coping techniques. Incorporating steady self-care habits that support well-being over time is essential. Just as ecosystems rely on continuous cycles—seasons replenishing nutrients, natural rhythms promoting stability—human resilience also depends on consistent patterns of rest, nourishment, movement, and daily structure. By attending to sleep, nutrition, exercise, and routine, individuals cultivate a foundation that can sustain emotional health through evolving environmental realities.

These pillars of self-care are not luxuries but fundamental needs. When external conditions provoke anxiety, the body's stress responses heighten, and missing basic physiological requirements can amplify distress. Adequate sleep, balanced meals, regular physical activity, and a stable daily rhythm collectively reduce vulnerability to panic, irritability, and hopelessness. They create a baseline of internal equilibrium so that when fear or sadness arises, it arises against a backdrop of physical steadiness rather than chronic depletion. Over time, meeting these basic needs fortifies mental stamina, enabling people to face threats with greater clarity, moral courage, and perseverance.

Sleep stands as one of the most critical elements of long-term resilience. Inadequate sleep impairs cognitive function, emotional regulation, and physical health. Without sufficient rest, the mind struggles to process complex information, making it harder to hold nuanced perspectives about climate issues or resist catastrophic thinking. Sleep supports memory consolidation, helping integrate facts, strategies, and emotional lessons into a coherent understanding. It also regulates mood, allowing the emotional brain—particularly the amygdala—to calm down overnight. By maintaining healthy sleep habits, individuals arrive at each new day more prepared to handle uncertainty without succumbing to panic or numbness.

Cultivating better sleep involves setting consistent bedtimes and wake-up times, avoiding excessive screens and alarming news late at night, and creating a restful environment. While these adjustments may sound mundane, their impact on resilience cannot be overstated. Imagine reading a dire climate projection after a good night's sleep versus after chronic sleep deprivation. With proper rest, the mind can process alarming information more rationally, balancing alarm with perspective. Without it, dread may swell uncontrollably. Over weeks and months, steady sleep patterns form an emotional shield, a protective layer that softens the edges of distress.

Nutrition also plays a profound role. The body and mind depend on nutrients to support neurotransmitter production, hormonal balance, and stable energy levels. Diets rich in whole foods—vegetables, fruits, whole grains, lean proteins—provide essential vitamins and minerals that influence mood and cognitive function. For instance, omega-3 fatty acids, found in fish and certain plant sources, help maintain brain health, potentially reducing mood swings and anxiety symptoms. Iron and B vitamins support steady energy and concentration, which matter when sifting through complex environmental data or making moral judgments under stress.

Emotional resilience can erode if the body runs on empty or fluctuates between sugar highs and crashes. Balanced meals modulate blood sugar, preventing sudden spikes in irritability or fatigue that might intensify anxious thoughts. Eating mindfully—tasting flavors, savoring textures—also doubles as a grounding exercise. Focusing attention on a meal can momentarily pull the mind from catastrophic futures into the tangible present. Over time, consistent, balanced nutrition fosters baseline emotional stability, making it easier to stay engaged with climate issues without burning out.

Exercise brings yet another dimension of resilience. Physical activity releases endorphins, improves cardiovascular health, and reduces muscle tension associated with chronic stress. Even moderate exercise—like brisk walking, yoga, cycling—can lift mood, sharpen focus, and counteract the body's stress chemistry. Environmental anxiety often tightens muscles, quickens breathing, and raises heart rate. Exercise counters these physiological markers, returning the body to a calmer baseline. Regular workouts also symbolize self-agency: if the world feels chaotic, the decision to move the body on purpose, to build strength or flexibility, is a reminder of personal power.

Beyond physiological benefits, exercise routines create opportunities to step away from screens and persistent alarming news. A walk in a park or a dance session at home offers a mental break, a time to process emotions subconsciously, and return refreshed. By committing to regular physical activity, individuals invest in their capacity to handle emotional strain. After a run or a yoga class, one might approach climate discussions with more equanimity, seeing challenges as something to grapple with thoughtfully rather than as an overwhelming storm. Exercise helps transform fear-laden energy into constructive vigor.

If sleep, nutrition, and exercise form the physical bedrock of resilience, routine adds the structural framework. In a world of shifting environmental predictions and policy uncertainties, personal routines provide a sense of predictability and stability. Knowing that certain tasks, rituals, or moments of rest recur daily or weekly provides anchors amid chaos. Routines need not be complex: a morning breathing exercise, a midday meal enjoyed without distraction, an evening walk, or a short journaling practice can create familiarity. This familiarity tells the nervous system that not everything is unpredictable, that life still contains patterns one can rely on.

Routines also reduce decision fatigue. When climate-related worries arise, it's easier to handle them if basic self-care practices are already baked into daily life. Instead of debating whether to exercise or go to bed on time, routine makes these choices habitual, freeing mental energy for more pressing concerns. Over time, routines weave a net beneath daily existence, catching individuals when anxiety tries to knock them off balance. Knowing that each morning includes a quiet moment to set intentions helps maintain focus on long-term goals, like supporting environmental adaptation or learning new skills for community resilience.

Together, sleep, nutrition, exercise, and routine form a comprehensive foundation for emotional stamina. When one area falters—perhaps a bad night's sleep—consistency in others—like a healthy breakfast or a midday walk—partially compensates, preventing a complete emotional nosedive. Over weeks and months, building these habits is like forging armor. The mind and body become more resistant to stressors, less reactive to bad news, and more capable of sustaining engagement with complex problems. Instead of crumbling after reading dire reports, individuals can process them more calmly, assess options, and choose meaningful actions aligned with their values.

These self-care foundations also interact positively with other coping techniques previously discussed—like reframing thoughts, practicing mindfulness, or using CBT and ACT tools. For instance, a well-rested, well-nourished brain finds it easier to identify and challenge cognitive distortions. When energy levels are stable, the mind can pause and think: "Is this thought catastrophic or balanced?" Similarly, mindfulness becomes more accessible when the body is not on edge from poor sleep or sugar crashes. ACT's emphasis on values-driven action resonates more when the individual is not battling physical exhaustion.

Long-term self-care also supports collective resilience. When activists, policy advocates, researchers, or community leaders practice healthy habits, they reduce burnout. This ensures that people invested in environmental solutions can continue their work over the long run rather than dropping out due to emotional fatigue. Teams benefit when each member maintains personal well-being—fewer mood swings, better concentration, improved morale. This fosters stable, constructive collaboration. Healthy individuals contribute to healthier networks, which in turn amplify their capacity to handle large-scale environmental challenges.

Addressing the basics of sleep, nutrition, exercise, and routine helps counter the narrative that self-care is selfish or trivial in the face of global problems. Just as well-maintained infrastructure supports societies through crises, well-maintained bodies and minds support individuals through emotional storms. This form of care does not replace activism, policy work, or education; it enables them. A person who invests in sleep and balanced meals does so not to escape reality, but to face it more effectively. Proper self-care is a strategic choice, ensuring long-term involvement rather than short-lived bursts of energy followed by collapse.

Admittedly, establishing these habits can be challenging amid personal and societal pressures. Some may struggle to find time for exercise or afford balanced meals. Limited sleep might stem from shift work, caregiving responsibilities, or stress-induced insomnia. Creating routines might clash with erratic schedules. These barriers suggest that while individual effort matters, systemic supports are also needed. Workplaces that respect work-life balance, policies that improve food access, community spaces for exercise, and public health education all contribute to a cultural environment that supports resilience. Self-care thrives best when not left solely to personal willpower but encouraged by social conditions that value well-being.

Adapting routines to personal preferences fosters sustainability. Not everyone enjoys early-morning runs; some may prefer evening yoga. One might appreciate gentle stretching before bed rather than high-intensity training. Similarly, nutrition need not follow strict diets or moralizing rules; finding enjoyable, wholesome foods that fit cultural traditions and personal tastes is key.

Sleep hygiene can start with small steps—reducing late-night screen time by ten minutes at first or experimenting with calming teas. Complexity is not reserved for climate scenarios alone; it also applies to self-care. Embracing an experimental approach—trying different sleep schedules, meal plans, or workout routines—helps discover what truly supports one's emotional equilibrium.

Integrating mindfulness into these foundational habits can enrich their benefits. Mindful eating transforms a meal into a grounding exercise. Savoring each bite, noting flavors and textures, anchors the mind in the present. Mindful movement—focusing on breathing and bodily sensations during exercise—reinforces awareness of the here and now. Mindful bedtime routines—like reading a soothing book or practicing breathwork—set the stage for restful sleep. By combining self-care basics with mindfulness, individuals weave a network of micro-habits that gently calm the nervous system and reaffirm agency.

As these habits solidify, individuals may notice that encountering negative climate news no longer triggers immediate panic or exhaustion. Instead, the body and mind have a reserve of stability. The mind can acknowledge fear while recalling that it got eight solid hours of sleep, ate nourishing meals, and followed a consistent routine. These memory traces reinforce confidence: "I've maintained my well-being before, I can handle this emotion now." Emotional reactions become more proportional, and the space between stimulus (bad news) and response (emotional collapse or destructive thinking) widens. In that space lies freedom to choose better responses.

Long-term self-care also encourages incremental improvements. Perhaps one starts by prioritizing sleep, knowing that exhaustion exacerbates anxiety. As sleep regularizes, energy allows more consistent exercise. With higher energy and mood, interest in cooking healthier meals grows. As all three stabilize, setting a regular daily routine—carving out time for reading climate reports calmly, reflecting on solutions, or engaging in local volunteer work—becomes easier. Each step supports the next, forging a virtuous cycle. Rather than feeling overwhelmed by overhauling everything at once, small, steady enhancements accumulate.

Social accountability can strengthen these habits. Finding a friend or colleague also aiming to improve sleep or establish an exercise routine can create mutual support. Checking in on each other's progress or attending a weekly class together turns solitary effort into shared commitment. Similarly, meal prepping with roommates or family ensures healthier eating becomes a team endeavor. Aligning personal well-being with community interaction fosters resilience at multiple levels—individual and collective. Everyone benefits when emotional steadiness reduces conflicts, miscommunications, and burnout within networks working on environmental solutions.

Reflecting on the moral dimension of self-care can also enhance motivation. Caring for one's physical and emotional health is not a distraction from moral duties but a foundation for fulfilling them. If anxiety threatens to derail engagement with climate activism, prioritizing rest and nourishment directly supports staying active in that cause. Balancing personal well-being with service to others prevents resentment and exhaustion from undermining moral aspirations. Recognizing that strength and clarity arise from a healthy body and mind makes self-care feel integral to the broader ethical project of protecting life and reducing harm.

For some, cultural traditions may already value these basics. Indigenous societies often incorporate communal meals, rituals that promote rest, and seasonal activities that keep bodies in tune with nature's rhythms. Traditional cuisines emphasize whole foods, and cultural celebrations often blend physical activities—dances, walks, pilgrimages—that integrate exercise into social life. Learning from these traditions can reinforce self-care as a natural part of human existence rather than a modern invention. Incorporating ancestral wisdom or cultural practices can give personal routines deeper meaning, linking one's well-being to generations of knowledge on harmonious living.

Technology can assist in maintaining these habits. Sleep-tracking apps, nutrition guides, online fitness classes, and productivity tools can support consistency. However, caution is needed. Relying excessively on gadgets can create stress if metrics become obsessions rather than guides. The aim is not perfection but balance. Using technology mindfully, one can schedule reminders to go to bed, find healthy meal ideas, follow gentle workout videos, or create timetables that reduce decision fatigue. Technology's role should be to facilitate a routine that feels nourishing, not to impose rigid standards that replicate all-or-nothing thinking in the realm of self-care.

Regularly reassessing these habits ensures they remain beneficial. Seasonal changes might require adapting exercise routines—perhaps indoor activities during harsh winters, outdoor cycling when weather improves. Dietary preferences can shift with changing availability of local produce or personal taste evolutions. Sleep patterns might need adjusting if work hours or family responsibilities alter. By approaching these habits as flexible and evolving, people prevent self-care from becoming another source of stress. Instead of forcing a rigid routine, embrace slight modifications to keep them relevant and supportive. Over time, this adaptability reflects the same ethos required to address climate challenges: ongoing learning, incremental adjustments, and resilience.

As internal foundations strengthen, individuals can better handle difficult conversations about climate futures. Instead of feeling emotionally drained by each new debate or piece of data, the body and mind carry reserves of calmness and physical well-being. More stable emotions allow for listening deeply, acknowledging concerns without getting defensive or defeated. This improved communication can influence how families, communities, and workplaces address environmental issues. Calm, attentive dialogue grounded in emotional resilience fosters cooperation, creative problem-solving, and empathy. Self-care thus ripples outward, enabling more effective collaboration.

Long-term self-care also improves the ability to celebrate partial victories or appreciate small joys amid troubling news. With balanced energy from proper sleep and nutrition, the mind is more open to noticing subtle improvements—a species recovering in one region, a city banning single-use plastics, a grassroots group planting community gardens. Enjoying a good meal with friends after a day of challenging climate work reminds that life holds beauty alongside struggle. Regular exercise endows a sense of strength and capability, reinforcing the notion that personal agency persists even under daunting circumstances. Such recognition of the good alongside the bad aligns with complexity-based thinking, sustaining hope without denying hardship.

In a sense, these habits form an internal climate of well-being. Just as planetary health depends on stable natural cycles, personal emotional health depends on stable cycles of rest, nourishment, and movement. By maintaining internal homeostasis, individuals better withstand external shocks. Anxiety fluctuates, but a well-rested body, well-fed brain, and a consistent daily structure moderate those fluctuations. One learns that storms of fear may blow through, yet they pass, and routine activities remain constants that provide safe harbor.

Such self-care also reduces reliance on maladaptive coping—like excessive caffeine, alcohol, or doomscrolling—that can intensify stress. Instead of reaching for quick fixes that worsen mood or fatigue, these foundational habits offer reliable, healthy alternatives. A stable routine ensures that when tension rises, returning to a known practice—like a short walk, a nutritious snack, or a quiet breathing session before bed—restores equilibrium without harmful side effects. Gradually, healthier self-care replaces harmful escapes, enhancing long-term resilience.

Encouraging others to adopt these habits fosters a supportive culture of well-being. If a friend struggles with eco-anxiety, gently suggesting improvements in sleep or offering to prepare a balanced meal together can help them find steadier ground. Families who jointly commit to regular exercise not only improve collective health but strengthen relationships, building trust and mutual support. Communities that promote access to healthy foods and safe places for recreation create conditions where more people can fortify their emotional stamina. This collective empowerment ensures that responding to climate uncertainty does not rest solely on heroic individuals but arises from collaborative networks of balanced, resourceful citizens.

Reflecting on progress after months of consistent self-care can validate the effort. Compare emotional reactions to bad news now versus when starting these habits. Are panic attacks less frequent? Is it easier to focus on constructive next steps rather than freezing in despair? Has empathy grown, thanks to stable energy and emotional reserves that allow listening to others' concerns without feeling overwhelmed? Observing these improvements builds confidence. Knowing that personal habits directly influence emotional resilience encourages further refinements and sustained commitment to self-care.

While no single habit guarantees immunity from anxiety, the combination of sleep, nutrition, exercise, and routine forms a strong protective net. Anxiety may still appear, but it may feel more manageable, shorter-lived, and less corrosive. This resilience does not deny the seriousness of climate challenges; it acknowledges them from a place of relative calm and moral steadiness. Over the long run, this approach ensures that engagement with environmental issues remains integrated into daily life rather than treated as an emergency that depletes all emotional reserves.

Finally, these self-care foundations integrate seamlessly with other strategies—reframing thoughts, accepting complexity, and using CBT or ACT tools. A well-rested mind more easily identifies cognitive distortions. A well-nourished brain can hold complexity without mental fatigue. A body supported by regular exercise responds to stress with less intensity, making it simpler to practice defusion or values-based action. A stable routine provides the time and consistency needed to incorporate mindfulness, journaling, or community engagement. These foundational habits thus serve as the bedrock upon which other resilience-building techniques rest, ensuring longevity and depth in emotional adaptability.

Embracing long-term self-care acknowledges that sustainable emotional health mirrors sustainable ecological health. Just as restoring ecosystems demands patience, ongoing maintenance, and respect for natural rhythms, nurturing personal well-being requires steady efforts that honor the body's and mind's inherent needs. Meeting these needs does not trivialize external challenges; it equips individuals to face them more effectively. By treating sleep, nutrition, exercise, and routine as integral components of resilience, people prepare themselves for the emotional journey ahead, ready to withstand uncertainty, learn from difficulties, and contribute thoughtfully to shaping a livable future.

Cultivating Gratitude, Hope, and "Micro-Moments" of Joy

When wrestling with the emotional weight of environmental uncertainty, many people find that fear, sadness, or anger tend to overshadow feelings of appreciation, optimism, and delight. It is natural for the mind to gravitate toward looming threats, catastrophic possibilities, and the urgency of what must be done. Yet, to remain engaged and not collapse under stress, nurturing positive emotions is equally important. Gratitude, hope, and small instances of joy act like emotional counterweights, balancing the moral gravity of the situation with the nourishment needed to persevere. Cultivating these qualities does not mean turning away from hard truths or indulging in denial. Instead, it recognizes that resilience emerges from a heart strengthened by understanding what remains precious, meaningful, and worth cherishing—even amid difficulty.

Gratitude often proves elusive in the face of grim forecasts. One might think, how can gratitude coexist with warnings of declining species, rising seas, or extreme weather events? Yet gratitude does not ignore these realities; it acknowledges them, then shifts focus to what remains stable, what people still value, and who helps carry burdens. Gratitude may arise from observing that communities come together after a disaster, that scientists share research freely, or that small policy wins accumulate over time. It can come from personal blessings: supportive friendships, a beloved natural place that still offers solace, the existence of art, music, and cultural traditions that enrich life. Recognizing these positives, even if modest, reframes emotional narratives from total despair to complexity, where loss coexists with treasures not yet destroyed.

This shift matters because persistent negative focus, while understandable, drains energy. Without occasional recognition of what's good, moral courage wanes. Gratitude, by highlighting reasons to care, nourishes moral conviction. For example, feeling grateful for ancestral knowledge that guided people through past hardships can reaffirm one's sense of continuity and purpose. Expressing gratitude for a community garden project that brings neighbors together underlines that collective action is not futile. Gratitude reminds individuals that while the future is uncertain, the present still holds value. Embracing gratitude helps maintain emotional balance, enabling engagement without drowning in sorrow.

To cultivate gratitude, one can start small. At day's end, reflect on three things that brought comfort or inspiration. Perhaps a conversation restored faith in human cooperation, or a sunrise revealed nature's beauty still intact. Maybe a policy debate showed some leaders willing to consider environmental justice. Writing these points down integrates them into memory. Over time, this practice creates an internal resource bank, a library of good experiences and achievements. When anxiety flares, recalling these moments counters one-sided pessimism. Such rituals transform gratitude from a fleeting emotion into a stable habit, a reliable anchor in emotional storms.

Another approach is to share gratitude with others. Thanking a friend who listened patiently to eco-anxiety confessions, acknowledging a local environmental educator for their work, or praising a volunteer who helps restore habitats amplifies positive emotions within networks. Collective gratitude builds solidarity—knowing that appreciation circulates among allies reduces feelings of isolation. As gratitude becomes a communal practice, people can draw strength from each other's recognition of efforts and values. Celebrating even small wins publicly—like a neighborhood's new recycling initiative—reaffirms that progress matters.

Hope, closely related to gratitude, also needs careful cultivation. In the climate context, hope can feel delicate, easily dismissed as naive or false. Yet hope is not blind optimism; it's the mindset that allows one to hold fear and possibility together. Hope acknowledges adversity but refuses to conclude that all futures must be dire. It rests on understanding that human creativity, policy changes, cultural shifts, and technological advances can still alter trajectories. Hope is not about denying the seriousness of challenges—it's about recognizing that uncertainty includes space for improvement and that moral agency persists.

Cultivating hope involves paying attention to signs of resilience. Notice when a species rebounds after conservation measures, or when a city adopts ambitious renewable targets. Observe that global cooperation, though imperfect, has produced treaties, alliances, and research exchanges that mitigate harm. Hope finds footing in these examples, seeing them not as trivial or insufficient but as evidence that change can occur. Over time, collecting such examples fortifies hope's foundation. Even if large-scale transformations are slow, small indicators of adaptability and problem-solving remind that worst-case scenarios are not destiny.

Another way to foster hope is through envisioning better futures. This means allowing the mind to imagine scenarios where policies tighten regulations, communities develop green infrastructure, youth activism shapes government agendas, and people adapt lifestyles to more sustainable patterns. These visions are not guaranteed outcomes, but they break the spell of inevitability that fear can cast. Visualizing better worlds provides direction and motivation. It encourages asking, "What steps lead toward that scenario?" and inspires concrete actions, reinforcing that efforts are meaningful. Hope thrives when individuals see their role in shaping outcomes, even if modestly.

Hope also emerges from connecting with communities of practice. Engaging in environmental groups, cultural organizations, or educational circles that share values and goals counters loneliness. Knowing others invest energy in solutions reduces despair. Seeing that one is not alone in hoping and working for improvement confirms that hope is not a personal whim but a collective endeavor. Over time, these networks build trust and resilience: when one person falters, others lend support, rekindling hope that might dim after a harsh news cycle. Hope, like gratitude, becomes woven into social fabrics, making it less susceptible to erosion.

While gratitude and hope provide emotional uplift, "micro-moments" of joy serve as sparks that brighten daily existence. These are brief instances of pleasure, wonder, or comfort that do not deny complex challenges but offer respite. A micro-moment might be admiring a flower blooming in a sidewalk crack, smiling at a child's laughter, tasting a ripe fruit, or noticing a bird singing against urban noise. These flashes of joy remind that life continues to produce beauty and that enjoyment is not frivolous even amid serious issues. Enjoying such moments does not detract from moral seriousness; instead, it refuels emotional reserves, preventing bitterness or cynicism from taking root.

Environmental anxiety can trick individuals into thinking that enjoying life is irresponsible, as if happiness betrays the cause. Yet sustainable engagement requires sustaining oneself. Joy is not betrayal; it's maintenance. Small joys help keep emotional aridity at bay, ensuring that dedication to solutions arises from love, not just anger or fear. If despair dominates, actions become grim and mechanical. In contrast, joy provides a sense of purpose—protecting what is cherished, celebrating that nature still offers surprises, and validating that human creativity can still produce beauty. These moments of joy strengthen moral commitments by grounding them in positive connections to the world rather than obligations alone.

One can intentionally seek micro-moments of joy. Paying attention to senses helps: savoring morning coffee, feeling warmth of sunlight, noticing intricate leaf patterns. Engaging in creative activities—writing poetry, playing an instrument, taking photographs—fosters joy that springs from self-expression and connection to the present. Exploring local green spaces, even if modest, reveals subtle ecologies thriving despite urban pressures. Each encounter with a butterfly, a resilient weed, a playful squirrel, or a fleeting rainbow counters the narrative that all is bleak. By accumulating these moments, the mind learns that complexity includes not only negative data but also instances of delight.

Another method is sharing these joys with others. Telling a friend about a beautiful sunset or a small community victory spreads positive energy. Social media, often a channel for alarming news, can also host brief celebrations: a post showing an urban garden in bloom or a local art piece inspired by environmental themes. Celebrating and amplifying positive micro-moments complements the serious work of advocating, protesting, and analyzing data. It ensures that emotional landscapes remain fertile, allowing hope and gratitude to flourish alongside alarm and determination.

Integrating gratitude, hope, and joy into daily routines stabilizes them as ongoing sources of strength. For example, start or end each day by recalling something you appreciate, articulating what you hope will improve, and noting a small pleasure experienced. Over time, this practice rewires the brain's attentional patterns. The mind no longer fixates solely on dire projections but acknowledges that reality is multidimensional. This does not dismiss severity; it places severity in context. Just as ecosystems thrive on biodiversity, emotional resilience thrives on diverse emotional states, including gratitude, hope, and joy.

Cultural narratives often highlight the importance of balancing distress with positive emotions. Many traditions encourage gratitude as a moral and spiritual practice, reaffirming that even in hardship, some aspects of life remain gifts. Others highlight hope as a virtue, sustaining efforts toward a just future. Joy, too, appears in rituals and communal celebrations, maintaining social cohesion despite adversity. Adopting these cultural lessons enriches personal resilience strategies. Instead of feeling guilty for seeking small joys or glimmers of hope, understand them as part of long-standing human wisdom on enduring hardships with grace.

These positive states also support moral agency. If anxiety feeds the idea that "nothing can be done," gratitude reminds that some things have been done and have helped. Hope suggests that more can be done if perseverance continues. Joy refuels the will to try. Together, they encourage acting on behalf of cherished values—protecting vulnerable species, advocating for clean energy, supporting policy reforms—despite lacking a guarantee of success. Positive emotions ensure that moral action arises not from desperation alone, but from a wellspring of care and recognition of what's worth saving.

Environmental educators, communicators, and leaders can incorporate these themes into their work. Emphasizing that complex problems coexist with partial victories and uplifting examples prevents audiences from concluding all is lost. Showcasing stories of local heroes, successful restoration efforts, technological breakthroughs, and cultural shifts fosters hope. Encouraging people to notice small wonders in their communities—even during difficult times—normalizes seeking joy. Reminding audiences to feel gratitude for the scientists, activists, and community members making a difference strengthens solidarity. By modeling these attitudes, leaders inspire more balanced emotional engagement, reducing burnout and motivating sustained action.

In activism, maintaining gratitude, hope, and joy improves relational dynamics. Movements avoid devolving into resentment or internal conflicts when they embrace positive emotions. Gratitude for supporters and allies helps maintain unity. Hope nurtures long-term planning rather than frantic short-term pushes. Micro-moments of joy at events—like singing uplifting songs, sharing nourishing meals, or admiring eco-art—enrich collective spirit. This creates a culture where participants feel valued and energized, more likely to stay involved for years rather than burning out after initial enthusiasm fades. Positive emotions become part of the movement's resilience strategy, not a distraction from its goals.

At the personal level, practicing these positive states does not always come easily. Anxiety and negativity bias can overshadow good news or small comforts. Persistence is key. Start by intentionally directing attention to something positive each day—maybe an inspiring climate initiative discovered online or a recent local improvement. Write it down. Over time, noticing the positive grows more automatic, balancing the mind's natural inclination to highlight threats. The brain's neuroplasticity allows it to become more adept at spotting hope-inspiring data points or joy-sparking scenes. Eventually, this skill transforms emotional landscapes, making it harder for despair to monopolize thinking.

Maintaining perspective is essential. Gratitude does not mean overlooking injustices or suffering. Hope does not mean ignoring that some impacts are irreversible. Joy does not mean trivializing harm. Each of these positive states coexists with awareness of pain. In fact, positivity gains authenticity when it stands alongside honest recognition of problems. It emerges as a deliberate choice to acknowledge complexity. Rather than demanding pure happiness or blind optimism, it means enriching the emotional palette so that sorrow, alarm, and anger do not define the entirety of one's experience.

ACT's concept of acceptance resonates here: acknowledging fear while acting on values aligns with embracing gratitude and hope despite distress. Similarly, CBT's reframing tools can spotlight the positive elements neglected by catastrophizing or all-or-nothing thinking. The synergy between these therapeutic frameworks and positive states is clear. Once distortions are managed, it becomes easier to see reasons for gratitude or hope. Once acceptance of discomfort is achieved, moments of joy feel more natural and less contradictory. Positive emotions and therapeutic tools reinforce each other in a virtuous cycle.

On a collective scale, if more individuals practice gratitude, hope, and joy, cultural narratives can shift. Public discourse might highlight not only problems but also solutions in progress, heroes on the frontlines, and ecosystems showing unexpected resilience. Hopeful storytelling encourages more people to join efforts, while gratitude for small policy advances supports patience during long negotiations. Micro-moments of joy shared publicly can inspire emulation. Over time, this cultural environment sustains a stable emotional climate conducive to long-term action. Instead of panic-driven mobilizations that fade, society can nurture steady, informed, value-driven involvement.

Gratitude also aids mental health in the face of harsh news cycles. By appreciating what remains functioning and supportive, people counter the sense of losing everything at once. Hope keeps future possibilities open, fostering what some call "hopeful realism," acknowledging difficulties but still pushing forward. Joy interrupts stress accumulation, preventing chronic anxiety from taking over. These emotions protect mental well-being from sliding into despair, making engagement more sustainable. Emotional health and sustained activism or policy work go hand in hand.

Finding personal symbols or reminders can reinforce these positive states. A photograph of a favorite natural landscape on a desk can prompt gratitude and joy during stressful moments. A journal dedicated to hopeful stories—emerging green technologies, local adaptation successes—becomes a resource to revisit when doubts surge. Even wearing a small token, like a pendant symbolizing nature's resilience or a bracelet gifted by a supportive friend, can trigger memory of these positive states. Over time, these tangible cues embed positive emotional habits into daily life.

One must also acknowledge that not everyone has equal access to experiences that foster these positive emotions. Inequalities and injustices mean some communities face immediate threats that make gratitude or hope harder to find. Still, even in harsh conditions, people have found ways to celebrate small victories and maintain cultural traditions that spark joy. Acknowledging these disparities and striving to spread resources, education, and opportunities more equitably can support everyone's capacity for positive emotions. Social justice reinforces emotional resilience by reducing chronic stressors and ensuring that hope and gratitude are not privileges reserved for the fortunate few.

Balancing alarm with gratitude, hope, and joy strengthens resolve. Alarm ensures vigilance, preventing complacency. Gratitude, hope, and joy ensure that vigilance does not curdle into nihilism. Together, they produce an emotional equilibrium suited for long-haul efforts. This balance also encourages creativity. When fear dominates, thinking narrows. With hope in the mix, minds open to innovative strategies. Gratitude for existing solutions inspires learning from them to scale up. Joy fosters curiosity and experimentation. Such openness is crucial because meeting climate challenges demands new ideas, adaptive policies, and cultural transformations.

Consistent practice cements these states. Initially, it may feel forced to seek a hopeful story after reading dire reports or to list reasons for gratitude when feeling overwhelmed. Yet perseverance yields fruit. Over time, positive emotions flow more naturally, requiring less effort. The mind learns that scanning for hope or recalling gratitude are normal responses to distress. Achieving this shift does not trivialize problems; it simply ensures that emotional energy is not monopolized by despair. Instead, energy can fuel long-term dedication, enabling moral depth, strategic thinking, and the resilience needed to navigate shifting environmental conditions.

Cultural artifacts—films, literature, music—can reinforce these emotional states. Watching a documentary that highlights both the severity of coral reef bleaching and the scientists working tirelessly on restoration projects exemplifies complexity balanced by hope. Reading a novel where characters find moments of joy while struggling for environmental justice humanizes the journey. Listening to songs celebrating nature's beauty while acknowledging its fragility can move hearts toward protective action. By choosing art that resonates with balanced emotional profiles, individuals internalize the idea that positivity and alarm can coexist harmoniously.

Role models who exemplify this emotional equilibrium further inspire others. Leaders who acknowledge risks honestly while maintaining genuine hope encourage followers to do the same. Activists who celebrate small wins publicly and express gratitude for supporters show that gratitude is not a sign of weakness, but a source of strength. Scientists who share both sobering data and positive developments in technology or policy encourage audiences to appreciate complexity. Such role models become living proof that it is possible to handle difficult information without losing moral traction.

In day-to-day life, cultivating positive emotions can be as simple as starting a meeting with a brief acknowledgment of recent achievements, or concluding a difficult conversation with a note of thanks for everyone's efforts. Incorporating short nature breaks or humor into long work sessions dealing with grim data can refresh minds. Celebrating progress—no matter how small—staves off cynicism. The principle is that steady, intentional infusion of gratitude, hope, and joy into routines and interactions builds an emotional climate supportive of sustained engagement.

When integrated into personal and collective practices, these positive states help ensure that environmental anxiety does not define the entire emotional palette. They restore agency, affirm moral responsibilities without allowing fear to cripple action, and keep ethical commitments tethered to experiences of worth, beauty, and love. Such integration produces a richer emotional landscape where people acknowledge alarm, learn from it, then pivot to appreciate what endures, imagine better futures, and find pleasure in small wonders. Far from complacency, this emotional richness fuels perseverance, ensuring that engagement with environmental issues lasts not just through one crisis, but across a lifetime of evolving challenges.

By embracing gratitude, hope, and micro-moments of joy, individuals and communities develop a reservoir of emotional resilience. They can return to these positive states whenever tension mounts or exhaustion threatens. Over time, these emotional skills form an essential layer of adaptive capacity, just as vital as technological solutions or policy reforms. Without emotional resilience, even the best policy faces public fatigue; without moral courage, even the greatest technology remains underused. Positive emotions are not decorative luxuries—they are functional assets, enabling the moral and intellectual strength required to navigate uncertain futures with dignity and resourcefulness.

As these practices become ingrained, individuals notice subtle shifts. Receiving alarming news might still hurt, but it no longer plunges them into despair. Instead, they remind themselves of progress made in other areas, recall that many people are working on solutions, and let themselves relish a peaceful moment outdoors. Hope resurfaces not as denial, but as recognition that no single data point seals fate. Gratitude emerges readily, tempering bitterness with appreciation for what still stands. Joy appears like a spark of light, renewing emotional energy when it begins to fade.

This transformation does not hinge on giant leaps. Small steps—writing one grateful thought per day, celebrating one incremental policy improvement, cherishing one tiny joy each afternoon—accumulate into a lasting emotional foundation. Over years, these habits reshape the mind, making it less susceptible to extremes, more tolerant of complexity, and better equipped for long-term moral engagement. In the same way ecosystems rely on myriad small interactions to maintain stability, emotional ecosystems rely on countless small moments of positivity to remain resilient under stress.

The quest for environmental resilience extends beyond infrastructures and policies into hearts and minds. While alarm ensures vigilance, gratitude recognizes what's worth protecting, hope fuels the resolve to try, and joy replenishes emotional vitality. Their interplay allows sustained moral effort: continuing to advocate, educate, adapt, and experiment despite unpredictability. Embracing these qualities fosters not naive optimism, but a profound moral steadiness grounded in values and lived experience. Individuals and communities equipped with gratitude, hope, and micro-moments of joy face the future with eyes open, hearts fortified, and the moral courage to keep walking forward through changing landscapes.

Embracing Imperfection and Letting Go of Eco-Perfectionism

As the urgency of environmental issues grows more apparent, a natural impulse may arise to strive for flawless eco-credentials—aiming to eliminate every trace of pollution from one's life, make exclusively sustainable choices, and never falter in moral responsibility. While the desire to do right by the planet is commendable, pursuing absolute environmental purity can become emotionally taxing and unrealistic. Embracing imperfection, on the other hand, acknowledges that while dedication matters, the messy realities of human life rarely fit neat ideals. Letting go of eco-perfectionism does not mean giving up or excusing harm; rather, it allows for steady engagement, moral growth, and sustained effort over the long term, even amid inevitable trade-offs and complexities.

The allure of perfectionism arises partly from the severity of ecological challenges. Facing threats to species, ecosystems, food security, and climate stability, individuals may feel that anything less than total virtue fails the planet. Guilt can accumulate if one falls short—using a plastic container occasionally, taking a necessary car trip, or supporting a policy that isn't fully green. This guilt intensifies pressure and anxiety, leading to emotional strain. Instead of feeling empowered by efforts to reduce harm, perfectionism casts every shortcoming as a moral failing, fueling despair and sometimes prompting people to abandon action entirely. Embracing imperfection means rejecting this all-or-nothing mindset and recognizing that partial progress, incremental improvements, and responsible choices within constraints still carry worth.

Just as no ecosystem is perfectly pristine—natural systems adapt, shift, and tolerate certain disturbances—human contributions to environmental well-being must navigate real-life conditions. Cultural, economic, and infrastructural factors limit the absolute control any individual can exert over their footprint. Even the most dedicated activists rely on technologies, goods, and services shaped by global supply chains. Complete insulation from environmental harm is unattainable. Accepting this reality reduces shame and fatalism, enabling people to focus on what can be done rather than fixating on what cannot be perfectly achieved.

Letting go of eco-perfectionism involves dismantling rigid standards that measure worth by unattainable benchmarks. It means acknowledging that riding a bike most days, but occasionally using a car, does not invalidate all the good done by cycling. A policy that cuts emissions partially may still represent a valuable step, despite not meeting every ambition. By appreciating increments, the mind transforms frustration into patience and moral perseverance. This shift frees emotional energy once consumed by self-reproach, allowing for more practical engagement, creativity, and coalition-building.

Recognizing that everyone lives within constraints helps. A person might lack access to locally sourced organic produce due to cost or geographical limitations. Another might need to drive for work or family commitments. Some communities lack robust recycling infrastructure or sufficient public transport. Blaming oneself entirely for these conditions ignores systemic factors. While individual actions matter, expecting perfection from oneself or others in an imperfect world leads to disappointment. Embracing imperfection invites advocating for better policies, technologies, and infrastructures that support more sustainable choices, acknowledging that personal virtue alone cannot resolve structural problems.

Moral complexity further justifies releasing perfectionism. Imagine someone who reduces personal carbon footprints significantly but must use air travel occasionally to visit aging relatives in distant places. Is this morally worthless? Hardly. It reflects balancing different moral claims—caring for family and limiting emissions. Similarly, choosing to eat mostly plant-based but occasionally consuming local fish might respect cultural traditions or nutritional needs. Such compromises do not erase environmental commitments; they demonstrate ethical discernment. Embracing imperfection encourages nuanced moral reasoning rather than moral absolutes that discount entire contexts and responsibilities.

Another insight arises from the risk of burnout. Eco-perfectionism demands constant vigilance, turning daily life into a series of moral litmus tests. Over time, this can lead to exhaustion, resentment, or cynicism. When perfection proves impossible, one might give up entirely, reasoning that if they cannot be faultless, why bother trying at all. Embracing imperfection breaks this cycle. Acknowledging that one can do good without doing everything perfectly sustains engagement. It allows for rest, forgiveness, and learning from mistakes, ensuring long-term involvement rather than short-lived intensity followed by withdrawal.

Embracing imperfection also humanizes the environmental movement. When advocates present themselves as flawless paragons, they may intimidate potential allies or inspire defensiveness rather than cooperation. If people perceive environmental efforts as an exclusive club requiring unattainable purity, many will opt out. On the other hand, acknowledging imperfection invites broader participation. It says: "We are all working on this, we all struggle with compromises, and every positive step helps." This inclusive attitude fosters solidarity, as people realize they need not be saints to contribute meaningfully. Instead, incremental improvements become collectively powerful, accumulating to influence markets, policies, and cultural norms.

To practice embracing imperfection, start by identifying unrealistic standards. Notice if a small slip—like using a disposable cup once—sparks disproportionate self-criticism. Ask: "Is it fair to dismiss all my previous efforts because of this one moment?" Recognizing such patterns allows reframing. Instead of "I failed," think, "I did well most times, and this time I encountered a constraint. I can learn from it or offset it in another way." Turning moral absolutes into moral trajectories—paths of gradual improvement rather than demands for instantaneous purity—fosters a kinder, more encouraging inner dialogue.

Compassion plays a key role. Treat oneself as a learner, not a culprit. Much like mastering a skill takes time and inevitable errors, becoming more sustainable also involves experimentation, setbacks, and gradual refinement. If an attempt to reduce plastic waste is partially successful, celebrate the reduction achieved rather than lament the unavoidable plastic straw used in an emergency. This self-compassion reduces shame, making it easier to persist and try new solutions. Over time, replacing harsh judgments with empathetic understanding builds emotional strength, vital for sustaining engagement amidst complex environmental realities.

Seeking community support helps reinforce these attitudes. In forums, discussion groups, or local environmental gatherings, people can share their struggles with imperfection. Hearing others admit they face limits or compromises normalizes the experience. Perhaps a fellow advocate explains how they must commute by car occasionally due to lack of public transport; rather than condemning them, the group acknowledges that systemic change is needed. This collective acceptance of imperfection builds trust and cohesion, galvanizing efforts to push for better infrastructures and policies that make sustainable choices more accessible. Compassionate communities highlight that everyone contributes differently, and moral worth is not contingent on flawless performance.

Integrating ACT principles can also facilitate letting go of perfection. Acceptance means acknowledging discomfort about not meeting ideal standards. Instead of fighting that discomfort, notice it, name it, and allow it without concluding moral failure. Defusion techniques help detach self-worth from the thought "I must be perfect or I'm a hypocrite." Seeing this thought as just a thought enables acting according to values—caring for the environment—despite imperfect adherence. Similarly, CBT tools help reframe rigid beliefs: if the mind insists that one must eliminate all waste to matter, question that assumption, seek counterexamples, and adopt a more flexible, supportive belief that partial efforts still matter significantly.

Recognizing that sustainable lifestyles unfold in evolving contexts aids releasing perfection. Just as ecosystems adapt to changing conditions, human behaviors adapt to new knowledge, resources, and technologies. An acceptable choice now might be replaced later as better options emerge. Rather than feeling guilty for not starting perfectly from day one, one can appreciate a growth mindset: learning continuously, updating choices, and acknowledging that even small shifts accumulate over time. Instead of fixating on a static ideal, celebrate each incremental improvement—fewer disposable items than last year, less food waste than before, more thoughtful product choices today than yesterday.

Embracing imperfection aligns well with complexity-based thinking. The environmental crisis resists simple narratives, and expecting personal purity tries to impose a false simplicity—either perfect or nothing. Complexity acknowledges that progress often emerges in fragmented, uneven patterns. One area improves, another lags behind. Instead of despairing at these discrepancies, complexity invites working strategically: use personal strengths to reduce environmental harm where possible, accept that some domains remain challenging, and trust that collective efforts will fill gaps. Imperfect actions form part of a larger mosaic that, over time, can influence significant shifts.

By accepting imperfection, individuals also alleviate the burden of constant guilt. Chronic guilt drains motivation and reduces emotional resilience. With a more balanced perspective, guilt transforms into a prompt for learning rather than a permanent mark of moral failure. Suppose someone buys a product discovered later to be less eco-friendly than advertised. Instead of wallowing in guilt, they can note the lesson—research brands more carefully next time—and move forward. With each iteration, efforts become wiser, more strategic, rather than stalling due to self-blame. This approach sustains morale and encourages continuous improvement.

Cultural traditions can provide insights. Many ethical systems acknowledge human fallibility and emphasize striving for betterment rather than attaining perfection instantly. Drawing on these traditions reassures that moral worth emerges from honest attempts, acknowledging human limitations, and caring enough to keep trying. Religious or spiritual frameworks that prioritize humility, forgiveness, and incremental good deeds support a mindset where ecological ethics can likewise evolve gradually. Similarly, humanistic or philosophical traditions may highlight that moral growth arises from wrestling with dilemmas rather than applying rigid rules.

In activism, embracing imperfection can shift strategy. Rather than criticizing those not fully meeting ideal standards, activists can invite them into the fold, praising partial steps and offering guidance to do more. This inclusive approach enlarges the movement's base, accelerates change, and avoids alienating potential allies by insisting on moral purity. When everyone feels their efforts are acknowledged, even if imperfect, they remain engaged and curious about learning more. This solidarity is crucial for large-scale transformations that require broad participation, not just a select few saints of sustainability.

From a psychological perspective, letting go of perfectionism frees cognitive resources. Energy once spent on self-criticism or hiding perceived moral failings can redirect toward problem-solving, advocacy, or outreach. Instead of ruminating over a single imperfection, the mind can focus on systemic improvements—like pushing for better waste management infrastructure. Emotional relief from perfectionism also enhances creative thinking, enabling innovators to propose partial solutions that complement others, building a network of overlapping efforts rather than searching in vain for a silver bullet.

Over time, practicing imperfection acceptance leads to a more stable emotional baseline. Reading alarming headlines or encountering resistance to green policies might still sting, but it no longer triggers existential despair about personal failings. Instead, one can say, "I'm doing what I can under present circumstances, and I can adjust as new opportunities arise." This stance encourages moral patience, recognizing that building sustainable societies is a generational endeavor. Imperfect steps taken today pave the way for future progress, and staying emotionally steady through that journey matters as much as any single accomplishment.

Micro-moments of joy and gratitude, previously discussed, dovetail well with imperfection acceptance. Enjoying a small positive environmental development—like noticing a local farmer's market grow—becomes easier when not demanding that everything be flawless. Partial successes feel meaningful when not overshadowed by unrealistic expectations. Joy finds room to breathe, acknowledging that beauty and resilience persist even in flawed conditions. Gratitude for what remains helps ward off frustration at what isn't perfect yet. Together, these positive emotions reinforce a mindset that cherishes improvement rather than lamenting incompleteness.

Role models who admit their imperfections and yet continue contributing to environmental solutions inspire others. When well-known activists or community leaders share stories of their own struggles—maybe their difficulty eliminating all single-use plastics or their reliance on non-ideal energy sources due to local constraints—they demonstrate honesty and relatability. Audiences see that moral dedication need not come wrapped in moral absolutism. This honesty reduces intimidation and guilt, encouraging more people to get involved. Imperfect role models expand the movement's ranks by illustrating that moral commitment coexists with human limitations.

Embracing imperfection also paves the way for learning from failures. If perfection is the goal, failure signifies total collapse. If improvement is the goal, failure provides lessons. For example, a pilot program for composting in a community might encounter contamination issues. Instead of branding it a pointless exercise, imperfection acceptance frames it as a learning step. The community analyzes what went wrong—lack of education on acceptable materials, insufficient supervision—and implements corrections next time. Gradually, systems refine. This iterative process, in which imperfection spurs adaptation, mirrors ecological patterns of trial, error, and resilience.

Another advantage of rejecting perfectionism is that it prevents moral elitism. If a few achieve near-perfect environmental lifestyles, treating them as the standard can alienate others who cannot. Embracing imperfection fosters egalitarian ethics: everyone can contribute improvements without needing to meet utopian ideals. This inclusive ethic harnesses the collective power of millions making moderate changes rather than relying on a tiny minority's purity. Democratically distributed effort can often outweigh the impact of a handful of individuals who attain near-zero footprints. Imperfect contributions scale across populations, driving systemic shifts.

In personal life, releasing eco-perfectionism eases daily stress. Instead of counting every carbon atom in despair, one can focus on manageable changes—like reducing meat consumption a few days a week, choosing local produce when feasible, or supporting community projects occasionally. Over time, these habits accumulate, and when conditions allow, one can further refine choices. Moral growth unfolds as a narrative of incremental adaptation rather than a singular test of moral purity. With this perspective, emotional resilience grows, making it more likely that individuals remain engaged for the long run instead of burning out in pursuit of unattainable ideals.

Mindfulness can enhance this process by noticing when perfectionist thoughts appear. Label the thought: "I'm insisting on perfect purity again." Recognize the associated tension, the urge to self-criticize. Then let that thought pass, returning focus to realistic goals and the next feasible action. Similarly, journaling can help process guilt or disappointment when not meeting a personal standard. Writing acknowledges the feeling, then reframes the experience as part of a moral journey. Over time, self-talk evolves from punishing to encouraging, reinforcing a healthy mindset that supports consistent effort despite imperfections.

One might wonder if letting go of perfection dilutes moral ambition. On the contrary, moral ambition thrives more sustainably when not burdened by impossible demands. By acknowledging imperfection, individuals free themselves to aim high without expecting overnight transformation. This paradoxically fosters greater total change over time. People feel free to attempt improvements without fear that partial outcomes discredit them. They may set ambitious goals—like significantly cutting household waste—knowing that achieving 70% of that reduction is still valuable progress. Without the perfectionist trap, they dare more often and persist longer.

Cultural narratives can shift when imperfection is embraced. Instead of media portraying environmental heroes as flawless saints, stories can highlight their struggles, compromises, and ongoing learning curves. This realism invites readers or viewers to relate, inspired to emulate their efforts rather than intimidated by unreachable standards. Over time, cultural acceptance of imperfection nurtures a collective identity where people feel part of a grand, evolving project. Each participant contributes a piece of the puzzle, acknowledging mistakes or shortcomings along the way, yet forging ahead together.

In political advocacy, admitting that policies are never perfect yet still worth pursuing can reduce polarization. Debates often stall because some demand flawless policies before acting, while others dismiss attempts as too flawed to matter. Embracing imperfection encourages iterative policymaking: implement partial measures, learn from results, refine further. This approach builds trust, as stakeholders see that political actors are not hiding imperfections but managing them transparently. Incremental success stories emerge, adding momentum to bigger reforms. Imperfection acceptance thus helps overcome policy paralysis fueled by unrealistic ideals.

Hope, previously discussed, also relates. Hope thrives when freed from perfectionist constraints. If hope depended on perfect outcomes, it would vanish easily. But if hope endures despite recognizing that progress is partial, messy, and incomplete, it becomes a resilient force. Hope under imperfection acknowledges that each step forward, however small, nourishes future possibilities. This kind of hope survives setbacks because it never hinged on unwavering purity or instant solutions. Instead, it rests on moral patience, trusting that collective incremental moves can bend the arc toward better environmental stewardship.

Over months and years, living with this mindset accumulates evidence of its worth. Instead of cycles of guilt and burnout, individuals experience steady engagement. They see their habits improve gradually: reducing single-use plastics by half over a year, improving dietary impacts incrementally, participating in local projects more frequently. Each small triumph, previously overshadowed by perfectionist disappointment, now shines as a meaningful achievement. This reinforces the cycle of moral effort and emotional resilience, building confidence that imperfect actions still carve out meaningful paths forward.

Sharing personal experiences of imperfection acceptance encourages others to adopt similar views. When someone admits, "I still use my car sometimes, but I've joined a carpool and reduced trips," they model attainable virtue. By highlighting not just final achievements but the journey, including stumbles and adaptations, they inspire empathy and mutual encouragement. The community learns that each person's imperfect contribution adds to a collective mosaic of improvements. This communal moral support weaves an emotional fabric that can withstand shocks, deter fatalism, and magnify small wins into cultural shifts.

This perspective also aligns with understanding nature itself. Ecosystems are not static paragons; they function despite pests, diseases, and disturbances, using resilience and adaptation. By mirroring nature's capacity to adjust rather than seek a static ideal, humans harmonize their ethical stance with ecological principles. Moral resilience, like ecological resilience, involves tolerating imperfection, learning from disruptions, and continuously evolving strategies. Embracing imperfection aligns human behavior with natural patterns, forging a more integrated approach to solving environmental problems in tune with how life thrives on Earth.

As these insights permeate personal and collective consciousness, expectations shift. Instead of evaluating oneself or others by strict green metrics, people appreciate progress made under constraints. Instead of collapsing into defeat at policy shortcomings, activists regroup, refine strategies, and try again. This long-term tenacity flows naturally once perfection ceases to loom as a requirement. With moral growth freed from absolute standards, creativity and innovation flourish. Advocates explore multiple paths—some partial, some complementary—knowing that piecemeal advances accumulate. The result is a more durable, adaptive movement capable of addressing evolving challenges as new data and technologies emerge.

At the end of each day, reflecting on choices made from this vantage point might yield greater peace. Perhaps not every action was optimal, but some were meaningful steps. Perhaps not every goal was reached today, but direction remains steady. This gentle, forgiving self-assessment maintains emotional equilibrium, ensuring that tomorrow's efforts continue rather than halt under the weight of self-imposed moral disappointment. Over time, such a practice becomes second nature, ingraining a habit of recognizing moral complexity, honoring effort amid imperfection, and walking calmly along a path of progressive improvements.

As environmental uncertainties persist, this mindset stands as a pillar of moral resilience. No one knows how all solutions will unfold, nor can anyone achieve total environmental purity. Yet everyone can contribute something valuable. Accepting imperfection and letting go of eco-perfectionism lays the groundwork for persevering through changing conditions. Instead of moral purity tests, shared endeavors emerge, guided by values and shaped by adaptation. Instead of infinite guilt, there is the grace to stumble, learn, and move forward again.

Over time, this ethic of imperfection may redefine environmental engagement from a test of personal flawlessness into a communal journey of moral growth. Through ongoing iteration, adjustments, and incremental betterment, the collective outcome may well surpass what any perfectionist blueprint could accomplish. Sustainability emerges not from isolated acts of purity, but from countless imperfect hands pulling together, each contribution a small stitch in the fabric of planetary care. Embracing imperfection unlocks the staying power needed to protect life, foster justice, and adapt with compassion as the future unfolds.

By caring for physical health over the long term, nurturing positive emotions, and releasing the burden of unattainable standards, individuals create an enduring framework for facing the complexities of environmental challenges. Adequate sleep, nourishing meals, regular exercise, and consistent routines build a stable physiological base that supports steadier emotions and sustained engagement. Alongside these habits, intentionally cultivating gratitude, hope, and moments of joy counters the weight of anxiety and sorrow, restoring a sense of meaning and moral purpose. Allowing imperfection acknowledges human limits, reframes personal efforts as part of an evolving journey, and encourages continuous learning rather than all-or-nothing thinking. Taken together, these approaches do not deny gravity or complexity; instead, they ensure that emotional strength, ethical commitment, and realistic expectations intertwine, making it possible to persist in caring, adapting, and striving for positive change in a world that demands resilience at every turn.

Chapter 8: Finding Meaning and Purpose

Identifying Personal Values and Why You Care About the Planet

In an era marked by shifting climates and ecological uncertainty, many find themselves questioning what truly matters and why. Stripped of old certainties, the future no longer appears a stable extension of the past, but rather a field of diverging possibilities. This upheaval can stir deep reflections on personal values. Understanding why one cares about the planet means probing the moral, emotional, cultural, and spiritual dimensions that anchor environmental concern. Identifying these personal values clarifies intentions, steadies moral compasses, and provides inner fuel to persist in protecting what is cherished, even as conditions change.

Environmental care does not spring from a single motive. For some, love for the planet emerges from simple admiration of nature's beauty: the grace of a heron stalking fish at dawn, the subtle textures of moss on an old tree stump, or the intricate patterns of insect life beneath fallen leaves. Others feel a moral duty—perhaps shaped by religious teachings emphasizing stewardship, philosophical principles highlighting justice, or cultural narratives venerating the Earth as a living relative. Some care out of concern for future generations, striving to safeguard a livable home for grandchildren yet unborn. Others feel empathy for vulnerable communities—rural farmers facing drought, coastal peoples losing land to rising seas, or low-income neighborhoods disproportionately hit by pollution. These diverse impulses form a tapestry of motivations, each thread reflecting a facet of human moral imagination.

Understanding personal values starts with introspection. Without clarity on what drives environmental engagement, efforts may feel hollow or forced. Perhaps one's care stems from a sense of interconnection—a recognition that humans do not stand apart from ecosystems but evolve within them, relying on soil, water, air, and biodiversity for sustenance and meaning. If this resonates, environmental care might align with a worldview that sees life as a web where harm done to one strand ripples through the whole. Another person might frame values in terms of fairness: it feels unjust that those who contributed least to emissions suffer most from climate impacts. Thus, caring about the planet aligns with a commitment to global equity and human rights. Identifying these core motivations infuses environmental action with moral purpose rather than mere obligation.

Emotional responses can guide the process. Observe what triggers sadness or outrage. If reading about disappearing coral reefs evokes grief, that sorrow points to valuing marine life's beauty, complexity, and vitality. If anger flares at corporate pollution, that fury reveals a value for honesty, responsibility, and the right of all beings to clean resources. Emotions serve as signposts: gratitude for local green spaces, tenderness towards pollinators, admiration for indigenous ecological knowledge, frustration at political inaction—these reactions highlight moral landscapes. Following emotional cues helps uncover deeper values. Why does witnessing deforestation hurt so much? Perhaps because it violates a personal principle that all species deserve respect or that humans must tread lightly on ancient forests.

It also helps to recall formative experiences. Memories of childhood outings—climbing a beloved tree, watching tadpoles transform, savoring orchard fruits in harvest season—can shape lifelong reverence for nature. Adolescents who learned about environmental challenges through documentaries or school projects may have integrated these lessons into their moral fabric. Adults who traveled to places under ecological stress, meeting communities adapting courageously, might carry a sense of solidarity. Reflecting on personal history illuminates how care for the planet did not appear from nowhere; it often grew from lived encounters. Understanding this origin story reaffirms authenticity, reassuring that environmental commitment is not a borrowed ideology but something embedded in personal narrative.

Cultural backgrounds offer another layer of insight. Some traditions revere certain landscapes, viewing them as sacred. Others emphasize cyclical relationships with nature—harvest festivals, rainmaking rites, respect for totemic animals. People raised in cultures that teach respect for elders and future generations may find their environmental values entwined with intergenerational responsibility. Identifying how cultural stories influenced current values can foster pride and continuity: one sees that caring about Earth aligns with longstanding traditions rather than a novel obsession. Alternatively, one might have grown up in a less nature-focused environment and consciously chosen a new value orientation after learning about environmental degradation. Recognizing this choice highlights moral agency, reminding that values can evolve and do not need external validation to remain meaningful.

Art, literature, and spiritual practices can also reveal personal values. The music that moves one to tears or paintings that evoke awe at natural grandeur often mirror deep aesthetic and moral appreciation for non-human life. Religious texts that frame humans as stewards or emphasize compassion for all creatures guide believers to defend creation. Philosophical essays discussing humans' place in the universe can inspire humility and reverence. Identifying which cultural products resonate most helps isolate key values: is it reverence for biodiversity's complexity, a sense of unity with all beings, or a thirst for beauty that transcends human constructs?

As values crystallize, the next step is articulating them. Simply naming them can be transformative. For example: "I care about the planet because I believe all living beings have intrinsic worth." Or, "I am committed to environmental protection because justice demands protecting vulnerable communities from climate harm." Another might say, "My motivation stems from awe at nature's creativity and the duty to preserve it for future minds to wonder at." Writing these statements clarifies purpose, turning vague feelings into guiding principles. Once expressed, values become touchstones. When anxiety or doubts arise, rereading or recalling these statements centers the mind, reestablishing moral direction.

Values often span multiple domains. One might value compassion, driving concern for climate refugees and species extinctions. Another might value ingenuity, seeing that humans can invent sustainable technologies if guided by ethical principles. Another might prize modesty, believing that human hubris caused many crises, and thus humility before nature's complexity is crucial. Values might also be relational: caring about the planet because caring about human health, cultural heritage, food traditions, or spiritual fulfillment all depend on healthy environments. Recognizing this interconnectedness allows values to reinforce one another, constructing a robust moral framework that supports resilience.

These identified values also influence how individuals navigate trade-offs. Environmental action often encounters dilemmas—should one prioritize emissions cuts even if it impacts certain industries? Should one accept some level of resource extraction if it reduces poverty? Values help guide such decisions. If equity ranks highly, then approaches that protect the poor or marginalized gain priority. If preserving biodiversity is paramount, then policies that safeguard habitats may win over short-term economic gains. Without clearly defined values, complexity can overwhelm. With them, complexity becomes manageable, as decisions align with moral anchors.

Caring about the planet is not static. As conditions evolve, values may shift in emphasis. Someone initially motivated by aesthetic appreciation might later highlight justice after witnessing environmental injustices. Another might start with a focus on future generations and, over time, broaden it to include empathy for non-human life. Reassessing values periodically ensures they remain relevant. This adaptive process mirrors ecological adaptation—moral ecosystems also respond to changing knowledge, challenges, and personal growth. Treating values as living entities that can mature prevents stagnation or dogma. Instead, values adapt, fueling sustained engagement.

Values are tested under stress. When frightening headlines appear or political stalemates persist, values serve as beacons. If discouragement whispers that nothing matters, recalling personal motives counters despair. "I care because I refuse to abandon future children to a barren world," or "I persist because protecting vulnerable communities is a matter of moral duty" may restore resolve. Without clear values, emotional storms scatter thoughts, making it easier to succumb to fatalism. With values articulated, emotional storms can be navigated: fear acknowledged, but determination preserved, since moral purpose outlasts transient doubts.

In communities and movements, shared values unite diverse participants. While each person's motivations vary, finding overlaps—like a common reverence for life, a shared belief in justice, or an agreed-upon principle of stewardship—builds cohesion. Such collective values bind people together, reinforcing solidarity even when tactics differ. Conflicts become constructive debates over how best to embody shared values rather than battles over incompatible goals. Identifying values publicly, discussing them in groups, and incorporating them into mission statements help keep activism morally grounded. Groups that know why they care about the planet can weather internal disagreements with grace, remembering ultimate purposes.

Values also shape communication with broader audiences. When explaining environmental issues, relating them to universal values—health, fairness, respect for the living world—reaches people's moral cores. Instead of presenting purely technical data, framing problems as matters of integrity, compassion, or preserving cultural legacies resonates emotionally. This resonance can invite more people to care, bridging ideological divides. Even those not initially invested in environmental causes may respond to appeals rooted in values they already hold—like wanting a safe future for children or believing in honesty and responsibility.

Reflecting on personal values may uncover hidden dimensions of care. One might realize that environmental protection aligns with love for poetry inspired by natural landscapes. Another might discover that a childhood experience rescuing a trapped animal instilled empathy for all creatures. Understanding that caring is not just intellectual but also emotional, aesthetic, and moral enriches motivations. This complexity of values ensures that when one aspect faces challenges—say, losing hope due to slow policy progress—other aspects sustain engagement, like beauty or compassion. Diversity of values within oneself forms an internal safety net against despair.

Values also inform priorities. Since no one can address every environmental issue simultaneously, personal values guide where to focus. Someone who values preserving wild habitats might volunteer for reforestation projects, while another, driven by social justice, might join campaigns to fund climate adaptation in vulnerable regions. Another, valuing innovation, might invest energy in clean technology start-ups. This selective emphasis is not avoidance; it's strategic realism. By aligning efforts with personal values, individuals find joy and meaning in their chosen niches, preventing emotional burnout from dispersing energy too thinly. Values thus help carve manageable pathways through moral complexity.

At times, doubts may arise: is it selfish to choose values that resonate personally rather than following the crowd? Yet genuine care thrives when rooted in authenticity. If a person feels most compelled by wildlife conservation, dedicating energy there yields more passion and perseverance than following a trend that doesn't align deeply. Authentic values ensure that caring is not performative or hollow but springs from true conviction. This authenticity fosters resilience because it does not depend on external approval. If others question one's focus, confidence arises from knowing these values are integral to personal moral identity.

Observing that values can coexist even if they occasionally conflict also enhances flexibility. For example, a love of technology and an appreciation for traditional ecological knowledge need not be mutually exclusive. One can value both innovation and respecting indigenous wisdom, seeking to merge these perspectives in problem-solving. Recognizing that values need not be singular or uniform enlarges moral imagination, allowing richer solutions that incorporate multiple dimensions of care. By embracing a spectrum of values, individuals become less susceptible to rigid thinking, more open to learning and adapting as new insights emerge.

Cultural traditions often highlight that caring about the Earth is not a new idea invented by modern environmentalism. Many ancient philosophies and spiritual teachings insisted on living harmoniously with nature, respecting its rhythms, and minimizing harm. Recognizing that current values align with ancestral wisdom gives moral depth and a sense of participating in a long lineage of ethical thought. This ancestral connection offers comfort: while modern challenges differ in scale, the moral impulse to cherish life's sources and cycles has deep roots, confirming that values formed now are not fleeting but part of humanity's enduring moral tapestry.

Values also bridge personal life with broader narratives. Considering why one cares about the planet may reveal that environmental protection resonates with the same principles guiding family care, friendship, or professional integrity. If honesty, compassion, and fairness guide personal relationships, it's logical they shape environmental ethics too. This integration reduces compartmentalization, ensuring that caring about climate or biodiversity is not a separate "activist identity" but a natural extension of existing moral frameworks. Aligning environmental engagement with other cherished roles—parent, teacher, artist—infuses daily activities with moral coherence, reducing internal conflicts and reinforcing identity consistency.

When hardships intensify—like natural disasters, failed policy negotiations, or witnessing ecosystem collapse—values act as a moral lighthouse. Even if immediate solutions are uncertain, values remind that caring is valuable in itself. Protecting what can be saved, bearing witness to loss with empathy, helping communities adapt—these actions gain moral significance from the values underpinning them. Without values, despair might argue that without guaranteed success, caring is pointless. With values, caring becomes a moral stance independent of outcomes. This stance yields dignity and purpose: one does right not only for results but because the underlying principles demand effort even in uncertainty.

In mentoring younger generations or educating students, highlighting the importance of identifying personal values can help them develop moral resilience early. Encouraging them to ask why they care about environmental issues nurtures a sense of ownership over their moral journey. If they realize their environmental concern aligns with respect for elders' legacies, love for animals, or protecting cultural diversity, this insight empowers them to face future challenges with confidence. Teaching values literacy ensures that future leaders and citizens do not just parrot environmental slogans but engage with depth and authenticity.

In activism, conflicts may arise between different approaches—direct action versus policy lobbying, incremental reforms versus radical transformations. Values mapping helps navigate such tensions. If both sides share fundamental values—compassion for the vulnerable, respect for non-human life—they can find common ground or at least respectful dialogue. Recognizing shared moral roots counters factionalism. Instead of seeing other strategies as betrayals, they become different interpretations of how to best embody shared ideals. This values-based understanding reduces infighting, channeling energy into constructive debate and synergy.

Values also encourage humility. Identifying personal values reveals personal biases. One might realize a strong aesthetic bias—protecting landscapes because they are beautiful—while underemphasizing justice concerns. Awareness of this bias invites learning: read about communities disproportionately affected by climate impacts, integrate fairness into personal motivations. This process refines values over time, pushing toward a more holistic ethic that accommodates multiple dimensions of environmental care. Embracing values as evolving guides rather than fixed dogmas encourages constant moral refinement, aligning with the principle that resilience means staying open to growth.

The interplay of values with emotions, reason, and culture creates a feedback loop. As values become clearer, emotional responses to events find context. Despair becomes less overwhelming when filtered through values that emphasize perseverance. Anger can channel into advocacy that aligns with moral goals rather than destructive impulses. Gratitude and hope root in understanding that these feelings support commitments. Similarly, rational analyses of policies and technologies gain significance when guided by moral compass points. Instead of juggling data without direction, values provide criteria to judge what matters and what to prioritize.

Over time, expressing values publicly can inspire others and attract allies. Stating, "I care about the planet because I believe in protecting the vulnerable and ensuring future prosperity" resonates with like-minded individuals, forming moral communities around shared principles. Such communities become moral support structures, where members reinforce each other's dedication when doubts surface. Knowing that others articulate similar values reassures individuals that their care is not idiosyncratic or naive. This collective affirmation bolsters resilience, making it harder for despair or cynicism to isolate or silence caring voices.

These personal values can also guide learning. Realizing one's emphasis on justice might motivate studying environmental laws, indigenous rights, or social equity frameworks. If beauty drives care, exploring ecological arts, landscape restoration aesthetics, or bio-inspired design deepens understanding. Values shape intellectual curiosity, leading one to seek information that strengthens and refines moral stances. This thirst for knowledge further empowers engagement—better-informed advocates argue more persuasively, choose more effective projects, and anticipate obstacles more strategically.

Eventually, identifying personal values and why one cares about the planet lays the foundation for moral resilience in the face of environmental uncertainty. Values distill motivations into coherent principles that endure beyond fluctuations in policy or technology. They provide direction when events are overwhelming, prevent emotional whiplash, and maintain moral coherence amid complexity. More than a shield against despair, values act as an engine driving constructive responses. By clarifying that care for the planet aligns with cherished principles—like compassion, fairness, beauty, integrity, stewardship—individuals and communities find lasting meaning and purpose in their environmental engagements. This meaning fuels sustained effort, ensuring that hope, integrity, and resilience persist despite ongoing uncertainty, forging a path of purposeful moral engagement.

Aligning Day-to-Day Actions with Deeper Moral or Spiritual Commitments

In a world where environmental challenges press on every side, many people wrestle with how to live their values in a tangible, consistent manner. High-level ideals—perhaps rooted in moral philosophies, religious teachings, cultural traditions, or personal spirituality—may feel lofty and inspiring, but translating them into everyday actions proves difficult. The gap between aspirations and daily habits can create tension, guilt, or confusion. If one treasures compassion, justice, reverence for life, or humility before nature, how can such principles shape the smallest decisions, from what to eat at breakfast to how to commute to work? Aligning day-to-day actions with deeper commitments involves bridging moral or spiritual convictions and practical routines, ensuring that abstract beliefs do not remain distant ideals but guide the living texture of existence.

This alignment does not require grandiose gestures. While large changes—shifting careers to environmental advocacy or moving off-grid—may appeal to some, they are not the sole path. Genuine alignment emerges in incremental shifts, conscious reflections, and repeated choices that better reflect cherished principles. Over time, these small changes accumulate, weaving moral commitments into the fabric of daily life. This process can feel humble and personal: selecting a locally grown fruit instead of an imported one, making time to recycle even when busy, or pausing to remember why driving less matters. These minor acts affirm that values matter and that implementing them need not wait for perfect conditions.

Before pursuing this alignment, it helps to identify core principles. Some may cherish compassion—caring about distant communities affected by climate shifts or empathizing with non-human creatures struggling to adapt. Others emphasize justice, feeling a duty to reduce harm disproportionately borne by marginalized groups. Still others hold spiritual beliefs that urge humility before nature's complexity, seeing every resource as a gift rather than a commodity. Another might value stewardship, interpreting human existence as entrusted with Earth's well-being. Clarifying which values resonate most provides a compass. Without a clear moral direction, daily choices remain reactive, swayed by convenience or habit. With principles defined, mundane decisions gain meaning, as each choice can be tested against the moral or spiritual lens that guides it.

Once these values are clear, consider starting with one area of daily life. Perhaps focusing on food choices first—aligning eating habits with respect for ecosystems and human dignity. For someone who values compassion towards animals, reducing meat intake or choosing sources with higher welfare standards expresses that principle on the plate. For someone driven by justice, supporting fair-trade, locally sourced produce reduces the exploitative chains hidden behind many global supplies. Rather than turning the kitchen into a moral battlefield, treat it as a creative laboratory where values inform culinary exploration. Over time, this realignment becomes normal, turning moral convictions into flavors and textures chosen thoughtfully.

Transportation choices also present opportunities. If humility before nature suggests minimizing personal environmental impact, evaluating commuting patterns might help. Perhaps cycling a few days a week or using public transport whenever possible. Even if not every trip can be sustainable due to logistics or time constraints, striving for incremental improvements displays loyalty to moral aims. Each ride on a bus, each short walk instead of driving, whispers: "I remember my values and act on them, however imperfectly." This gentle persistence undermines cynicism and reaffirms that caring is not a hollow claim but something lived.

Aligning actions with values also involves recognizing trade-offs. Suppose one values both efficiency and reducing emissions. Driving might be quicker in personal scheduling terms, but biking or carpooling might lower one's carbon footprint. Balancing these considerations is tricky, but moral alignment encourages leaning towards choices that honor core convictions. Not every instance will yield a perfect solution—sometimes necessity wins, and one must drive to meet urgent obligations. The key is to avoid despair or self-flagellation when imperfection arises. Instead, treat each compromise as a learning moment: could future planning reduce the need for that car trip, or could new infrastructure support greener transport later?

Spiritual frameworks can deepen this process. For those whose faith traditions emphasize stewardship, daily rituals reinforcing gratitude for nature can inspire small acts. Saying a short prayer before meals, acknowledging the labor of farmers, the fertility of soil, and the grace of rain reminds that food is not merely a commodity but a sacred gift. This reverence can translate into reducing food waste, choosing ethically sourced ingredients, and sharing surplus with neighbors. Spiritual practice becomes not just private contemplation but a catalyst for practical ethics.

Cultural narratives also guide alignment. Traditional festivals tied to harvests or seasonal changes highlight nature's cycles. Participating in these ceremonies or remembering cultural stories about respectful use of resources can inspire changes in daily routines. If a cultural tale warns against overfishing, recall it when buying seafood, supporting sustainable fisheries. If another story praises frugality and careful resource management, apply it by fixing appliances rather than discarding them. Aligning day-to-day actions with cultural values creates continuity, ensuring that ancient wisdom lives through modern practices.

Another dimension involves communicating these changes to others. Without preaching, sharing the reasoning behind certain choices can invite collective growth. Explaining to friends why choosing a vegetarian dish resonates with compassion or justice can spark conversations. When family members see that one's decision to reduce plastic aligns with a deeply held moral principle rather than trend-chasing, they may respect and perhaps emulate the effort. Such dialogues humanize moral commitments, showing that they are not abstract dogmas but sources of meaningful life direction.

Challenges inevitably surface. Modern lifestyles often present convenience and affordability at odds with sustainability. Aligning day-to-day actions with moral commitments may mean investing more time researching products, learning about supply chains, or planning meals ahead. Initially, these efforts may feel cumbersome. Yet, as one's mindset shifts from viewing these steps as burdens to seeing them as expressions of faithfulness to values, the inconvenience lessens. These extra steps become acts of devotion, strengthening resolve and pride in living authentically.

When discouragement arises—seeing ongoing environmental degradation despite personal efforts—moral grounding helps maintain perspective. The purpose is not to singlehandedly save the planet through individual actions but to embody principles that matter. Each eco-friendly choice, no matter how modest, resists apathy. Aligning daily life with values is a moral stand, signaling that exploitation is not the inevitable default. When large-scale impacts feel distant, knowing that one's routines consistently reflect moral depth provides emotional resilience. Even if systemic changes lag, personal integrity holds firm. Over time, as more individuals integrate principles into behavior, collective momentum may influence markets, policies, and cultural norms.

Variety in approaches keeps the process flexible. If one area proves difficult—maybe responsible electronics consumption is hard due to complex supply chains—another area can offer more accessible changes, like energy conservation at home. The point is not flawless purity but a pattern of ongoing moral effort. Each improvement reinforces the idea that principles are not decorative but operational. Over months and years, incremental shifts accumulate, leaving a lifestyle noticeably gentler on ecosystems and more respectful of life. This steady transformation demonstrates that values are not static beliefs tucked in the mind's corner, but dynamic forces shaping daily existence.

Another helpful tactic is setting intentions before beginning activities. For instance, upon entering a grocery store, remind oneself: "I will choose items that align with compassion and justice." With this intention set, product labels are read more carefully, local and organic produce are preferred, and less sustainable options are considered last resorts. Similarly, starting a workday with a mental note—"Today I will reduce paper waste and remember my commitment to stewardship"—anchors moral awareness amid routine tasks. These small rituals keep principles active in the mind, preventing them from fading under daily pressures.

In some cases, engaging in community activities accelerates alignment. Joining a local sustainability group, volunteering at an urban farm, or attending workshops on zero-waste living exposes one to practical tips, supportive peers, and shared moral frameworks. Observing how others navigate trade-offs, handle setbacks, and celebrate improvements can inspire and instruct. Mutual encouragement transforms solitary struggles into collaborative learning. Communities that celebrate partial wins, empathize with difficulties, and pool knowledge help individuals stay aligned without feeling isolated.

Over time, this alignment can influence career choices or educational directions. If one's values highlight justice and sustainability, seeking work at organizations that uphold these principles might become appealing. Whether shifting into environmental education, green business strategies, policy research, or staying in a current field but pushing for more ethical practices, values guide professional development. Even within a conventional job, advocating eco-friendly measures, supporting ethical sourcing, or reducing resource consumption integrate moral vision into professional spheres.

Care must be taken to avoid moral arrogance. Just because one aligns actions with values does not mean others who struggle or prioritize differently are less moral. Everyone faces unique constraints—financial hardship, cultural pressures, limited information. Humility reminds that moral alignment is a personal journey. Instead of judging others harshly, share successes and setbacks honestly, allowing empathy and dialogue. Moral alignment as a personal contribution rather than a measure of moral superiority preserves unity and openness.

Aligning with deeper commitments also encourages moral consistency. Without deliberate effort, people compartmentalize: acting ethically in one domain but ignoring principles in another. Integrating values across daily activities reduces this fragmentation. If respecting life guides dietary choices, it might influence product sourcing or household chemicals too. Each domain touched by values harmonizes moral identity, reducing internal conflict and reinforcing a coherent ethical character.

Values evolve through feedback loops. Implementing changes aligned with principles can yield positive emotional responses: a sense of peace, coherence, or pride. Tangible outcomes—reduced waste, fewer single-use plastics, lowered energy bills—add evidence that moral alignment is not futile. These reinforcements encourage advancing further. Instead of seeing small wins as trivial, treat them as proof that integrity can shape lived reality.

Emotional resilience increases when daily life and moral identity coincide. Facing troubling climate reports, one can say, "Though I cannot fix everything, I have chosen a path consistent with what I hold sacred." This reassurance counters despair. By living out values, fewer internal conflicts arise—no guilt for enjoying a comfort moderated by careful choices, less shame in occasional compromises. Instead of feeling defeated by imperfections, each step forward embodies sincerity and growth.

Periodically reassessing progress keeps alignment fresh. Has meal sourcing improved since last year? Has commuting become greener? Are interactions with nature more mindful and reverent? If progress stalls, consider new strategies or resources. Adjusting values emphasis or learning about alternative solutions prevents stagnation. Alignment aims for dynamic equilibrium rather than rigid constancy, inviting continuous improvement and adaptability.

Cultural and media influences might press in opposite directions. Strong value awareness helps filter these pressures. If cherished principles discourage consumerism, flashy ads fail to captivate. If honesty and respect matter, greenwashing tactics appear hollow. Moral clarity frees one from automatic conformity. Even when compromises occur, they do so knowingly, minimizing moral dissonance.

For those who draw on spiritual traditions, prayer or meditation can deepen alignment. Before making choices, brief reflections—"May this decision honor life and justice"—focus intentions. Spiritual practice transforms mundane tasks into moral opportunities, reinforcing that even small actions hold spiritual significance.

Values also inform responses to setbacks. If a policy effort fails, moral grounding interprets it as a lesson rather than a reason to quit. The principle of perseverance encourages trying again or shifting tactics. Compassion suggests supporting allies feeling discouraged. Moral alignment transforms defeats into learning experiences, fueling resilience.

As alignment solidifies, everyday activities cease feeling neutral. Preparing meals, commuting, shopping—all become moral expressions. Values seep into all corners of life, humanizing environmental care and making it a natural extension of personal character. Instead of performing sustainable actions reluctantly, one embraces them as coherent aspects of a life story guided by principles.

This integration can quietly inspire others. Observers might notice thoughtful consumption patterns or respectful relationships with nature. Curiosity arises: "Why this effort?" Explaining that these actions flow from moral or spiritual commitments resonates more than abstract preaching. Leading by example demonstrates that caring is practical and accessible, not reserved for rare saints. Over time, cultural norms shift as more people find this model appealing.

In crises, pre-established moral alignment provides stability. Instead of scrambling to find meaning after a disaster, one meets challenges with a moral foundation already woven into daily life. Even under pressure, principles guide actions—helping neighbors, sharing resources, and advocating for responsible rebuilding. Moral grounding anchors emotional response amid chaos, preserving dignity and identity.

Aligning actions with deeper commitments means no longer treating environmental care as an external obligation. Instead, it becomes integral to who one is. Values become lenses for interpreting reality, and actions become the language expressing them. Instead of feeling forced, sustainable choices arise from internal coherence. Emotional tension eases as moral aspirations find real outlets in daily acts.

As patterns cement, one's sense of moral coherence grows. Instead of compartmentalizing ethics and routine, life flows holistically. Doubts or temptations lose force against established habits that reflect chosen values. This inner consistency supports long-term resilience and adaptability. Even as knowledge evolves and conditions shift, the capacity to realign actions with principles endures.

Such moral integration invites ongoing learning. Values remain stable anchors, yet new insights refine strategies. Better products, community initiatives, or policy opportunities appear. Embrace them, guided by moral compasses that remain steady even when embracing novelties. Flexibility ensures moral alignment never becomes static righteousness, but a living practice open to better solutions.

In these ways, aligning day-to-day actions with deeper moral or spiritual commitments transforms caring for the planet from a theoretical stance into lived reality. Over time, small changes accumulate, expressing reverence, justice, compassion, or stewardship in myriad details. The outcome is a meaningful existence where one's principles guide the hand at the grocery shelf, the choice of transportation, the handling of waste, and the engagement with community. The result is both a gentler ecological footprint and a richer moral life, forging resilience and purpose on the path toward collective environmental healing.

Turning Worry into Motivation to Protect What You Love

As environmental uncertainties intensify, many people find themselves caught in cycles of worry and distress. The worry might begin as a quiet unease upon hearing that glaciers are retreating faster than expected, or it could flare into full-blown anxiety after reading reports of species loss, extreme weather, or stalled policy reforms. Yet beneath the surface of these fears often lies a profound care for the planet and its inhabitants. Worry, while uncomfortable, can serve as a gateway to moral awakening rather than a trap. With conscious effort, it can be transformed from a source of paralysis into a driving force that energizes actions aligned with values. Turning worry into motivation means recognizing that the emotional charge sparked by ecological threats can fuel purposeful engagement, sustained learning, and persistent advocacy.

An initial step is to acknowledge worry as a signal rather than a flaw. Instead of berating oneself for feeling anxious, understand that worry arises because something cherished appears threatened. Concern over coral reefs declining is not a random phobia; it reflects love for marine biodiversity, awe at the intricate relationships among ocean species, or moral outrage at destructive human activities. Similarly, worry about rising seas jeopardizing coastal communities might express empathy for those who could lose homes, livelihoods, or cultural heritage. By seeing worry as evidence of caring deeply, one reframes anxiety from a personal failing into a moral indicator. This shift reduces shame and invites curiosity: what does this worry reveal about what truly matters?

Once worry is recognized as rooted in care, the next step involves directing its energy outward. Without guidance, worry can cycle through one's mind, amplifying fear. But with intention, the emotional energy can be channeled into seeking information, connecting with others, and identifying meaningful actions. If dread arises over deforestation, let that dread prompt research: learn about which forests are most endangered, discover local and global initiatives restoring habitats, or find out how consumer choices affect logging practices. Information transforms amorphous fear into structured knowledge. Knowledge, in turn, identifies entry points for action—contacting organizations working on reforestation, adjusting personal purchases to avoid products linked to forest destruction, or supporting policy campaigns that protect old-growth woodlands. Each piece of information reframes worry from helplessness to engagement.

Sometimes worry seems too vast, overwhelming the mind. Breaking down big worries into manageable parts can help. Instead of stressing about "the planet's doom," specify what precisely triggers fear. Is it melting polar ice? Then focus on learning about polar ecosystems, and see who is working to mitigate threats. Is it water scarcity? Research where droughts hit hardest and what irrigation methods or community adaptations are being tested. Each narrowed focus clarifies pathways forward. By pinpointing a particular issue, worry becomes more concrete, and acting on that issue feels more feasible. Over time, addressing small facets of a massive problem builds confidence and trust in one's ability to contribute.

Engaging moral imagination also helps. Instead of letting worry feed catastrophic visions of inevitable collapse, envision scenarios where efforts reduce harm, adaptations bear fruit, and communities find creative solutions. This does not deny severity; it acknowledges that futures are not fully predetermined. Imagining what success might look like—healthier coral reefs after local restoration projects, cleaner rivers following stricter pollution controls, or stable food supplies after regenerative farming takes root—recruits hope. When worry pushes one toward despair, these envisioned successes become mental counterweights. They remind that efforts matter, that choices shape outcomes, and that worry's energy can be channeled to help approach desired scenarios.

Worry often grows more intense when people feel isolated. Connecting with others who share concerns transforms private anxiety into collective momentum. Join local environmental groups or online forums where people discuss challenges and solutions. Hearing how others cope, what actions they take, and what small victories they celebrate reveals that worry is widespread but not paralyzing everyone. Instead, many translate their anxieties into volunteering, political advocacy, educational campaigns, or community resilience projects. Witnessing these transformations provides tangible role models. If they can turn worry into constructive action, why not follow suit? Emulating successful examples builds confidence and reduces feelings of helplessness.

Conversations with friends and family also help. Expressing worry to trusted individuals may result in empathy, reassurance, or even practical suggestions. Sometimes those less directly involved in environmental issues can offer fresh perspectives. They might remind that incremental changes do add up, or propose simpler first steps. Just articulating anxieties out loud can reveal that certain fears, while large, are not entirely beyond influence. These dialogues deepen understanding of why one cares—uncovering values embedded in concern—and clarify how personal strengths or resources could best serve protective efforts.

Another powerful approach involves pairing anxiety management techniques with purposeful action. Grounding exercises, mindfulness, or reframing thoughts can stabilize emotions sufficiently to choose deliberate responses. Once calm, ask: "What action can I take today that aligns my concern with a positive outcome?" Maybe it's writing a letter to a representative, donating to a conservation nonprofit, or attending a workshop on sustainable gardening. The act itself, however small, channels emotional tension into productive energy. Each time worry surfaces, repeat this process: calm the mind, recall core values, and choose an action step. Over time, worry becomes a prompt for engagement rather than a trigger for panic.

Aligning personal strengths with chosen actions boosts motivation. If one is a skilled communicator, writing articles or talking to neighbors about environmental issues can feel natural and impactful. If someone has organizational abilities, planning community clean-ups or fundraisers might fit their talents. Artistic souls could create music, art, or poetry that inspires empathy for threatened ecosystems. By linking worry-driven motivations to innate skills, efforts feel less forced and more authentic. This authenticity sustains long-term involvement. Instead of seeing environmental efforts as burdensome chores, they become meaningful expressions of identity and moral purpose.

Another dimension involves recognizing that worry sometimes springs from witnessing damage done to what one loves. Perhaps a beloved childhood lake now suffers from pollution, or a cherished tradition tied to a stable climate becomes harder to maintain. These losses hurt because they threaten sources of meaning and comfort. Realizing this emotional root clarifies that protecting these cherished elements is not just an abstract good but a personal imperative. Worry then transforms into determination: "I love this place, these traditions, these species, and I refuse to lose them without a fight." This personalized moral drive ignites sustained motivation, fueled by love rather than fear alone.

In some cases, spiritual or philosophical teachings can guide the transformation of worry. If a faith tradition encourages viewing adversity as a call to moral action, then worry about climate impacts can be seen as an invitation to fulfill stewardship duties or practice compassion. If a philosophy emphasizes life's interconnectedness, worry about collapsing ecosystems reaffirms one's role in protecting the larger web. These teachings provide frameworks that dignify worry, making it not a weakness but a catalyst for deeper ethical engagement. Prayer, meditation, or reflective reading can help process anxious feelings and emerge with renewed resolve to serve life's well-being.

Cultural narratives also support this shift. Many stories, myths, or historical accounts show communities facing daunting challenges and responding creatively. Drawing on such narratives—heroes who restore damaged lands, communal efforts that overcome resource scarcity—provides role models and moral confidence. When worry arises, recalling these stories offers a roadmap: others before have turned fears into proactive responses. Cultural memory thus refutes the notion that anxiety must lead to helplessness. It proves that worry can signal moral readiness, prompting a community or individual to muster strength, empathy, and innovation.

Implementing a practice of inquiry whenever worry surfaces strengthens motivation. Instead of passively enduring fear, actively question it: "What specifically worries me now?" "Which values does this fear highlight?" "What can I do that aligns with these values?" "Who can I collaborate with?" Answering these questions transforms worry into a problem-solving prompt. The mind shifts from rumination to planning. By repeatedly guiding anxious thoughts through this constructive channel, worry's energy moves away from draining emotional reserves and toward fueling purposeful steps. Over time, this process becomes habitual, ensuring that worry no longer derails but instead recalibrates engagement.

One must be cautious not to expect instant gratification. Turning worry into motivation is gradual. Early on, actions may feel small and unsatisfying. Planting a tree when entire forests vanish might seem trivial, yet it is a start. Donating to an environmental group may not yield visible results overnight, but it supports a network pushing for long-term change. Over months, cumulative efforts generate tangible impacts—restored habitats, policy shifts, cultural dialogues—and one can look back to see how initial worry, once a source of distress, became a steady current carrying forth moral engagement and incremental improvements.

Consistency in channeling worry into action also builds resilience. The next time alarming news emerges, the established habit of responding with considered steps and moral clarity reduces emotional shock. Instead of sinking deeper into despair, the mind recalls previous experiences of transforming worry, reinforcing confidence. Gradually, fear's grip weakens because it no longer dictates behavior; the person has proven their capacity to respond morally and constructively. Confidence nurtures a feedback loop: the more often worry leads to action, the less threatening worry feels, allowing even more energy for solutions.

In group contexts, encouraging each other to harness worry's energy can galvanize collective efforts. When someone expresses anxiety about coral bleaching, the group can ask: "What can we do together?" Maybe they organize a beach clean-up, fundraise for a marine research station, or write joint letters advocating for marine protected areas. Turning individual worry into group initiatives amplifies impact. Seeing anxiety shared and transformed builds solidarity. Members learn that worry is not something to hide but to articulate, transforming negative emotions into communal motivation that drives forward momentum.

Leadership that acknowledges fear openly and channels it into action can inspire others. If environmental advocates admit their own worries—about political gridlock, rising temperatures, or ecosystem collapse—but show how these worries motivate them to try harder, to learn more, to unite people, then followers see vulnerability as strength. Transparency about anxiety humanizes leaders, making them relatable. It also demonstrates that worry is normal and can be productive. Moral authority arises not from pretending fear does not exist but from modeling how to use it as fuel for moral bravery and creative problem-solving.

Embracing flexibility in the face of worry enhances adaptability. If one strategy fails, anxiety might flare again. Instead of discouraging, treat it as a nudge to adjust tactics. Worry can prompt reevaluation, leading to new methods or alliances. For instance, if lobbying at a certain political level proves ineffective, anxiety about time running out might prompt shifting focus to grassroots organizing or youth empowerment. In this sense, worry acts as a compass: when one direction hits a dead end, anxiety signals a need to chart a new path. This agile response prevents stagnation and fosters continuous learning.

Another key element is celebrating successes that result from worry-inspired actions. If anxiety about urban air quality led to joining a local clean-air coalition, and that coalition eventually convinces the city to introduce greener public transport, acknowledge how worry initiated this journey. Reflecting on successes reminds that worry, when harnessed, can produce real-world improvements. This memory strengthens motivation for future endeavors. Every time alarm arises, recalling past transformations of fear into progress assures that such conversions are possible again, encouraging perseverance.

Hope also intertwines with worry in this process. While worry highlights what could go wrong, hope identifies what could go right. Balancing the two leads to constructive energy. Worry without hope can immobilize; hope without worry can breed complacency. Combined, they create a dynamic tension: concern for potential disaster motivates vigilance, while hope for better outcomes encourages proactive measures. Recognizing this interplay helps manage emotional states. Instead of trying to eliminate worry, pair it with a hopeful vision, ensuring that action is guided both by caution and aspiration.

Connecting worry to personal values is essential. If anxiety flares when reading about melting ice sheets, identify which principle is threatened—maybe justice for future generations or love for fragile polar ecosystems. By naming the value implicated, the mind realizes worry is not random fear but care distilled into anxiety. This clarifies moral obligations. To honor love for future generations, take steps that secure long-term stability—supporting educational programs about climate adaptation, helping build local resilience. Worry transforms into a moral reminder, a call to live up to chosen principles. In this way, anxiety becomes an ally in staying true to moral commitments.

Spiritual or reflective practices can strengthen this transformation. When feeling overwhelmed, some may pray, asking for courage to turn fear into compassion-driven action. Others meditate, observing worry as a passing mental state and visualizing how to channel its energy into one constructive deed. These inner dialogues calm the nervous system, replace panic with focus, and reaffirm that anxiety is not the endpoint but a stepping stone. Spiritual insights might frame worry as a prompt to fulfill one's role as a guardian of creation, a collaborator in healing rather than a passive observer of destruction.

Artistic expression can also facilitate this shift. Writing poetry about fears, painting images that reflect both threat and resilience, or composing music inspired by uncertain futures all reframe worry's emotional intensity into creative output. This creative release eases tension while offering new perspectives. Art can embody the desired transformation: worry portrayed not as a dark pit but as fertile soil where seeds of action sprout. Later, sharing this art with others who feel similar anxieties might inspire them to follow suit, building a cultural narrative that normalizes turning fear into moral fuel.

At times, fear might escalate into moments of panic. Grounding techniques can prevent panic from overwhelming motivation. After calming the body and mind, revisit the question: "What caused this fear? How can I respond constructively?" Maybe reading about an extinct species triggered sorrow. That sorrow can now guide sending support to a habitat restoration initiative or contacting local representatives about stronger conservation laws. Even intense episodes become chances to reaffirm that emotions, while strong, need not rule passively. Instead, they guide attention toward moral obligations not yet met.

Diversifying one's engagement channels worry's energy more effectively. Instead of fixating on a single policy or a single behavioral change, branching out into multiple approaches—educational outreach, supporting climate justice organizations, adopting greener personal habits—ensures that if one avenue hits obstacles, others remain viable. This variety reduces the risk that setbacks in one area destroy morale. Maintaining multiple streams of action prevents anxiety from crushing motivation. If a global treaty stalls, local community gardening might still flourish. If personal lifestyle changes feel too incremental, involvement in a social movement may yield broader impact. Each domain of action supports the next, weaving a safety net against despair.

Reflecting on personal growth over time reinforces the transformation. Early on, worry might have induced withdrawal or numbness. After consistently directing anxiety toward learning, adapting habits, or joining initiatives, compare past and present states. Perhaps now worry prompts emailing an environmental journalist or volunteering once a month, whereas before it led to isolation. This progress confirms that fear can evolve into courage. Knowing that transformation is possible, individuals gain confidence that they can continue refining this skill. Anxiety's role changes from tormentor to motivator, albeit one that must be guided with patience and discernment.

Value-driven perseverance also matters. If love for certain ecosystems motivates action, remind oneself regularly of that love. Return to a favored natural spot, observe its living richness, and reconnect with the initial source of caring. Such rejuvenation prevents moral fatigue. Instead of dreading new climate warnings, these reminders reinforce commitment. Nature itself, even damaged, can inspire. Witnessing recovery in small pockets—regenerated wetlands, thriving pollinator gardens—validates efforts, confirming that worry-inspired work is not futile. Observing concrete improvements counteracts cynicism, fueling further motivation to protect what still remains beloved.

As skills in channeling worry grow, consider mentoring others. If friends or colleagues express anxiety about environmental issues, share experiences of converting fear into action. Suggest small first steps, offer resources, and empathize with their emotional struggles. Demonstrating that it's normal to feel worried yet possible to turn that worry into positive motion encourages others to break free from paralysis. Over time, a supportive community of people who treat worry as an activation energy rather than a shutdown response emerges. This collective mindset can influence broader cultural attitudes, making fear-driven stagnation less common and collective problem-solving more likely.

A sense of moral continuity also results from this transformation. Instead of viewing environmental crises as periodic reasons for alarm, see them as ongoing calls to moral vigilance. Worry becomes predictable: whenever alarming data appears, it stirs care once again. Anticipating this cycle prepares the mind, reducing shock and allowing quicker transitions into action. This reliability gives emotional stability. Knowing that anxiety will surface regularly, one learns to welcome it as a reminder of values and respond accordingly. Over time, this normalized cycle diffuses anxiety's power to destabilize, integrating concern into a balanced moral life.

Moral humility accompanies these changes. One might acknowledge that although worry inspires action, single efforts cannot solve entire crises. But accepting these limits does not negate significance. Worry prompts doing one's part—improving local conditions, influencing some policy, educating certain groups, inspiring a handful of people—and these contributions join countless others. Instead of craving omnipotent impact, humility embraces partial influence as meaningful. This perspective is liberating: anxiety no longer demands heroic feats, merely sustained engagement according to ability and resources. Freed from unrealistic burdens, motivation thrives as a steady flame, not a brief, explosive spark.

Communication styles may shift as well. Speaking about environmental issues while acknowledging personal anxieties humanizes the message. Instead of broadcasting doom or scolding others, one can say, "I've felt worried about species loss, and that worry led me to support a local habitat restoration project. It makes me feel more hopeful to do something tangible." This narrative avoids shaming and invites empathy. Others see that fear can yield constructive outcomes. This disarms the notion that concern must be repressed or that emotional responses are unhelpful. Instead, emotional honesty combined with moral action models a healthy pattern others can replicate.

Long-term integration occurs as one's identity evolves. Initially, transforming worry into motivation might feel like a deliberate technique. After repeated cycles, it becomes intuitive. Observing alarming news prompts a near-automatic response: identify relevance, recall core values, select a helpful action. Anxiety transforms from a hindrance into a prompt. The difference is stark: where once fear caused distress and retreat, now it triggers moral creativity and determined problem-solving. Over years, individuals who practice this skill become resilient advocates capable of adapting strategies, supporting communities, and sustaining hope even under relentless environmental challenges.

Cultural narratives play a role here too. If communities celebrate stories of people who overcame fear by acting for the common good, these stories reinforce the idea that worry need not corrode motivation. For example, historical accounts of individuals responding to local environmental disasters with organized clean-ups, improved protections, or educational campaigns highlight that anxiety can be a starting point. Cultural memory thereby encourages approaching climate anxieties as opportunities to rally, not reasons to surrender. Tradition, myth, and collective memory all contribute to normalizing fear-to-action transitions, making it part of a shared moral language.

As more people internalize this approach, collective emotional states shift. Instead of societies overwhelmed by gloom, we may see communities that acknowledge emotional strains but quickly convert them into resolute efforts. This does not guarantee easy success, but it creates a psychological environment more conducive to sustained involvement. Markets and policies may respond as citizens actively voice demands, influence consumption patterns, and support leaders who take environmental responsibilities seriously. Anxiety thus no longer remains confined to personal turmoil; it moves through social channels, sparking reforms and innovations.

Confidence in this method comes not from blind hope, but from observing incremental results. Suppose anxiety about ocean plastics led someone to reduce single-use plastics, and later their city passed regulations curbing them. Although not solely caused by that individual's shift, knowing that personal effort contributed to broader pressure reinforces belief in moral agency. Next time worry flares—about microplastics in soil or endangered pollinators—confidence encourages tackling these issues, expecting that collective action can emerge likewise. The virtuous cycle strengthens as each worry converted into action and incremental improvement validates the approach.

This emotional-muscular memory, so to speak, makes worry predictable and manageable. Each new environmental report no longer catches one off guard. The mind says: "Yes, this is alarming. I know how to handle this feeling. I will research, connect with allies, try a small reform, and keep going." Over time, this mental repertoire grows varied and flexible, offering multiple ways to respond: direct activism, donation, lifestyle adjustment, conversation with policymakers, or creating supportive art. Instead of fixation on despair, one sees a toolbox filled with morally guided actions prepared to handle any fresh dose of anxiety.

In sum, turning worry into motivation involves acknowledging that fear emerges from care, channeling that care into information-seeking and incremental improvements, connecting with others to magnify impact, and consistently choosing moral action over despair. This ongoing practice changes the emotional climate of one's life, making worry less threatening and more generative. By treating anxiety as a signpost pointing toward what one cherishes, each fear-laden episode becomes an opportunity to affirm values through deeds. Such moral alchemy transforms anxiety's corrosive energy into a sustainable, guiding force that continuously renews engagement, deepens moral character, and supports the collective endeavor to heal and protect our shared home.

Part III: Strength in Numbers—Connecting with Others

Chapter 9: Breaking Isolation Through Community

The Importance of Not Facing Eco-Anxiety Alone

As environmental threats intensify, many people find themselves grappling with a persistent undercurrent of anxiety. This fear can stem from dire headlines about climate shifts, coral reef die-offs, melting glaciers, or mass extinctions. It can arise from encountering polluted streams near home, observing unseasonal weather patterns, or discovering that one's favorite wooded spot has turned into a sprawling development. However it emerges, eco-anxiety is rarely a calm, isolated feeling. Instead, it often cascades into questions about moral responsibility, the fate of future generations, and what it means to live ethically in times of uncertainty. Facing such emotional weight alone can be overwhelming. Yet, there is strength to be found in community, in seeking out others who share concerns, in exchanging experiences, insights, and encouragement. Not facing eco-anxiety alone transforms distress from a private burden into a shared catalyst for resilience, empathy, and collective action.

Isolation magnifies fears. When one sits with anxiety in silence, it can balloon into a sense that nobody else cares or that the issues are too vast for any individual to influence. Without others to talk to, it becomes easier to doubt one's own emotional responses—perhaps feeling silly for worrying so deeply, or wondering if these fears are simply overreactions. Over time, isolation can erode hope, leaving a person feeling powerless and misunderstood. By contrast, connecting with others who experience similar feelings normalizes the emotional struggle. Realizing that many grapple with eco-anxiety affirms that concern is a natural, human response to unprecedented global challenges. In community, individuals discover that they are not the only ones losing sleep over rising seas or dwindling pollinators. This recognition alone can provide immense relief, assuring that anxiety is not an odd personal failing but evidence of moral engagement and empathy.

The first step toward community is acknowledging a desire for connection. Social norms often discourage openly expressing fears about large-scale problems. People might fear appearing too emotional, too grim, or too political. Yet eco-anxiety thrives in shadows. By voicing these emotions—whether in a casual conversation with a friend, at a community meeting, or in an online forum—one starts to break the silence. Often, the response is surprisingly supportive. Friends might admit they too have worried silently, relieved that someone else named the elephant in the room. Strangers in environmental discussion groups reveal parallel struggles, instantly erasing the notion that one is alone. In these shared spaces, fear transforms into solidarity. What felt like a private torment emerges as a communal concern, and the burden disperses among many shoulders.

Communities that acknowledge eco-anxiety vary widely. Some form organically: a group of neighbors discussing concerns about water shortages decides to meet regularly to share news and coping strategies. Others arise in structured environments, like workshops, support groups, or local chapters of environmental organizations that include emotional support in their mission. Online platforms also serve as gathering spaces, connecting individuals across continents who, despite geographic distance, share similar fears and hopes. Religious congregations, cultural clubs, professional networks, and educational institutions can all provide forums for discussing environmental anxieties, weaving moral, cultural, and intellectual dimensions into the conversation. The common thread is that these environments hold space for honest emotion, affirming that it's okay to feel overwhelmed, sad, or frightened.

In community, mutual validation emerges. Hearing another person say, "I feel scared too, and sometimes I don't know what to do about it," releases tension. People realize their emotions are not weaknesses but indications of caring deeply. Such validation interrupts self-blame. Instead of judging oneself for anxiety, individuals learn to accept these feelings and explore them together. This collective acceptance reduces shame, making emotional energy available for constructive thinking. Instead of spinning alone in dread, the anxious mind sees paths forward—perhaps small steps like joining a local clean-up, writing to policymakers, or attending a climate rally. Knowing others stand ready to take similar steps encourages one to move from paralyzing worry into action.

Beyond emotional support, communities offer diverse perspectives. Alone, a person might feel trapped in a narrow range of responses—maybe just recycling more or feeling guilty about driving. Within a group, people share coping strategies, practical tips, and success stories. Someone might introduce local sustainable food initiatives, another describes how they convinced their workplace to adopt green policies, while another shares a personal mindfulness practice that eases nighttime worries. This exchange transforms anxiety into learning. Instead of feeling stuck in fear, community members expand their toolkit for responding. Multiple viewpoints also highlight different value systems—some driven by justice, others by reverence for nature's beauty, others by future-oriented care for children. Exposure to this moral diversity enriches understanding, affirming that there are countless ways to translate concern into meaningful contribution.

Connections also nurture long-term resilience. Environmental crises will not resolve overnight, and staying engaged requires endurance. Alone, stamina wanes easily under relentless grim news. Together, people can take turns uplifting each other. On days when one person feels hopeless, another might bring fresh optimism—perhaps they read about a new conservation project succeeding against odds. Next week, roles reverse, and the once-optimistic person might feel despair, while another member steps in with encouraging news or a humorous anecdote to lighten the mood. This cyclical support ensures emotional resilience. No single individual must remain strong all the time; strength rotates through the community, maintaining collective equilibrium.

Moreover, by not facing eco-anxiety alone, individuals gain practical confidence in collective action. Environmental problems often loom too large for solitary endeavors. Joining forces with others demonstrates that collective efforts multiply impact. A solo letter to a senator might feel futile; a stack of letters or a well-attended town hall meeting shows that many care, increasing the likelihood of policy shifts. On a local scale, working together might mean sharing gardening tips to promote pollinator habitats, organizing carpool systems, starting a community solar initiative, or launching environmental education programs in schools. Witnessing tangible outcomes from collective efforts validates concern. Instead of futile hand-wringing, anxiety channels into real-world changes. Successes, however small, chip away at the idea that fear is pointless. They confirm that emotional distress can be harnessed to build rather than tear down.

Community engagement also refines moral discernment. Solitary anxiety can lead to all-or-nothing thinking: either total despair or desperate hope for miraculous solutions. In group settings, discussions untangle complexities. People debate strategies, compare notes on effective activism, weigh short-term sacrifices against long-term gains. Such dialogues produce nuanced understanding—maybe perfect solutions don't exist, but incremental improvements matter. Recognizing incremental wins reduces anxiety's intensity, replacing futile longing for perfection with patient commitment to stepwise progress. This moral patience stabilizes emotions. The group's collective wisdom guides emotional responses, preventing catastrophic fears from overshadowing the incremental nature of real-world progress.

Cultural narratives within communities matter as well. Some groups draw on religious teachings about stewardship, framing eco-anxiety as a spiritual call to care for creation. Others reference indigenous traditions emphasizing balance and reciprocity, providing a moral story that affirms healing relationships with land. Secular groups might highlight democratic principles—environmental integrity as a matter of justice and fairness. By embedding personal fears into larger cultural or moral narratives, individuals find belonging. They realize their anxiety is part of a broader human struggle to align society with ethical principles. Understanding that one's emotional turmoil resonates with deep cultural currents relieves isolation and confers dignity on emotional responses.

Artistic and creative expressions in community settings also help. Workshops where participants draw, sing, or write poetry about their eco-anxieties transform raw fear into shared cultural artifacts. Viewing another's art depicting melting ice caps or drought-stricken fields evokes empathy. Recognizing these emotions in others' creations validates personal feelings. Creative collaborations channel anxiety into beauty, anger into calls for justice, sadness into elegies that honor what's lost. These cultural productions become communal emotional vessels, reminding participants that they sail together on the same uncertain seas, steering by shared moral lights.

The presence of role models and mentors in community spaces further alleviates isolation. Seasoned activists, educators, or elders who endured past ecological struggles can reassure novices that moral engagement persists despite setbacks. Hearing how previous generations faced acid rain, ozone depletion, or local pollution crises and pushed for reforms gives context. It shows that environmental problems have long been tackled collectively, often with successes that once seemed unimaginable. This historical continuity counters the despair that arises from feeling one's era is uniquely doomed. It also encourages intergenerational solidarity—young people learning from elders, elders inspired by youthful energy. These relationships weave moral fabrics strong enough to carry multiple generations forward.

As communities form, it's essential to ensure inclusivity. Eco-anxiety affects everyone differently. Some marginalized communities already bear disproportionate ecological burdens and might approach environmental fears from the vantage of injustice endured for decades. Ensuring that all voices—rural, urban, indigenous, immigrant, low-income—are heard enriches understanding. Diverse communities help members see that eco-anxiety can unite people across backgrounds who share a desire to protect life's foundations. Realizing that worry traverses lines of class, race, religion, or nationality dissolves the idea that caring is a niche concern. Instead, it appears as a universal moral sentiment rooted in common human vulnerability and hope.

Inclusivity also ensures that solutions proposed within communities address multiple dimensions—technological fixes, policy reforms, cultural revitalization, educational initiatives. When members bring varied perspectives, the group sees that no single tactic suffices. This moral pluralism, nurtured by collective dialogue, prevents narrow thinking. It encourages moral empathy. If someone struggles with eco-anxiety about deforestation in distant rainforests, another facing polluted water in their neighborhood can highlight local actions. This cross-pollination of issues and responses strengthens emotional resilience, proving that while each problem is unique, strategies and moral courage can be shared.

Communities can also establish rituals that anchor emotional well-being. Regular meetings where members take turns expressing what worries them and what inspires them create an emotional rhythm. Just as religious congregations have rituals for solace and reflection, environmental communities can adapt similar practices—lighting candles for threatened species, holding silent reflections for ecosystems under stress, celebrating small victories with a shared meal. Rituals provide structure to emotional expression, ensuring that anxiety doesn't erupt chaotically but finds channels that acknowledge pain while pointing towards healing and action. Over time, these shared practices build trust, friendship, and a sense of collective identity.

Technology can facilitate these connections, especially when geographic barriers make face-to-face gatherings difficult. Virtual groups, social media communities, video calls, and online webinars connect people from different regions. These digital spaces can become safe harbors for emotional honesty. Moderators can encourage respectful listening, guide discussions away from despairing spirals, and prompt solution-oriented thinking. While online interactions sometimes lack the warmth of in-person contact, they can still break isolation effectively. They allow people in remote areas or underrepresented communities to find peers. For instance, a person in a rural locale might discover others scattered worldwide who share their eco-anxieties and their cultural perspectives, forging bonds that transcend distance.

Yet community alone isn't a magic cure. Like any support network, it requires nurturing. Members must commit to listening, respecting differences, and moving beyond venting to constructive problem-solving. Encouragement to act is important—merely complaining together won't reduce anxiety long-term. Celebrating incremental steps taken by individuals or subgroups—a school project completed, a letter-writing campaign launched—reminds everyone that the group's purpose is not just emotional comfort but moral empowerment. Achieving a balanced dynamic—validating fears without getting stuck in them—fosters long-term resilience. Over time, communities become adept at guiding emotional energy from worry toward action, and from despair toward strategic hope.

Another subtle benefit of community is learning to rely on others during emotional dips. In solitary anxiety, a bad day can feel like a confirmation that nothing matters. In a group, one can say, "I'm feeling so discouraged today," and others might respond, "We've been there. Here's how we coped." Hearing coping strategies normalizes emotional fluctuations. As trust builds, members feel safe revealing vulnerabilities. This mutual vulnerability tightens bonds, enabling deeper empathy and reinforcing the notion that no one must shoulder eco-anxiety alone. The community's memory stores countless stories of members who overcame low points, creating a collective narrative that resilience is possible.

Communities can also reach beyond their own membership. By organizing public events—workshops, film screenings, discussion panels—they invite the wider public into the conversation. This outreach reduces isolation at a societal level. As more people engage, eco-anxiety becomes recognized not as a fringe emotion but as a reasonable response to real dangers. Widespread acknowledgment reduces stigma. People who once buried their worries might openly join dialogues, discovering supportive networks. Eventually, a cultural shift can occur: caring about environmental issues and feeling anxious about them is seen as normal and commendable, not odd or alarmist.

This cultural normalization has ripple effects. When eco-anxiety is shared openly and channeled into community efforts, policymakers and leaders cannot dismiss environmental concern as niche. Public pressure intensifies. More voices demand sustainable policies, corporate accountability, and fair adaptation measures. The moral urgency emanating from connected communities influences media narratives, academic research priorities, and political agendas. Over time, this collective moral pressure can alter systems. Feeling anxious alone inspires few. Feeling anxious together, openly and constructively, can sway institutions to respond ethically.

Engaging with communities also refines communication skills. Initially, some may struggle to articulate their fears without sounding overly pessimistic. In supportive groups, learning to express complexity—"I'm scared, but I'm also hopeful," or "I feel powerless, yet I believe small steps matter"—becomes easier. This nuanced communication resonates better in public forums and can persuade more effectively. Improving communication reduces misunderstandings, bridging ideological divides. Thus, community nurtures not only emotional resilience but also the ability to engage diverse audiences, mobilize allies, and find common ground with those who may differ in approach but share underlying moral concerns.

In personal life, reducing isolation through community brings tangible mental health benefits. Anxiety becomes less overwhelming, allowing individuals to function better in daily tasks. Knowing one can attend a monthly meeting to share worries and hear updates transforms anxiety from a lurking dread into a manageable emotion. One can say, "I'll save these concerns to discuss with my group," or "Next week I'll learn if anyone tried this solution." Anticipation of collective reflection relaxes the mind. Over time, these interactions form emotional anchor points—a monthly check-in, a weekly online chat, an annual gathering—infusing life with rhythms that process and relieve stress, preventing emotional buildup.

Communities need not remain static. Some focus on local issues—saving a nearby wetland—then evolve to address broader climate questions. Others start online and grow into in-person volunteer teams. This dynamism mirrors ecological adaptability. Just as species in ecosystems form changing alliances, communities adapt their focus, scale, or membership as conditions evolve. Such adaptability ensures long-term relevance. If policy windows open, the community pivots to advocacy; if a natural disaster strikes, it shifts to emergency relief and emotional support. Maintaining moral purpose and flexibility helps communities navigate evolving crises, consistently offering refuge from isolation.

Finding community might require some effort: searching for environmental clubs at universities, attending public lectures at nature centers or local cultural groups, exploring faith-based organizations embracing environmental stewardship, or joining digital platforms dedicated to eco-concerns. Attempting multiple avenues increases the chance of discovering a group that matches personal values and communication styles. Not every space will feel right—some might be too policy-driven, others too emotional, still others too narrowly focused. The key is to remain patient and open-minded, understanding that forging meaningful connections takes time. Over months, friendships blossom, trust deepens, and what began as a loose network transforms into a true community.

Accepting help from others and offering help in return is integral. When feeling anxious, seeking reassurance from group members, sharing struggles, or asking for suggestions acknowledges that it's okay not to have all answers. Similarly, when feeling stable, one can support another member going through a rough patch. This reciprocal dynamic reinforces that community is not a passive resource but an active moral relationship—everyone contributes, everyone benefits. Over time, these relationships transcend environmental issues, forming bonds of friendship, trust, and shared ethical journeys that endure beyond any single crisis.

Children and youth benefit immensely from community support. Growing up in a time of climate instability can imprint fear early on. If young people find spaces where adults and peers respectfully discuss eco-anxiety, explain complexities, and model proactive responses, they learn resilience from the start. Instead of inheriting isolation and despair, they inherit networks that show caring is normal, action possible, and fear manageable. This early formation of moral and emotional literacy equips them to become future stewards, leaders, and educators, passing on the tradition of connected resilience to subsequent generations.

Ultimately, recognizing the importance of not facing eco-anxiety alone underscores that moral courage and emotional health flourish in community. Individual anxiety, no matter how intense, finds pathways to expression, validation, and transformation when shared. Instead of paralyzing the heart, fear energizes collective effort. Instead of sowing despair, it cultivates empathy and mutual support. Communities, large or small, local or global, in-person or virtual, become places where worry reveals care, care sparks conversation, conversation leads to solidarity, and solidarity channels into constructive action. Over time, this cycle fosters moral resolve strong enough to confront daunting environmental challenges, ensuring that no one must bear the weight of eco-anxiety in solitude.

Finding Support Groups, Online Communities, Local Networks, and Faith-Based Circles

As people discover the strength that comes from sharing eco-anxiety with others, the natural question becomes: where to find these supportive communities? Different personalities, cultural backgrounds, and logistical factors mean that no single type of group fits everyone. Fortunately, a wide variety of settings, from structured support groups to informal local networks or faith-based circles, cater to diverse needs. Understanding the spectrum of options makes it easier to identify the right match. Whether one seeks a therapeutic environment focused on emotional well-being, a bustling online forum rich in global perspectives, a hands-on local initiative that blends action with connection, or a spiritual community offering moral guidance, possibilities abound. Each type of gathering transforms isolation into companionship, nurturing a sense that eco-anxiety is manageable and, more importantly, that caring about the planet is a shared moral endeavor rather than a solitary burden.

Formal support groups dedicated to environmental distress have emerged as professionals recognize eco-anxiety as a legitimate emotional challenge. Such groups often resemble therapeutic circles, guided by mental health professionals, counselors, or trained facilitators who create a safe space for open discussion. Participants share fears, sadness, or anger related to ecological threats, and the facilitator introduces coping strategies—breathing techniques, cognitive reframing, or mindfulness practices—to help process emotions. The advantage of a formal support group is structure: sessions meet regularly, follow an agenda, and ensure confidentiality. This framework can be comforting for individuals wary of sharing personal feelings in unstructured contexts. Over time, attendees gain emotional tools to handle distress, feeling less overwhelmed by grim headlines. Moreover, hearing others' journeys reveals that their struggles are not unique. While these groups might charge fees or require registration, many organizations or non-profits run free or low-cost sessions, especially in communities hard-hit by environmental changes.

Therapeutically oriented groups help break down complex emotions into understandable parts. A participant might say, "I feel paralyzed by guilt because I'm not doing enough," and the facilitator can guide the discussion toward acceptance and incremental action steps. Such an environment encourages vulnerability, reducing stigma around anxiety. The presence of a professional ensures that if intense emotions arise, they are met with expertise and sensitivity. Additionally, these support groups sometimes form alliances with environmental psychologists or eco-therapists who specialize in this domain, refining the conversation into deeper understanding of how personal identity, moral values, and environmental realities intersect. Such insights not only alleviate distress but also clarify motivations for civic engagement, activism, or personal lifestyle changes.

If formal groups feel too clinical or require resources not easily accessible, online communities present a more flexible option. On social media platforms, discussion boards, or dedicated websites, people can join forums or groups centered on environmental concerns. These digital spaces break geographical barriers, connecting someone in a rural town with individuals in coastal cities, tropical regions, or even overseas. The diversity of online membership means encountering a broad range of perspectives. A person worried about Arctic ice melt might learn from someone living in the Arctic region, who can share firsthand experiences and cultural insights. Another struggling with eco-anxiety about agricultural collapse might find solace in farmers discussing regenerative practices. The sheer variety of participants fosters empathy and encourages comprehensive understanding, reminding that while each location faces distinct challenges, all are interconnected in a global tapestry of environmental change.

Online communities often permit anonymity, lowering the threshold for participation. Shy individuals or those afraid of judgment can initially lurk, reading others' stories before chiming in. Over time, they may gain confidence to share their own experiences. Moderators in many online groups maintain respectful norms, preventing trolls or cynics from hijacking the conversation. Rules that encourage constructive dialogue ensure that anxiety is met with supportive responses rather than dismissal. This environment can be especially comforting for people in areas where eco-concerns are not widely discussed. They realize that even if local friends or family show little interest, global allies exist who understand their fears.

However, online spaces, while accessible, can become overwhelming if not navigated carefully. The influx of news links, doom-laden forecasts, and passionate debates may intensify anxiety. Selecting well-moderated, solution-oriented platforms helps balance emotional expression with hope. Some online groups feature weekly "solutions threads" focusing on positive developments, volunteer opportunities, or policy wins, thus offsetting the heavy emotional load of negative news. Over time, repeated engagement teaches participants to filter information, draw strength from collective wisdom, and refine their voice in discussions. Online interaction also can lead to offline connections—members who discover they live nearby might organize local meetups, blending digital solidarity with face-to-face community-building.

Local networks form another vital avenue. These might include neighborhood associations, environmental clubs at schools, gardening circles, or community centers hosting workshops on sustainability. Local networks excel at transforming abstract concerns into tangible action. If worry about plastic pollution plagues someone, joining a neighborhood clean-up not only addresses the problem physically but also provides camaraderie. In these hands-on settings, eco-anxiety dissipates as participants focus energy on improving their immediate environment. Seeing others enthusiastically replant trees, construct pollinator habitats, or set up compost systems reassures that people care and can enact positive changes, even within constraints. Real-world progress, shared laughter during a community event, and the mutual encouragement of neighbors who say, "We're in this together," produce emotional grounding hard to replicate virtually.

Local networks have the advantage of place-based identity. Participants can rally around a shared landscape—a local river, forest, or community garden—that symbolizes collective stewardship. The emotional resonance of caring for common ground reinforces bonds. Unlike abstract global issues, local efforts yield visible outcomes: a restored wetland, a successful farmers' market offering sustainable produce, or a school orchard bearing fruit. Witnessing these successes firsthand counters despair. Instead of feeling like a powerless observer of distant catastrophes, one becomes an engaged citizen shaping one's immediate environment. Over time, these networks cultivate empowerment, confidence, and resilience against eco-anxiety. Facing challenges shoulder-to-shoulder with neighbors, individuals learn that no one must navigate fear alone; each contributes skills, ideas, and moral support to the group's evolving efforts.

Faith-based circles add another dimension. Many religious traditions contain teachings that encourage caring for creation, viewing nature as sacred, or emphasizing the moral duty to protect life. In these communities, eco-anxiety can be framed spiritually. Prayer services asking for guidance in stewardship, sermons on ethics of consumption, or religious festivals celebrating harvest cycles remind participants that environmental engagement has deep moral roots. In faith-based settings, discussing anxiety about endangered species or climate refugees weaves moral concern with spiritual purpose. Members understand that their emotional pain over ecosystem destruction isn't just psychological distress but a moral and spiritual call to act compassionately.

Faith-based circles often provide long-standing rituals and symbolic actions that comfort believers. Lighting candles for threatened habitats, dedicating prayers to climate victims, or interpreting sacred texts through an ecological lens infuses the conversation with transcendent meaning. Participants realize that their worries about pollution or deforestation resonate with ancestral wisdom and divine imperatives. This moral gravitas can transform anxiety into reverent commitment. Additionally, these circles frequently possess established social structures—committees, fundraising events, educational programs—offering multiple ways to translate concern into deeds. Whether organizing an interfaith climate vigil, advocating for green policies within religious bodies, or partnering with environmental charities, these spiritual communities channel eco-anxiety into guided, meaningful service.

In all these settings—support groups, online communities, local networks, faith-based circles—one theme stands out: the move from passive worry to active engagement. Fear by itself can stagnate. Coupled with others' support, it flows into purposeful action. The chosen community type might depend on personality. A more introverted person might start online, building confidence before seeking in-person groups. Another, craving hands-on tasks, heads straight to local gardening clubs. Someone drawn to moral frameworks finds faith-based gatherings appealing. Yet, mixing different communities can also help. A person might attend a monthly therapeutic support group, participate weekly in a local river clean-up, and join an online forum for global updates. Each environment nurtures different aspects of emotional resilience.

Cultural fit also matters. In some regions, environmental associations form around indigenous leadership, reinforcing values of ancestral land care. Others might focus on youth voices, driven by passionate teenagers who demand intergenerational responsibility. Some groups emphasize policy advocacy and research, suitable for those who find solace in informed debate and evidence-based strategies. Others highlight creative expressions—art installations, storytelling festivals, eco-poetry readings—to process anxiety through aesthetic experience. Exploring multiple avenues, attending a few initial meetings, and sensing whether the group's ethos aligns with one's values is wise. If the first attempt feels mismatched—perhaps too formal or not action-oriented enough—trying another setting can eventually yield a supportive circle that feels like home.

It's essential to remember that these communities, while supportive, are not places to offload all responsibility. Shared eco-anxiety encourages empathy, but constantly dwelling on fear without working towards solutions risks creating cycles of despair. Moderators or facilitators in good groups balance open emotional expression with encouraging problem-solving. They might say, "We understand the fear. What small step can we take today?" or "Let's name three positive developments this month." By continually steering the conversation from mere venting to forward-looking approaches, these communities maintain hope and prevent emotional stagnation. Accountability among members ensures that the group's purpose—transforming fear into constructive moral force—remains clear.

Over time, membership in a supportive community can change how individuals relate to environmental news. Instead of absorbing dire data in solitude, they might think, "I'll bring this up with my group and see what they suggest." Anticipation of collective response tempers panic. Hearing varied interpretations—one member sees an opportunity for renewed activism, another highlights adaptive measures—broadens understanding. Difficult headlines become catalysts for collective learning rather than triggers for isolated despair. This shift reduces stress and cultivates a healthier relationship with environmental information. People learn to contextualize challenges within a network of caring minds, distributing emotional weight across many shoulders.

Another advantage of these communities is discovering allies outside one's comfort zone. An online forum might introduce someone to a policy analyst from another country who highlights successful green legislation. A local group might include a teacher who can bring environmental lessons into classrooms, inspiring youth engagement. A faith circle could contain a small business owner willing to experiment with sustainable supply chains. These unexpected connections expand the range of possible interventions. Instead of feeling restricted by personal limitations—lack of expertise, funds, or authority—participants realize that within their network someone has complementary skills, resources, or influence. Collective strength emerges not just from empathy but from synergy, leveraging diverse talents to address complex problems more effectively than any lone individual could.

As relationships deepen, trust grows. Members begin confiding in each other about not only eco-anxieties but also broader life challenges. Solidarity extends beyond environmental concerns, forging friendships grounded in shared moral commitments. This camaraderie reveals that ecological care is not a narrow issue; it intertwines with social justice, personal well-being, and cultural identity. Knowing that the same people who help navigate eco-anxiety also care if you're feeling isolated at work or worried about family health integrates environmental efforts into a wider moral and emotional landscape. Such holistic support further reduces the temptation to retreat into isolation. Instead, environmental anxiety is seen as one of many interconnected life struggles that communities can help address.

Practical advice on engaging these communities can help first-timers. Start by researching local environmental meetups—websites, community bulletin boards, or green organizations' newsletters often list upcoming events. Attend a workshop or a film screening and strike up conversations. Ask about ongoing projects. If shy, consider bringing a friend who shares your concerns. Online, search for keywords like "eco-anxiety support," "climate grief groups," or "environmental mental health forums." Social media platforms often host private groups where members must answer brief questions to join, ensuring earnest participation. For faith-based engagement, talk to clergy or lay leaders, explain your eco-anxiety, and inquire if the congregation has environmental initiatives or interest groups. Many faith communities welcome suggestions to integrate ecological care into their activities.

Once part of a community, consider small steps to strengthen bonds. Offer to help organize an event, contribute a skill—maybe you're good at graphic design, writing, or public speaking. Sharing your talents and time builds reciprocity and trust. If the group lacks direction, gently propose a format—a monthly discussion topic, a rotating responsibility to research good news, or a shared resource library. By proactively shaping the community's structure, you align it more closely with the goal of mutual support and moral action. Initiative also fosters a sense of investment: the community is not just a service you consume but a living entity you help sustain.

In digital forums, respect the rules, engage kindly, and appreciate diverse opinions. If a new member expresses intense fear about climate feedback loops, respond compassionately, offer factual reassurance, and mention steps you or others took to cope. Positive interactions encourage newcomers to stay and contribute. Over time, your empathetic presence could inspire others to adopt a similar tone, reinforcing a culture of constructive dialogue. Avoid sensationalism; if you share troubling news, balance it with possible solutions or reputable sources. This approach nurtures an atmosphere where eco-anxiety is acknowledged but not exploited for emotional shock. Instead, it's consistently channeled towards informed, morally grounded responses.

In faith-based circles, if no structured environmental group exists, consider initiating one. Speak to leaders about hosting a study session focusing on scriptural, philosophical, or cultural teachings related to environmental stewardship. Encourage incorporating nature walks, blessing ceremonies for local ecosystems, or prayers for affected communities. Such steps legitimize environmental care within spiritual life, granting it a stable place in the community's moral calendar. Over time, these practices normalize conversations about eco-anxiety as part of spiritual practice—lamenting losses, seeking guidance for wise stewardship, and finding solace in shared purpose.

Local networks often appreciate fresh energy. If you join a community garden collective, for example, you might suggest a monthly "story circle" where everyone shares how they feel about recent environmental news. Placing emotional reflection alongside gardening tasks integrates mental well-being with practical solutions. Similarly, proposing a climate book club or documentary night after a tree-planting session weaves moral inquiry and fellowship into action. Such initiatives remind participants that environmental involvement isn't just about chores—it's also about supporting each other emotionally, learning collectively, and strengthening moral convictions.

Communities should remain open to adaptation. If a group's focus becomes too narrow—e.g., exclusively policy talk with no space for emotional support—suggest adding a short "emotional check-in" segment. Conversely, if discussions become too emotional and seldom move towards action, propose planning a small campaign or educational outreach. This balance ensures that the community doesn't stall. Members continuously refine their approach, guided by shared moral goals: turning fear into constructive agency, complexity into collaborative understanding, and sorrow into empathy-fueled action.

As relationships mature within these groups, a sense of collective identity forms. Eco-anxiety shifts from a private malaise to a sign of moral citizenship, an emotional compass pointing to what's at stake. Members learn that their values align, whether centered on protecting future generations, ensuring environmental justice, or honoring the sacredness of Earth. Over time, this shared identity transcends initial fears. Anxiety no longer isolates—it becomes a bond linking people who care deeply. In this moral community, differences in tactics or emphases can be navigated through dialogue and compromise, sustained by trust that everyone shares a baseline moral concern: safeguarding life's integrity.

At a societal level, as communities multiply and interlink—support groups connecting with local networks, faith-based circles interfacing with online forums—an informal but powerful infrastructure forms. This infrastructure spreads norms of emotional support, moral reflection, and active problem-solving. The ripple effects are cultural: people witnessing positive outcomes in one community replicate similar initiatives elsewhere. Environmental organizations collaborate with mental health advocates, religious leaders partner with local environmental educators, online platforms direct participants to offline volunteer opportunities. Over time, this interconnected web reduces fragmentation. No single community stands alone; each becomes part of a larger moral ecosystem that consistently converts eco-anxiety into communal strength.

For individuals, involvement in multiple communities can provide layered support. Someone might rely on a monthly support group for emotional processing, a local citizens' assembly to tackle policy issues, a faith circle for spiritual grounding, and an online forum for quick updates and global perspectives. Each community addresses a different dimension of environmental engagement. Emotional well-being finds refuge in supportive groups, moral courage grows through shared value discussions, intellectual clarity emerges from diverse conversations, and practical skills develop through community-based projects. This holistic approach ensures that eco-anxiety never festers in isolation; it's always matched by a chorus of voices guiding each other toward wiser, kinder, more sustainable paths.

As the world continues changing, new community forms may arise. Virtual reality gatherings might simulate shared experiences in fragile ecosystems, galvanizing participants worldwide. Cultural exchanges between distant communities facing different climate impacts could occur, each inspiring the other. Faith traditions might host interfaith summits integrating ecological ethics at their core. Each innovation further erodes isolation, making it harder for eco-anxiety to lurk unaddressed in private corners. Instead, anxiety becomes a recognized starting point for connection and moral awakening. With every step, people learn that caring for the planet is not a solo quest. The planet's challenges belong to everyone, and the moral courage required to address them is easier to sustain when woven through networks of understanding.

Recognizing that effective communities embrace imperfect participation also helps. Some members engage deeply, attending every meeting and initiating projects. Others come sporadically, dropping in when anxiety spikes or when they have time. The community should remain welcoming, not pressuring individuals to prove their commitment constantly. Understanding that people's emotional and logistical capacities vary—some have caregiving duties, health constraints, or shifting work schedules—upholds compassion. A flexible, low-pressure environment encourages return visits and long-term bonding. When members know they can step back without losing respect, they remain connected at least on some level, returning when they're ready.

Likewise, communities can find ways to acknowledge and celebrate emotional growth. Perhaps after a season of meetings, the group notes how members who once hesitated to speak now share openly, or how someone who arrived anxious and pessimistic now expresses cautious hope.

Recognizing these emotional changes affirms that community support works, not just externally but internally. It shows that courage is contagious and that collective empathy can heal emotional wounds. Such milestones encourage newcomers, proving that by investing in these circles, one can gradually transform private anguish into shared moral vigor.

By maintaining a forward-looking stance, communities remain resilient through inevitable challenges. If certain members struggle with differences in opinions—say, over whether to focus on systemic activism or personal lifestyle changes—the group can facilitate respectful dialogue, reminding everyone that these differences arise from caring. By returning to shared values, communities navigate conflicts constructively. Over time, negotiating differences with goodwill further consolidates trust. The community not only soothes eco-anxiety but also demonstrates democratic moral practice: listening, compromising, and reaffirming commitments without fracturing.

In conclusion, finding support groups, online communities, local networks, and faith-based circles is essential for not facing eco-anxiety alone. Each type of community offers unique benefits. Formal support groups provide therapeutic structure, online platforms connect distant allies, local networks foster place-based action, and faith-based circles integrate spiritual guidance. People can combine these resources, tailoring support to their personal style and cultural context. Together, these diverse communities form a societal safety net against emotional isolation. Instead of being crushed by fear, individuals learn to hold and share their anxieties, rediscovering courage and hope in collective moral engagement. Over time, these networks grow strong enough to influence cultural norms, policies, and collective imaginations. Eco-anxiety, once a harbinger of lonely despair, becomes a catalyst for unity, compassion, and sustained efforts toward a more just and sustainable world.

Sharing Stories to Feel Seen, Understood, and Less Alone

The act of sharing personal experiences—telling stories about what it feels like to worry about the planet's future, grieve disappearing habitats, or grapple with moral responsibilities—transforms isolation into connection. Stories convey more than information; they carry emotional truths that resonate deeply with others. In communities built around eco-anxiety support, the exchange of narratives becomes a powerful tool, allowing members to see themselves reflected in each other's journeys. Suddenly, distressing feelings are no longer private burdens but points of empathy linking hearts and minds across backgrounds and geographies. By sharing stories, individuals not only feel seen and understood; they realize their emotions are valid, their concerns are natural, and their hopes and fears are part of a collective moral tapestry rather than personal flaws.

This storytelling fosters understanding on multiple levels. On a personal note, narrating one's emotions—"I felt helpless after reading about coral bleaching," or "I panicked when I considered how drought might affect my hometown"—articulates feelings that might otherwise remain tangled. Speaking or writing these sentiments externalizes them, reducing their intensity. As listeners nod in recognition or respond, "I've felt that too," fear and shame diminish. Confirmed by another's empathy, anxiety loses its edge. Telling one's story clarifies its contours: is the worry rooted in moral concern for future generations, sadness over losing beloved landscapes, or frustration at political inertia? As the narrative unfolds, the storyteller gains self-awareness, understanding their emotional triggers, values, and motivations more clearly.

For listeners, hearing stories builds solidarity. When someone recounts their eco-anxiety—perhaps describing how a childhood forest now reduced by logging provokes both anger and mourning—listeners see parallels in their own experiences. Even if they never visited that forest, they understand the universal sting of loss. Feeling recognized in another's tale reassures them they are not odd or oversensitive for feeling similarly distressed. This mutual validation counteracts the cultural silence that often downplays environmental emotions. No longer must one guess if others care or worry, for stories confirm it directly. The resulting relief is palpable: one's anxieties belong to a shared human condition responding to planetary instability, not a private defect.

Cultural differences enrich storytelling's impact. In a diverse group, stories flow from many places and perspectives. Someone from an island community threatened by rising seas might share how her family debates relocating ancestral graves, grappling with the spiritual trauma of losing ancestral lands. Another, living in a Northern city, might detail anxiety sparked by increasingly unpredictable winters and the confusion of adjusting traditions once anchored to stable seasons. Listening to these varied accounts broadens everyone's understanding of ecological impacts. Each narrative reveals another facet of the global crisis, expanding empathy beyond one's immediate horizon. Realizing that anxiety is felt globally yet in context-specific ways promotes a moral cosmopolitanism—caring not only about one's backyard but about distant ecosystems and cultures too.

These narratives also humanize abstract data. Reading that a certain species declines by a certain percentage can trigger sadness, but it remains somewhat detached. Hearing a group member describe memories of that species—perhaps how as a child they watched monarch butterflies migrating each autumn—makes the loss vivid and personal. Such stories transform statistics into lived experiences, deepening emotional engagement. This deepened engagement strengthens moral resolve: if people find themselves touched by another's heartbreak, they are more likely to support efforts to restore butterfly habitats or advocate for policies that protect pollinators. Stories thus bridge knowledge and moral action, converting abstract concern into heartfelt impetus.

Communities often incorporate storytelling sessions as structured activities. In support groups, a facilitator might designate a portion of each meeting for personal reflections. Online forums might have "story threads" where members post essays or videos describing their emotional landscapes. Local networks might host story circles or storytelling workshops, teaching skills to convey experiences more effectively. By valuing storytelling as a communal practice, these groups reaffirm that emotional honesty is welcome, not burdensome. Over time, participants become more comfortable sharing vulnerable feelings, knowing others will respond compassionately. This openness encourages ongoing dialogue that refines understanding and continually reassures members that they are not alone.

Telling stories also invites solutions. After someone shares their eco-anxiety narrative, others may suggest strategies that helped them cope. If a storyteller expresses feeling stuck—perhaps immobilized by fear of climate collapse—listeners who overcame similar paralysis might share what actions restored their sense of agency. One might say, "I felt the same despair until I joined a local reforestation effort; replanting seedlings gave me hope." Another might add, "I started writing letters to policymakers; even small advocacy steps lessened my anxiety." These story-informed suggestions resonate more than abstract advice because they emerge from lived experiences. The storyteller sees that transformation is possible—someone else stood in similar shoes and found a way forward.

As trust grows, more nuanced stories emerge. Instead of focusing solely on fear and sadness, members reveal complexities—moments of joy discovered amid worry, instances when anxiety spurred creativity, or evolving relationships with nature inspired by moral convictions. A person might recount how their initial dread over melting glaciers led them to study environmental law, now channeling fear into systemic reforms. Another could describe how anxiety pushed them to form a community garden, discovering new friendships and small victories feeding collective resilience. These narratives illuminate anxiety's potential for moral awakening: fear can mobilize compassion, empathy, and determination previously dormant.

Storytelling also challenges stereotypes. Sometimes environmental concerns are dismissed as "privileged worries" or "overreactions." Hearing from someone whose community faces immediate harm—like saltwater intrusion ruining farmland or heatwaves affecting vulnerable neighbors—discredits trivializations. Another might share that their eco-anxiety is less about luxury fears and more about losing cultural traditions reliant on stable ecosystems. Stories undermine simplistic narratives, reminding everyone that eco-anxiety is not a niche phenomenon but arises wherever humans cherish ecosystems, depend on natural resources, or sense injustice in uneven climate burdens. This expanded perspective fosters inclusive solidarity, uniting people from varied socioeconomic, ethnic, and age backgrounds under a shared moral umbrella.

Artistic storytelling intensifies these effects. Poems, songs, short films, or illustrations capturing eco-anxiety's emotional texture convey truths that plain speech sometimes misses. Visual metaphors—like a painting of a tree shedding leaves under smoky skies—tap into feelings beyond words. When community members share artistic creations, they enrich collective emotional vocabularies, offering new ways to discuss eco-anxiety. Artistic storytelling grants permission to explore complexity: one might depict anxiety as a river carrying both toxic debris and seeds of regeneration. Another might write a poem blending loss and hope, each stanza shifting tone. Experiencing these artistic expressions together deepens empathy and binds participants through shared aesthetic encounters. The resulting emotional resonance transcends language barriers and intellectual analysis, forging a bond that encourages collective perseverance.

Digital storytelling is equally potent. A short personal essay posted in an online forum may prompt dozens of replies within hours, transforming solitary fear into a communal conversation. Podcasts featuring interviews with people processing eco-anxiety amplify voices, allowing listeners to feel a speaker's emotional cadence. Webinars or live streams can host storytelling events spanning continents, with participants sharing experiences from different time zones. Technology democratizes storytelling: anyone can record a short video from their phone, narrating how local deforestation changed their sense of home, and upload it to a platform where others respond in kind. This global tapestry of stories fosters moral imagination, ensuring that eco-anxiety no longer isolates individuals in silence. Instead, it sparks a continuous exchange of narratives, each adding depth and understanding.

Family and intergenerational storytelling plays an important role. Older relatives might recall times of ecological abundance, seasonal predictability, or traditional practices now endangered. Younger family members may share their anxiety for the future they'll inherit. By listening to elders' memories and grandchildren's worries side by side, families weave a longer moral thread, connecting past, present, and future. These intimate exchanges underscore that eco-anxiety is not just about current troubles but also a legacy affecting those yet unborn. Caring spans generations, and stories anchor these connections in lived experience. Family storytelling can thus transform eco-anxiety from a solitary weight into a shared family narrative, inspiring intergenerational cooperation—elders passing wisdom, youth bringing fresh energy, and all collaborating to sustain what they hold dear.

In faith-based or spiritual circles, storytelling can integrate eco-anxiety into moral teachings. Members might share personal testimonies about how environmental changes tested their faith or prompted reevaluation of spiritual duties. A worshipper might describe how walking through a drought-stricken field redefined their understanding of stewardship, pushing them beyond ritual into concrete care for that land. Others may recount visions or dreams highlighting interdependence. Such stories blend moral reflection with lived faith, confirming that ecological concern is not a diversion from spiritual life but an integral dimension. Community members discovering similar spiritual resonances find comfort in knowing that their spiritual path accommodates environmental anguish, guiding them from despair towards service and reverence.

When stories highlight not only anxiety but also resilience and adaptation, they remind participants that emotional states can evolve. One member might recall a time they felt hopeless, then recount how joining a local climate justice group or adopting more sustainable habits gradually rekindled hope. These narratives demonstrate transformation, proving that eco-anxiety need not be permanent paralysis. Listeners realize they too can find ways to cope and flourish amid uncertainty. Encouraged by these success stories, individuals dare to try small steps—reaching out to neighbors, contacting a policy representative, or starting an educational blog—believing now that change is possible.

Occasionally, disagreements arise. Different experiences might lead to varied interpretations—some prefer technological optimism, others champion radical systemic change. Storytelling can mediate these tensions by focusing first on shared emotional truths before debating solutions. If everyone recognizes that underlying their differing strategies is a similar grief for lost species or fear for coastal towns, they approach disagreements with empathy rather than hostility. Stories humanize the interlocutors, reducing them from ideological opponents to fellow humans navigating moral complexity. Constructive dialogue emerges, grounded in mutual understanding that all care, though they express care differently. This empathetic foundation allows negotiations and compromises that might have been impossible if no emotional common ground existed.

For newcomers, storytelling eases entry into these communities. Arriving fearful and unsure, a newcomer listening to ongoing narratives soon realizes that everyone once stood where they stand—uncertain, overwhelmed, but hopeful for solutions. By hearing how others overcame initial inertia, learned coping methods, or found meaning amid pain, the newcomer gains a roadmap. Feeling less alienated, they start to share their own perspective. Perhaps they describe feeling helpless upon reading climate reports, and in return, a long-time member recounts how similar despair prompted them to organize a local green festival. This exchange shapes a welcome: worry is greeted with understanding, and fear meets examples of growth, turning apprehension into budding trust.

As communities mature, storytelling traditions evolve. They might record oral histories, preserving accounts of how the group confronted environmental crises year after year. Over time, these recorded narratives form a collective memory, accessible to future members. Such archives help future generations understand that today's eco-anxiety is neither new nor uncharted territory. Past members faced challenges, found ways forward, and documented their emotional journeys. This historical continuity reassures newcomers that eco-anxiety can be integrated into a long, ongoing moral effort, producing inspiration and continuity rather than feeling like a novel catastrophe leaving everyone adrift.

Recognizing the healing power of storytelling, some communities host special events—a storytelling night dedicated to exploring eco-grief, poetry slams focused on climate themes, or a documentary screening followed by an open mic session where viewers share emotional reactions. These rituals celebrate emotional honesty and moral reflection, placing eco-anxiety at the center of communal cultural activities. Instead of treating fear as something shameful or hidden, events showcase it as part of the moral fabric connecting everyone. By normalizing public discussion of environmental emotions, storytelling challenges cultural norms that expect stoicism or detachment. Gradually, society learns that eco-anxiety is not weakness but a sign of moral engagement, and that sharing it can uplift rather than discourage.

Such collective vulnerability forges a resilient community identity. When crises intensify—extreme weather hitting local fields, sudden policy reversals threatening environmental gains—members return to storytelling. They process fresh wounds by recalling past hurdles survived and lessons learned. Circling together, they tell how they once felt cornered but emerged stronger. This narrative continuity forms a bulwark against despair. Each new challenge becomes an episode in a larger saga of moral courage. Participants no longer fear that fear itself will isolate them; they know they can translate it into stories that unite them deeper. Like a chorus, each voice adds volume to collective determination.

In educational contexts, storytelling transfers moral wisdom to younger generations. Teachers might invite elders or community activists to share their ecological struggles and triumphs with students, blending environmental science with personal testimony. Students confronted with dire climate data find hope in learning that anxiety is normal and that previous generations also navigated uncertainties, sometimes successfully. This moral inheritance, passed through stories rather than sterile lectures, creates emotional scaffolding. Young people realize they are not forging meaning from scratch; they inherit narratives that demonstrate resilience, adaptability, and moral striving amid complexity.

Policy advocates can also use storytelling strategically. When lobbying for environmental protections, they might share personal accounts of those affected by pollution or climate impacts. These stories humanize issues that lawmakers might otherwise treat as abstract numbers. Hearing a mother's fear for her children's health due to contaminated water, or a farmer's heartbreak as drought withers her fields, moves decision-makers beyond intellectual debate. Storytelling cuts through political rhetoric, appealing to moral empathy and urging responsive policies. In this way, the communities that foster emotional storytelling indirectly influence policy landscapes, ensuring that data and arguments are complemented by narratives that clarify what's morally at stake.

Even globally, storytelling can forge transnational solidarity. Online forums or cultural exchange programs allow people in one region to share experiences with another. A coastal villager in the Pacific describing rising seas may inspire activists in a European city to amplify the call for climate justice. Stories from indigenous communities restoring habitats might motivate urban dwellers elsewhere to support Indigenous-led initiatives. These exchanges weave a global moral community, each story adding a thread of compassion linking distant lives. Recognizing that eco-anxiety is felt worldwide shatters illusions of isolation and encourages a universal moral mandate: to protect life and fairness across borders.

To keep storytelling vibrant, communities must ensure everyone can speak. Facilitation techniques help quieter members share narratives: encouraging equal speaking turns, using prompts like "Describe a moment when you felt hopeful despite your anxiety," or "Recall a memory of nature that shaped your values." These prompts guide participants toward introspection and expression. Some might prefer written forms—journals, letters read aloud—while others are comfortable speaking spontaneously. Accessibility matters too—translating stories into multiple languages if the group is diverse, or accommodating different forms of communication. By designing inclusive storytelling practices, communities guarantee that voices from all walks of life enrich the collective narrative.

The synergy of storytelling with other community elements is powerful. Stories complement intellectual discussions of solutions, grounding them in emotional experience. They augment problem-solving sessions by reminding participants of what's at stake—real lives, cherished traditions, beloved places. They enliven action-oriented gatherings, reminding everyone that each volunteer hour or policy effort arises not just from duty but from heartfelt care. Storytelling breaks the monotony of technical talk, reintroducing passion and purpose. Supported by the trust and empathy cultivated elsewhere in the community, these narratives deepen relational bonds, ensuring that members relate not solely as co-activists but as moral companions united by shared vulnerability and shared aspiration.

The transformative nature of storytelling emerges over time. Participants may recall their early days in the community, arriving anxious and skeptical. After hearing multiple stories and eventually telling their own, they integrate into the moral fabric of the group. Anxiety no longer isolates them; it connects them. They realize their voice matters, their fears are listened to, and their perspective contributes to collective wisdom. Observing others' growth—from fear to confidence, from despair to renewed resolve—encourages the belief that they too can transform.

Through iterative storytelling, communities continuously reaffirm that eco-anxiety is not an endpoint but a catalyst for moral engagement, learning, and collaborative resilience.

As more communities adopt these storytelling practices, a broader cultural shift takes shape. Media outlets, for instance, might highlight personal testimonies of climate-impacted individuals, humanizing news reports. Educational curricula might integrate oral histories from local elders who witnessed environmental changes over decades, enriching students' moral imagination. Artistic collectives might produce anthologies of eco-narratives, spreading moral insights to wider audiences. These cumulative efforts embed a storytelling ethic into societal responses to climate challenges, ensuring that technocratic solutions never overshadow the moral and emotional dimensions of planetary care.

At the personal level, participating in these storytelling exchanges encourages ongoing growth. Each time one shares a story, they refine understanding of their own motivations, clarify their moral stance, and rediscover inner resources. Hearing others' stories reveals previously unseen moral angles, teaching empathy and broadening moral horizons. Over months and years, individuals become adept at translating anxiety into narrative form, turning raw fear into structured insight. This skill extends beyond eco-anxiety—improving emotional communication in general, strengthening relationships, and enhancing one's ability to navigate life's complexities with honesty and compassion.

In essence, sharing stories within eco-anxiety communities ensures that no one must face the moral weight of environmental threats alone. Personal narratives link individuals into a supportive network where each voice finds validation and inspiration. These stories transform fear into empathy, isolation into belonging, and uncertainty into moral clarity. As members witness each other's emotional journeys, they internalize the lesson that while environmental futures remain uncertain, moral resilience and collective effort persist. Together, stories weave a social fabric strong enough to bear heavy burdens, stable enough to explore difficult truths, and flexible enough to adapt as conditions evolve.

This collective narrative building transcends any single moment. It represents a long-term moral project: forging a culture where eco-anxiety prompts caring action rather than despair. Over time, as these communities share stories, their emotional intelligence grows, their moral imagination expands, and their capacity to face shifting conditions increases. In these shared stories, individuals and groups find not just solace, but motivation to push forward. Their voices, once solitary cries of worry, merge into a chorus affirming that to care about the planet is to be part of a grand, ongoing human story—one of striving, learning, cooperating, and hoping amid the immense task of protecting our only home.

Bringing people together transforms private fears into sources of shared moral strength. Instead of wrestling alone with eco-anxiety, individuals find that seeking companionship, exchanging insights, and telling personal stories builds a supportive environment where isolation gives way to empathy and encouragement. By acknowledging that caring about the planet and feeling deeply troubled by its fate is a natural, human response, communities offer reassurance, diversity of perspectives, and pathways to action. In such groups—whether formal support circles, online forums, local networks, or faith-based gatherings—worry becomes a bridge linking hearts that recognize common values and long for meaningful solutions. This collective engagement not only softens the emotional burden but also ignites a renewed sense of purpose, inspiring participants to translate concern into practical efforts for healing and protecting what they love. Over time, working together nurtures moral resilience, affirming that the human capacity for understanding, cooperation, and hopeful action runs deeper than fear, and that facing global challenges side by side can make all the difference.

Chapter 10: Honest, Compassionate Communication

Talking About Climate Worries with Family, Friends, and Colleagues Without Alienation

As climate-related fears intensify, many people feel torn between the urgency of their concerns and the fear of pushing away those they care about by speaking openly. Voicing eco-anxiety to family members, close friends, or colleagues can be challenging. Some worry about appearing too negative, while others fear judgment or dismissal. Yet honest, compassionate communication about climate worries can open the door to deeper understanding, stronger relationships, and cooperative problem-solving. Instead of driving wedges between people, conversations can inspire empathy, invite collaboration, and broaden everyone's moral perspectives. The key lies in approaching these talks with tact, patience, and humility, ensuring that what emerges is not alienation but closeness grounded in shared human vulnerability and hope.

For many, the first barrier is the perception that climate issues are too big or too politicized to discuss comfortably. While it's true that environmental topics often spark debate, it's also true that they connect to values almost everyone holds dear: health, security, fairness, and the well-being of future generations. Recognizing these universal links can shift the focus from confrontation to common ground. Instead of starting a conversation with dire scientific projections or policy critiques, begin with personal stories, feelings, or memories. Describing how you felt upon seeing a local river polluted or a once-stable season now unpredictable can humanize the issue. Such authenticity makes others more receptive. They see you not as a "climate preacher" but as a fellow human being facing uncertainty and caring deeply about shared resources and cherished places.

Emphasizing common values before delving into specifics matters. If speaking to a parent, recall a childhood tradition—visiting a local lake, enjoying seasonal harvests, or marveling at wildlife. Show how your concern stems from wanting to preserve these experiences for others. With friends, highlight that you're worried because you care about everyone's long-term security, not because you wish to judge their choices. With colleagues, frame your eco-anxiety as a moral response to challenges that could affect everyone's professional future or community stability. This approach positions climate-related fears not as niche obsessions but as heartfelt reactions rooted in love, responsibility, and respect. It's harder to reject someone's feelings when you see that they arise from values you yourself consider important.

It helps to prepare mentally before starting these conversations. Consider what outcome you'd find beneficial. Is it simply to be understood and not feel alone in your worries? Or do you hope to motivate cooperative actions like reducing office waste, supporting local environmental initiatives, or exploring more sustainable household habits? Clarifying your goal guides how you present the topic. If you only seek emotional validation, let that be known: "I've been anxious about these environmental changes and just need someone to understand that I'm not overreacting." If you hope for action, specify small, concrete steps that feel achievable: "I'm thinking of cutting down on single-use plastics. Would you consider doing it with me?" Breaking down big problems into manageable requests eases tension. People become less defensive when they see you're not demanding radical life overhauls overnight, just exploring ways to align actions with shared concerns.

Tone matters. Start gently and acknowledge complexity. Instead of insisting, "We must do X right now, or we're doomed," say, "I've been reading about these changes in weather patterns and feeling uneasy. It makes me wonder if there are small things we could do together that might help us feel less helpless." This openness invites dialogue rather than confrontation. It signals that you respect the other person's intelligence and agency. By avoiding moral superiority or alarmist extremes, you keep the door open for honest exchange. This approach fosters trust, encouraging loved ones or peers to share their own thoughts, even if they differ.

Listening is crucial. Honest communication isn't just about presenting your fears; it's also about understanding others' emotional landscapes and concerns. After expressing why you feel anxious, pause to hear their reaction. Perhaps your friend nods sympathetically, or maybe they hesitate and say, "I understand, but I'm not sure how bad it really is." Resist the urge to argue immediately. Instead, ask questions: "What makes you feel uncertain?" or "What parts of the issue seem unclear?" By validating their perspective—acknowledging that it's normal to feel overwhelmed, skeptical, or unsure—you reduce defensiveness. Demonstrating respect for their viewpoint increases the chance they'll reciprocate, acknowledging that your anxiety is not baseless or exaggerated.

Empathy goes both ways. If a family member dismisses your fears, consider their background and motivations. Are they downplaying the problem to shield themselves from worry or maintain a sense of normalcy? Maybe they feel powerless and resort to denial for emotional self-preservation. Understanding this can prevent alienation. Instead of labeling them as uncaring, recognize that everyone grapples with uncertainty differently. Gently express that you're not trying to rob them of hope or comfort, but rather invite them into a more honest understanding of shared challenges. For example: "I get that it's scary and easier to think it's not a big deal. I sometimes wish I could brush it off too. I'm just feeling this worry deeply and hope we can at least talk about it openly."

Humor can help soften tense moments. Light-hearted jokes about media exaggerations or acknowledging your own emotional intensity can release pressure. For instance, "I know I might sound like a doom-and-gloom reporter, but I'm not trying to audition for a disaster documentary. I just care a lot!" Infusing gentle humor signals that you're aware these topics can be heavy. It shows humility, assuring others you don't see yourself as morally superior or humorless. Humor can humanize the conversation, reminding all participants that they're on the same team, navigating tough realities together rather than battling each other.

Timing and setting matter too. Bringing up serious anxieties at a busy family dinner or a rushed coffee break at work might backfire. Choose a calm, private setting where people have time to reflect. A relaxed weekend afternoon walk, a quiet evening chat, or a dedicated lunch break conversation can create better conditions for meaningful exchange. Starting with a lighter topic or something positive—a recent nature documentary that impressed you, a small environmental success story—can ease into deeper concerns without feeling abrupt or overwhelming. The environment shapes mood, so a peaceful context encourages open-mindedness and patience.

Be mindful of emotional pacing. If discussing dire predictions or personal fears, pause occasionally to check in: "How are you feeling hearing this?" or "Am I overwhelming you with too much information?" Offering these breaks respects the other person's emotional limits. It conveys that you care not only about the planet but also about the emotional well-being of those you're talking to. This consideration prevents flooding them with grim data, which might trigger defensiveness or shut-down. Instead, co-manage the emotional temperature of the conversation, ensuring both you and they remain comfortable enough to engage productively.

When disagreements arise, avoid framing differences as moral failures. If a colleague doubts the severity of climate models, resisting the urge to call them ignorant or uncaring is key. Instead, try: "I see we have different interpretations of the data. I'd love to understand what sources shape your view. Maybe I can share what convinced me." Framing it as mutual curiosity rather than a battle of right versus wrong encourages constructive dialogue. Likewise, if your sibling believes individual actions don't matter compared to corporate emissions, acknowledge their point and suggest exploring complementary approaches: "I agree systemic changes are crucial. But maybe personal habits can support broader shifts. How about we talk about ways both levels matter?" By incorporating their perspective into a larger, more nuanced understanding, you invite them to join you in exploring complexity rather than defending fixed positions.

Another useful strategy is storytelling. Instead of bombarding relatives with charts, share a personal moment that affected you: "Last summer, the creek near our childhood home had almost no water. It reminded me how precious these places are to us." Emotions conveyed through anecdotes can bypass intellectual resistance. People relate to personal experiences more readily than abstract arguments. When they see climate worries grounded in vivid memories, moral outrage, or love for particular places, it's harder to dismiss them as unfounded hysteria. Stories also open the door for them to share their own experiences—maybe they recall strange weather that disrupted a cherished tradition. Mutual storytelling nurtures empathy and trust.

If you encounter strong resistance, be patient. Not everyone is ready to confront these issues. Avoid pressuring them into agreement right away. Sometimes planting a seed is enough: "I understand this might not feel urgent to you now. But I hope you'll remember this conversation if you notice changes that start to worry you later." Trust that repeated gentle exposure to honest, compassionate communication may shift attitudes over time. Just as environmental changes unfold gradually, changing hearts and minds also requires patience, persistence, and understanding that people process information at different paces. Pressing too hard too soon can backfire, causing them to dig in their heels. Instead, leave the door open for future exchanges.

Compassionate communication also acknowledges that eco-anxiety can manifest in different forms. Some express anger at political failures, while others sink into grief over what's lost. Recognizing these variations allows you to tailor your approach. If a friend is always angry about corporate pollution, acknowledge their anger and show you respect their feelings: "I see why you're furious; it's unfair. I feel that too, sometimes. Let's think about how we can channel that anger into productive action." By validating emotions before suggesting solutions, you establish solidarity. In contrast, if someone is more numb or indifferent, try gently awakening their empathy by asking how they'd feel if their favorite pastime or hometown tradition disappeared due to environmental changes. Posing questions that evoke personal connection can soften indifference without attacking their stance.

Role modeling honesty and vulnerability encourages others to reciprocate. By admitting uncertainty—"I don't have all the answers, and I'm still learning how to cope"—you show that you're not positioning yourself as an expert but as a concerned individual willing to grow. This humility lowers defenses. Others may feel safer admitting their own lack of understanding or confusion. Shared humility fosters a sense of collective moral exploration rather than a debate over who's right. Over time, these dynamics foster an environment where everyone can become more informed, conscientious, and supportive.

If a conversation goes awry and becomes contentious, don't be discouraged. Not every attempt will succeed immediately. Sometimes stepping back, acknowledging the tension, and offering to revisit the topic later allows emotions to settle. You might say, "I can tell this is getting heated, and I don't want us to argue. Maybe we can pause and try again another time." Respecting boundaries and emotional states shows your commitment to a healthy relationship. It reassures them that you value the person more than winning a debate. This respect often resonates over time, making them more open to future dialogues.

In workplaces, where differences in opinion or hierarchical relationships complicate discussions, focus on shared professional interests. For example, if colleagues worry that environmental talk distracts from company goals, frame sustainability as risk management or brand reputation enhancement. Emphasize that caring about resource efficiency, stable supply chains, and avoiding reputational damage aligns with business interests. Adapting your message to their priorities can reduce alienation. Later, as trust grows, you can introduce the moral and emotional aspects more comfortably. The key is to find a starting point that resonates with their worldview, gradually paving the way for deeper conversations.

Over time, consistent honest communication can create supportive micro-communities within families, friend circles, or workplaces. People become accustomed to open exchanges about complex feelings, acknowledging that environmental stressors affect everyone differently. Gradually, a culture develops where it's normal to say, "I'm feeling uneasy about those wildfire reports" and receive understanding, not dismissal. As these networks expand, they reduce isolation at a structural level. People begin to see eco-anxiety as a shared emotional response that warrants collective moral consideration. Instead of tiptoeing around the subject, they address it head-on, solving misunderstandings and fostering empathy.

Cultural narratives that depict honest, compassionate environmental communication—films, articles, podcasts—also help normalize this mode of discourse. When people encounter stories of families discussing eco-anxiety constructively or workplaces implementing green initiatives after thoughtful dialogue, they realize these outcomes are achievable. Real-life success stories reinforce the idea that it's possible to engage in meaningful, non-alienating conversations. They provide templates for how to speak, listen, and respond. As these narratives spread, fewer people perceive talking about climate worries as socially risky; instead, it becomes a mark of moral responsibility and relational sincerity.

In educational settings, teaching students from a young age to discuss environmental issues openly and empathically lays a foundation for future honest communication. Role-playing exercises, classroom debates with emphasis on respect and compassion, and assignments encouraging students to interview family members about environmental changes cultivate these skills. Children raised with these norms grow into adults comfortable sharing climate concerns without fear of alienation. Over generations, this cultural shift can strongly influence how societies handle environmental crises—through dialogue, empathy, and collaborative problem-solving rather than silence, denial, or polarization.

If at any point you feel conversations stagnate in repetitive patterns—one party always defensive, the other always alarmed—consider bringing in external resources. Suggest reading a balanced article, watching a documentary that carefully explains issues, or attending a neutral workshop together. External content can act as a mediator, helping both sides find common understanding without feeling personally targeted. Sometimes introducing a well-researched report, a speech by a respected moral leader, or data from a trusted institution reframes the discussion in less personal, more collective terms, allowing everyone to focus on content rather than perceived personal attacks.

Reflect on progress periodically. After several conversations, notice changes in tone. Are your family members less dismissive now? Do colleagues show more interest in sustainable practices? Acknowledge small improvements and celebrate them. Positive reinforcement encourages continued openness. Even if they haven't fully embraced your viewpoint, subtle shifts—like fewer defensive comments or a willingness to try an eco-friendly product—indicate progress. Recognize these increments and express appreciation for their openness, reinforcing the idea that mutual understanding grows through patience and good will.

In time, honest, compassionate communication can deepen trust, not just about climate issues but also in other areas. Demonstrating that you can broach tough topics respectfully and handle conflicts calmly builds relational confidence. Loved ones or coworkers see that discussing difficult truths doesn't destroy bonds; it can strengthen them. This relational resilience means future challenges—environmental or otherwise—can be navigated more collaboratively. Instead of avoiding problems, everyone learns to face them together, leveraging empathy, patience, and honesty as collective tools.

Ultimately, discussing climate worries with family, friends, and colleagues without alienation is possible when rooted in empathy, shared values, careful framing, listening, and humility. Not every conversation will produce immediate agreement or action. But by consistently using respectful, compassionate approaches, you create conditions where misunderstandings can ease, barriers can soften, and moral understanding can flourish. Over time, these dialogues not only reduce personal eco-anxiety by securing emotional support but also lay the groundwork for broader cultural change. As more people communicate openly and kindly about their fears and hopes, a collective moral awakening emerges, strengthening society's capacity to confront environmental challenges with unity, wisdom, and sustained moral courage.

Empathy, Listening Skills, and Nonjudgmental Dialogue

Talking about environmental anxieties and moral uncertainties can be fraught with misunderstanding. People come to these conversations carrying diverse values, emotional states, and cultural backgrounds. Without careful attention to how we communicate, attempts to discuss climate worries might quickly devolve into debate, denial, or defensiveness. Empathy, listening skills, and nonjudgmental dialogue offer a way forward. They transform a potential battlefield of clashing opinions into a shared space for honest moral exploration, understanding, and growth. By practicing empathetic engagement, we create an environment where even difficult truths can be acknowledged without crushing hope, where differences can be navigated without alienation, and where moral courage can flourish rather than wither.

Empathy lies at the heart of effective communication about eco-anxiety. Rather than striving to "win" a conversation or prove one's point, empathic engagement involves stepping into the other person's emotional and intellectual space. This doesn't mean agreeing with everything they say; it means recognizing their perspective as valid to them and respecting the emotional reality behind their stance. If a friend dismisses fears about rising seas or extreme weather events, empathizing doesn't require endorsing their dismissal. Instead, it entails acknowledging why they might feel resistant—perhaps it's protecting a cherished worldview, a sense of stability, or coping with uncertainty by minimizing threats. Understanding this emotional need, rather than labeling them as ignorant or uncaring, opens doors for connection.

Similarly, if someone expresses deep sadness or anger over environmental damage, empathetic listening acknowledges that these emotions flow from moral concern, rather than dismissing them as exaggerations. Imagine a coworker passionately lamenting how climate changes threaten future generations. Instead of responding with facts or counterarguments, start by reflecting their feelings: "I hear how upset and worried you are about what the future holds for your kids. It sounds heartbreaking and overwhelming." This reflection assures them that you see their emotional reality, not just their words. Empathy validates their experience, reducing defensiveness and making them feel understood, which sets the stage for deeper, more productive dialogue.

Developing empathy begins with self-awareness. Recognize your emotional triggers. Do you become frustrated when someone seems apathetic or uninformed? Acknowledge these feelings privately before entering the conversation. Remind yourself that everyone's journey to understanding environmental issues differs. By managing your internal reactions, you approach the discussion with a calm, open demeanor. This internal preparation allows empathy to flow naturally. You're not faking compassion; you're genuinely striving to see the world through another's eyes, even if their viewpoint challenges your own.

Active listening skills complement empathy. Active listening means focusing fully on the speaker, giving them your undivided attention, and conveying through body language, eye contact, and verbal acknowledgments that you are hearing them. In face-to-face conversations, lean slightly forward, maintain gentle eye contact, and nod occasionally. Verbal cues like "I see," "mm-hmm," or "that makes sense" indicate engagement. Online, active listening involves reading carefully, acknowledging their points, and responding thoughtfully rather than skimming or replying hastily. When they finish speaking, summarize what they've said before expressing your viewpoint. For instance, if a family member worries that addressing climate issues might jeopardize jobs, restate their concern: "You're worried that rapid changes to reduce emissions might cause some people to lose work, which feels unfair and risky." Hearing their thoughts reflected back accurately assures them you truly listened.

Active listening also entails asking clarifying questions. If someone says, "I just don't see how we can trust these climate models," instead of immediate rebuttal, ask: "Can you tell me more about why the models feel unreliable to you?" Such questions invite them to elaborate, revealing underlying fears or uncertainties. Perhaps they encountered conflicting information or feel overwhelmed by scientific complexity. By understanding the root of their doubt, you can respond more effectively, addressing their actual concern rather than assuming ignorance or bad faith. In this way, active listening transforms potential confrontation into a joint search for understanding.

Nonjudgmental dialogue is the next essential component. Nonjudgmental doesn't mean abandoning moral convictions or pretending all views are equally valid. Rather, it means suspending immediate value judgments about the person you're speaking with. Refrain from labeling them as "selfish," "stubborn," or "clueless." This stance creates a safe emotional climate. People sense when they're being judged and often respond with defensiveness or withdrawal. By withholding judgment, you invite them to open up more honestly. You communicate that you respect their dignity as a person, even if you disagree with their perspective.

Nonjudgmental dialogue also involves framing differences as opportunities for growth, not as moral failings. Suppose a colleague insists that individual actions are pointless without system-wide changes. Instead of calling them cynical, acknowledge their point: "You feel that personal efforts don't count much unless big players change. That's an important consideration." Now you've recognized their reasoning without disparaging them. From there, you can share how individual and collective actions can intersect, explaining that personal choices often influence cultural norms, consumer demand, and eventually policy—thus working hand-in-hand with systemic reforms. By treating their viewpoint as a piece of a larger puzzle rather than a flawed stance, you avoid labeling them and keep the conversation constructive.

Another key aspect of nonjudgmental dialogue is being open about your own uncertainties. If asked a challenging question—such as how exactly a certain policy will affect rural communities—admit if you don't know. Honesty about gaps in your knowledge humanizes you and reduces any impression of moral high-handedness. It shows that you're not there to dominate the conversation with supposedly complete answers but to navigate complexity together. This humility can inspire the other person to lower their defenses and share their own doubts or areas they'd like to learn more about. By meeting in a zone of mutual exploration, both parties can move from combative stances to collaborative inquiry.

Balancing empathy, listening, and nonjudgmental dialogue can feel tricky in practice. One helpful tactic is the "OARS" approach borrowed from motivational interviewing techniques: Open-ended questions, Affirmations, Reflective listening, and Summarizing. Open-ended questions ("What concerns you most about climate policies?") invite fuller responses. Affirmations acknowledge the person's efforts or emotional courage ("I appreciate that you're sharing these worries with me"), building trust. Reflective listening involves paraphrasing their statements to show understanding, while Summarizing periodically checks for clarity and agreement. Though originally developed for counseling, OARS aligns well with moral conversations about eco-anxiety, guiding them towards empathy and understanding rather than conflict.

Cultural factors also shape how empathy, listening, and nonjudgmental dialogue unfold. In some cultures, direct emotional expression may be uncommon, so gentle approaches or indirect conversation starters might be more appropriate. In others, strong opinions are welcomed as signs of passion, and the challenge is to channel that passion without personal attack. Recognize these cultural nuances. If someone from a community that values emotional restraint appears distant or reserved, interpret that not as disinterest but as adherence to communication norms. Adjusting your style—perhaps by starting with factual data before sharing feelings—can help bridge cultural differences. By showing respect for their communication traditions, you reduce misunderstandings and encourage them to reciprocate.

Another dimension is the role of shared activities. Sometimes discussing eco-anxiety directly feels heavy. Instead, engage in a shared task first—cooking a meal with locally sourced ingredients, walking through a park, or watching a documentary about climate adaptation. These activities create a relaxed atmosphere, allow mutual appreciation of nature or sustainability practices, and make conversation flow more naturally. As you both handle tangible elements—preparing food or marveling at wildlife—the abstract concern of climate change becomes more relatable. The activity can prompt comments like, "Seeing these birds thriving here gives me hope, even though I'm worried about other habitats," opening a window for honest exchange. When spoken in a shared context, fears feel less isolated and more part of a joint reality.

When you encounter someone who strongly denies environmental issues or becomes hostile, empathy and nonjudgmental dialogue become even more critical. While you cannot force them to see reason, you can maintain composure and continue practicing good listening. Acknowledge their frustration: "I can sense this topic upsets you, and you're feeling skeptical about the entire narrative. That's understandable." This doesn't concede the factual ground but grants them emotional legitimacy. If possible, find a subtle common point—maybe they value economic stability and you can frame environmental policies as long-term investments in prosperity. Even if they remain unconvinced, your respectful demeanor keeps communication channels open. Over time, your steady approach may soften their stance, or at least lay groundwork for future conversations when circumstances change.

Empathy and nonjudgmental dialogue also apply to self-talk. Before and after challenging discussions, treat yourself with kindness. If a conversation didn't go well, avoid harsh self-criticism. Reflect: "I tried to understand their viewpoint and shared my feelings honestly. It's okay if we didn't reach agreement today." Acknowledge the difficulty of these conversations; they involve complex moral dimensions, personal fears, and often inadequate societal discourse. Self-empathy prevents burnout. It reminds you that progress in moral understanding is incremental and that every attempt, even if bumpy, contributes to a larger cultural shift towards open, ethical conversations about our planet's future.

Over time, consistently applying empathy, listening, and nonjudgmental dialogue changes the emotional tone of environmental conversations in your circles. Friends might start initiating these chats, asking how you feel about recent environmental news because they know the topic is safe to discuss without judgment. Family members may approach you after reading a troubling article, confident that you'll listen patiently and respond kindly. Colleagues might share ideas for office sustainability, trusting that you'll value their input rather than dismissing it. As trust builds, conversations become more substantive and morally productive. People dare to ask harder questions—about the ethics of consumption, the fairness of global resource distribution, the balance of economic progress and ecological integrity—knowing that these inquiries will be met with understanding rather than accusation.

In workplaces, leadership can set a tone of empathic communication. If a manager encourages open dialogue, acknowledges that people worry about environmental implications for the company's future, and demonstrates nonjudgmental listening skills, employees feel safer raising concerns. Over time, this may lead to innovative green initiatives, improved morale, and a sense that the company recognizes moral dimensions in its operations. Similarly, educators who model empathic listening when students express eco-anxiety teach the next generation that moral reflection is not a side issue but integral to learning and ethical reasoning. Students grow up internalizing respectful discourse, likely becoming adults who handle eco-anxiety conversations with similar grace.

As networks of empathetic communicators expand, larger communities and even societies can shift from polarized shouting matches to more constructive dialogues about climate challenges. Instead of entrenched camps hurling accusations, a culture of empathic inquiry emerges. Public forums, town halls, and media debates may incorporate moderators trained in active listening and nonjudgmental facilitation. Imagine a political debate where candidates are prompted to acknowledge the emotional weight of climate impacts before discussing policies. Such framing grounds the conversation in shared humanity, reducing toxic polarization. While perfect civility is never guaranteed, incremental improvements in communication norms can mitigate the worst excesses of hostility and denial.

Global implications arise, too. If empathy, listening, and nonjudgmental dialogue become international norms, negotiations on climate treaties or resource management might proceed more smoothly. Diplomatic delegations might sincerely acknowledge each other's fears and constraints, moving beyond mere positional bargaining. Such moral atmosphere could help bridge gaps between nations with disparate interests, fostering compromise and long-term cooperation. While this vision may seem idealistic, it's rooted in practical communication skills that anyone can practice in everyday life. Summed across countless interactions, these shifts can influence policy climates and cultural attitudes at scale.

For individuals dedicated to moral engagement, refining empathy, listening skills, and nonjudgmental dialogue is a form of moral self-improvement. Consider these qualities as moral muscles that strengthen with use. Over months and years, as you practice, emotional intelligence grows. You become more adept at reading cues, detecting fears hidden beneath bravado, distinguishing honest confusion from willful ignorance, and knowing when to push for clarity or when to yield space for reflection. Such moral finesse helps navigate moral gray areas inherent in environmental issues. Recognizing these skills as part of your ethical toolkit inspires confidence that, even amid uncertainty, you can facilitate dialogues that foster understanding rather than deepen conflict.

In essence, empathy, listening, and nonjudgmental dialogue form the bedrock of honest, compassionate communication. They guide us away from monologues toward true dialogues, away from defensive stances toward shared moral inquiry. By seeing others as moral agents grappling with complex truths rather than adversaries, we encourage them to open up and reciprocate empathy. By listening fully, we learn the contours of their fears and values, enabling us to respond precisely and kindly. By withholding judgment, we create safe space for all to evolve their thinking. Over time, this practice not only reduces eco-anxiety isolation but also builds moral communities capable of addressing environmental challenges with wisdom, cooperation, and unwavering respect for our shared human dignity.

While these approaches require patience, they pay dividends in emotional resilience and collective readiness. As more people adopt empathic communication, the ecosystem of dialogue improves. Conversations about climate worries transition from draining arguments to meaningful exchanges that illuminate moral perspectives, spark new ideas, and reinforce a sense of common purpose. In this transformed conversational landscape, eco-anxiety isn't a barrier to understanding but a catalyst for deeper moral connections and collaborative efforts to protect the Earth we share.

Turning Contentious Discussions into Opportunities for Mutual Understanding

Even with empathy, listening, and nonjudgmental dialogue as guiding principles, discussions about climate change and environmental responsibilities may still become contentious. Moral stakes are high, future outcomes uncertain, and people often cling to entrenched beliefs. Facing disagreements can be intimidating—who wants to risk a heated argument with a friend, family member, or colleague? Yet it's precisely in these tense moments that opportunities arise to deepen moral understanding and strengthen communal resilience. Contentious discussions, rather than signaling failure, can become transformative if approached with courage, humility, and a willingness to learn. By reframing disputes as shared moral inquiries rather than zero-sum battles, participants can find common ground, expand their moral imaginations, and emerge with greater respect for each other's perspectives.

The first step in transforming contention into mutual understanding is recognizing conflict as a sign that something meaningful is at stake. People rarely become passionate about topics they find irrelevant. If someone argues fiercely about the effectiveness of renewable energy, it signals that they care—about economic livelihoods, energy security, or ethical resource management. Acknowledging that everyone involved cares deeply about something important, even if they define importance differently, sets the stage for constructive engagement. This reframing moves the conversation away from "us-versus-them" dynamics and into a zone of shared moral investment. Instead of labeling the other side as obstructionist or naive, see them as fellow moral agents grappling with complex dilemmas under uncertain conditions.

Adopting a stance of curiosity helps. When confronted with a contentious assertion—like skepticism about climate science or dismissal of individual efforts—ask sincere questions. "What leads you to believe that individual actions don't matter?" or "Can you share what concerns you about relying on climate models?" These questions convey a desire to understand, not trap. Curious inquiries invite the other person to elaborate, often revealing underlying values or fears.

Perhaps their skepticism about climate models stems from past experiences of misleading media coverage, or their dismissal of personal efforts arises from frustration that big corporations evade responsibility. Unearthing these layers clarifies that beneath heated opinions lie legitimate anxieties, moral intuitions, or intellectual uncertainties.

As the other person responds, maintain the empathetic and listening posture established earlier. Reflect their points back accurately: "I hear that you worry about job losses if we move away from certain industries, and that feels unfair to those workers." Acknowledging their concerns as real and morally pertinent diffuses tension. This approach prevents the conversation from becoming a debate over who's right and who's wrong. Instead, it becomes an exploration of how to address both environmental imperatives and social justice. Once they feel understood, they might soften their tone and become more receptive. Being genuinely curious and reflecting their concerns without judgment cultivates trust, encouraging them to grant the same courtesy when you present your perspective.

Another tactic involves identifying and highlighting shared values. Even when positions seem polar opposite—one person demands immediate decarbonization, while another insists on slower transitions—both might value stability, fairness, and a habitable future. If you discover these common values, state them explicitly: "We both want a secure future for our children, and we both dislike seeing anyone unfairly burdened by changes." Emphasizing shared values reminds participants that they stand on the same moral field, even if they differ on strategies. This reminder counters the perception that disagreement means moral estrangement. Instead, it shows differences as variations in how to achieve a commonly desired moral outcome, shifting the narrative from confrontation to collaboration.

In some cases, acknowledging complexity helps. Environmental issues rarely lend themselves to one-size-fits-all solutions. By admitting complexity, you signal that you're not pushing an oversimplified agenda. For example, if someone resists aggressive emissions cuts out of fear for local industries, concede that transitions are indeed complex. "You're right; changing energy sources can't happen overnight without affecting people's jobs. The question is, how do we manage that change responsibly, so workers aren't left behind?" This acknowledgment validates their concern, demonstrating sensitivity to trade-offs. Once they sense that you respect the complexity and don't view them as barriers to progress, but rather as important voices ensuring fairness, they may relax their defensive stance.

Using neutral language and avoiding inflammatory terms also reduces defensiveness. If you say, "It's absurd that you think personal efforts are pointless," you immediately put them on the defensive. Instead, try: "I'm curious why you feel personal efforts don't contribute enough." Similarly, avoid labeling their viewpoint as "denial" or "ignorance." Instead, frame it as uncertainty or skepticism: "I understand you're skeptical about climate projections." Neutral language focuses on the idea rather than attacking the person's character, preserving dignity on all sides. Over time, this respectful tone encourages reciprocity, prompting others to address your views without resorting to insults or dismissive language.

When it's your turn to present a different viewpoint, do so with humility. Instead of stating, "Here are the facts that prove you wrong," try, "I've read studies suggesting that personal choices can influence market demand and eventually shape policy. I'd be interested in your thoughts on that." This phrasing shares information while inviting dialogue rather than announcing a verdict. Humility acknowledges that no one has a monopoly on truth. By presenting evidence as something to consider, rather than a weapon to wield, you grant the other person space to integrate new information without feeling coerced.

Listening for emotional cues is crucial in contentious exchanges. Sometimes a person clings to a certain stance not out of pure logic but because they feel threatened—threatened by change, by loss of identity, or by moral condemnation. If their voice rises or they become agitated, consider the emotional subtext. Are they afraid that environmental policies will harm their cultural traditions? Are they angry at perceived hypocrisy from elites who preach sustainability without making personal sacrifices? Naming these emotions can help: "It sounds like you're frustrated by what seems like hypocrisy in how environmental issues are addressed. That frustration makes sense." Once these emotions are acknowledged, the conversation can shift from clashing arguments to addressing emotional needs. If you can empathize with their frustration, you might then find ways to meet those emotional needs, like exploring policies that ensure fairness and transparency.

Time and pacing matter. If a conversation grows too heated, suggest taking a break. "This topic is really intense. Maybe we can pause and revisit it another time," communicates respect for emotional limits. Returning to the discussion after a cooldown can prevent permanent relational damage and allow both parties to reflect. Sometimes, sending a follow-up message the next day, referencing a point they made and reiterating appreciation for the conversation, helps keep channels open and more amicable. Over multiple conversations, trust builds, and previously contentious points become opportunities for joint learning rather than confrontation.

Consider using metaphors or personal stories to communicate your viewpoint. If discussing global warming's complexity, share a personal anecdote—how noticing earlier spring blooms in your hometown made you realize the local climate is changing. This humanizes your argument, replacing abstract data with lived experience. Metaphors can also bridge understanding: comparing the climate system to a complex machine we barely understand fully might help someone skeptical of climate modeling see why uncertainty doesn't equate to unreliability. Analogies resonate emotionally, fostering mutual understanding. Just ensure your metaphors don't oversimplify to the point of misrepresentation; the goal is to clarify, not distort, the complexities at hand.

Another approach is to seek common actions that both sides can endorse. If you can't agree on the severity of climate predictions, perhaps you can agree on steps that improve local air quality, reduce waste, or increase community resilience. Suggesting small, manageable initiatives—planting trees in the neighborhood, encouraging a local school to install solar panels—shifts focus from abstract disagreements to tangible cooperation. Working side by side on a communal project bonds people, reduces hostility, and might later create a more receptive atmosphere for discussing bigger differences. Over time, experiencing the positive results of joint efforts can break down initial resistance to wider climate measures.

Humor, when used gently and respectfully, can defuse tension. A light-hearted remark acknowledging the complexity—"Well, if saving the planet were easy, I guess we'd all be superheroes by now!"—can ease the pressure. Humor should never mock or belittle the other person, but it can illuminate the shared difficulty of these moral problems. Laughter, even mild, creates a moment of humanity amidst intellectual sparring. This moment might prompt both sides to remember they're human beings trying their best under tough conditions, and that agreement in some form may still be possible.

If multiple people are involved in a contentious discussion—like a group meeting or a family gathering—structured facilitation helps. Set ground rules at the start: everyone gets equal speaking time, no interrupting, and no personal insults. Agree to summarize each speaker's main point before anyone responds. This formal structure may feel artificial, but it ensures fairness and respect. The result is that contentious points are examined more carefully, not drowned out by emotional outbursts. As participants see that the process values fairness and clarity, they become more willing to engage sincerely. Over time, such structured dialogues can evolve into norms of ethical conversation that persist even in less formal encounters.

Remember that not every contentious discussion will yield a breakthrough. Some disagreements run deep, rooted in identity, ideology, or misinformation resistant to quick resolution. Accepting this reality reduces frustration. The goal is not always to change minds instantly, but to improve mutual understanding and maintain relationships. By treating contentious moments as stepping stones rather than dead ends, you preserve the possibility of future progress. People remember being treated respectfully, even if they didn't shift their stance immediately. Months or years later, they might revisit the topic with fresh eyes, or new circumstances might make them more open to considering alternative views. The memory of a calm, empathetic conversation can linger, planting a seed that might later germinate.

When frustration surfaces, practice self-care. If you feel emotionally drained, take time to recover. Remind yourself that moral progress is iterative. Applaud your effort to maintain empathy and nonjudgmental dialogue despite difficulties. Each attempt at constructive engagement refines your skills and moral character. Learning from each encounter—what worked, what triggered defensiveness, how humor or stories helped—enables you to approach future contentious discussions with greater nuance and patience.

Over time, as you apply these strategies consistently, your ability to handle contentious discussions improves. Conflicts that once felt unbearable become more manageable, even enlightening. You realize that contention often highlights areas of moral uncertainty that deserve thoughtful exploration. Instead of seeing disagreements as obstacles, you embrace them as chances to hone moral reasoning, clarify values, and identify shared goals. The emotional resilience and moral intelligence developed in these exchanges can ripple outward, influencing how others approach tough conversations. Observers may note your calm approach and decide to emulate it, gradually raising the standard of discourse in your community.

This shift in perspective—viewing contentiousness as opportunity—is itself morally significant. It reflects a commitment to treating others as moral equals capable of growth, rather than as adversaries to be defeated. It expresses faith in the human capacity for learning and moral adaptation. Such faith is crucial in an age where environmental challenges require unprecedented cooperation. Without trust that people can change and reach understanding through conversation, it's easy to fall into cynicism. By engaging skillfully even when tension rises, you testify to the value of persistent moral effort, demonstrating that navigating complexity and emotion can yield better outcomes than polarized silence.

Within communities that integrate these approaches, contentious discussions become less feared. Members know they have tools—curiosity, empathy, active listening, reframing—to prevent meltdown and nurture comprehension. The community's moral cohesion strengthens, equipping it to address environmental issues with unity rather than division. Disagreements remain but transform into productive dialectics that refine collective strategies and ensure no perspective is ignored.

Applied globally, widespread adoption of empathic, nonjudgmental methods for handling contention could profoundly improve the climate discourse. Politicians, activists, educators, and ordinary citizens who internalize these principles might find it easier to forge compromises, experiment with hybrid solutions, and accept incremental progress. Instead of stalemating over ideological differences, societies could learn to manage disagreements as normal features of democratic moral life, ensuring that no moral voice is drowned out by anger or fear. This ethical evolution in communication could accelerate the search for wise and equitable responses to environmental challenges.

Ultimately, turning contentious discussions into opportunities for mutual understanding is about moral courage. It's about choosing dialogue over dismissal, patience over immediate victory, and respect over ridicule. By approaching disputes with humility, empathy, and a focus on shared values, you become a moral bridge-builder. Over time, these efforts pay off: hearts soften, minds open, and once-inflammatory topics become catalysts for joint learning. Even if not every conversation ends harmoniously, the aggregate effect moves the moral landscape closer to a culture where environmental anxieties are no longer wedges dividing people, but rather sparks igniting collective moral imagination. In that culture, contentious discussions aren't dreaded—they're embraced as vital crucibles forging stronger, more understanding communities poised to tackle the greatest moral challenge of our era.

As these principles become second nature, conversations about climate and its challenges shift from perilous territory into meaningful exchanges that strengthen rather than strain connections. The willingness to speak honestly about fears, guided by empathy and a genuine effort to listen, helps ensure that climate anxiety does not fester in silence or erupt into bitter disputes. Instead, it becomes a catalyst for richer moral understanding and collaborative problem-solving. Approaching even the most difficult topics with humility and kindness shows that acknowledging complexity need not alienate, and that even in moments of contention, common ground can be found. By consistently practicing respectful communication, individuals and communities develop the emotional and moral resilience needed to navigate environmental uncertainties together. Each carefully chosen word and thoughtful pause contributes to a culture of dialogue that encourages growth, fosters trust, and preserves the integrity of relationships in the face of daunting global responsibilities.

Chapter 11: Turning Anxiety into Collective Action

Channeling Fear into Meaningful Engagement: Volunteering, Advocacy, Civic Participation

As global environmental challenges intensify, many people find themselves navigating a complex emotional landscape. Concerns about climate disruption, biodiversity loss, and resource scarcity can provoke deep anxiety, occasionally bordering on despair. Yet within these same fears lie the seeds of moral courage and collective resilience. The key lies in channeling that anxiety not into paralysis, but into meaningful engagement—whether that means joining a local volunteer initiative, raising one's voice in civic forums, or participating in advocacy efforts that push institutions toward more sustainable practices. By translating fear into constructive action, individuals affirm their moral agency, foster hope, and contribute to lasting change. Instead of remaining passive witnesses to environmental degradation, they become active co-creators of a better future, each action reinforcing personal well-being and communal strength.

Recognizing Anxiety as a Catalyst for Action

For many, eco-anxiety can feel like an emotional burden, a nagging voice that points to the fragility of ecosystems and the uncertainty of tomorrow's world. Yet anxiety, while uncomfortable, also signals moral awakening. It emerges because we sense something profound at stake. If apathy were the norm, no one would lose sleep over melting glaciers or polluted coastlines. Worry arises precisely because we value life, justice, and continuity. This realization reframes anxiety from weakness into a sign of care. Instead of condemning oneself for feeling anxious, it's more constructive to ask: "What does this anxiety tell me about what I care about? How can I channel this care into action that reduces harm and builds resilience?"

This shift in perspective transforms anxiety from an inward stressor into an outward prompt. Like an alarm bell alerting us to danger, environmental worry calls us to respond. Acknowledge the discomfort, but do not linger in it. Focus on the underlying moral impetus—the knowledge that something precious needs protection. From there, identify a domain where your energy, skills, or resources can make a difference. Fear can thus be transmuted into resolve, confusion into learning, and uncertainty into experimentation.

Starting with Small Steps: Local Volunteering

One of the most accessible pathways from anxiety to action is local volunteering. Close to home, opportunities abound. Community gardens, wildlife rehabilitation centers, beach clean-ups, reforestation projects, and energy-efficiency initiatives all welcome helping hands. Volunteering doesn't require special expertise or long-term commitments. Even a few hours per month can yield tangible benefits. Pulling invasive weeds in a local park, for instance, directly supports biodiversity, making the environment healthier for native plants and pollinators. Planting trees or restoring wetlands contributes to carbon sequestration and ecosystem balance. Sorting recyclables at a community event or helping organize a zero-waste fair can educate neighbors and shift norms.

These hands-on activities not only address environmental issues but also soothe anxiety. Engaging physically with nature—feeling soil beneath your fingers, observing a restored habitat thriving—reinforces that positive change is possible. Instead of feeling helpless before global crises, you see immediate outcomes: seedlings growing stronger, cleaner riverbanks, neighbors learning something new. Witnessing progress, however modest, instills hope. Over time, regular volunteering can become a grounding ritual, a reminder that while no single effort solves everything, collective incremental actions add up.

Volunteering also connects you with like-minded individuals. Fellow volunteers share similar values and anxieties, forging camaraderie. Conversations flow naturally: "What brought you here?" "How does this work help you cope with worry?" Exchanging experiences normalizes your concerns and reveals that many people respond to environmental distress by doing something constructive. These peer relationships form emotional support networks, confirming that no one stands alone in caring about the planet's well-being.

Joining Environmental Organizations and Advocacy Groups
Beyond local volunteering, joining organized groups magnifies your impact. Environmental non-profits, advocacy networks, and community coalitions channel collective strength into targeted campaigns. They may focus on preserving a particular habitat, influencing municipal policies, advocating for renewable energy incentives, or campaigning for stronger environmental regulations. Aligning with an organization whose mission resonates with your values means adding your voice and labor to an ongoing effort with established strategies, resources, and expertise.

This structured involvement can alleviate anxiety by providing direction. When overwhelmed by complex global challenges, it's reassuring to work within a group that has honed its approach. Seasoned activists, policy experts, and community leaders can guide newcomers—suggesting effective lobbying techniques, navigating bureaucratic hurdles, or refining messaging to gain public support. Instead of reinventing the wheel, you contribute to collective momentum, learning from those who've walked similar paths before.

These groups also transform moral concern into measurable progress. Tracking milestones—like passing a city ordinance restricting single-use plastics or convincing a local business to adopt sustainable packaging—demonstrates that action leads to results. Celebrating each victory strengthens resolve. Facing setbacks, members find solace and encouragement in each other's stories of persistence. Advocacy groups also link local efforts to broader movements, helping volunteers see how their local clean-up or policy push fits into a global tapestry of environmental justice, climate action, and conservation initiatives. This sense of interconnectedness counteracts despair by highlighting that millions worldwide are pushing in the same direction.

Harnessing the Power of Civic Participation
Anxiety can sap a sense of control. Yet civic participation—voting, attending public hearings, contacting representatives, or petitioning lawmakers—reclaims agency. Democracy offers tools to influence policies and resource allocation. While systemic changes often move slowly, ignoring civic channels surrenders the battlefield of decision-making to those uninterested in long-term environmental stewardship. Instead, channel eco-anxiety into a determination to have your voice heard. Reach out to elected officials, share your concerns respectfully but firmly, and urge them to prioritize sustainability. Support candidates who foreground climate resilience or

environmental justice. When ballot measures or referendums address ecological issues, vote accordingly.

Each civic action, from signing a petition to providing testimony at a council meeting, transforms worry into concrete advocacy. Even if outcomes remain uncertain, knowing you participated in shaping decisions can reduce anxiety. You're not a passive spectator of unfolding crises but an active moral agent trying to steer society toward responsible choices. Over time, these civic engagements foster political literacy and confidence. As you learn how proposals become laws, how budgets reflect priorities, and how community input influences urban planning, your anxiety dissipates in the face of understanding. Knowledge replaces helplessness, and moral commitment finds institutional channels.

Collaborating Across Social Divides
Environmental issues transcend cultural, political, and economic divisions. While anxiety might tempt us to rally only with those who share our worldview, bridging differences can amplify impact. Anxiety about climate displacement, for example, may motivate reaching out to groups concerned about immigration, human rights, or poverty alleviation. Similarly, farmers worried about shifting rainfall patterns can partner with environmental scientists or technology startups innovating drought-resistant crops. By forging alliances with stakeholders who approach the problem from various angles, you broaden the constituency for change.

Such coalitions reduce isolation and demonstrate that shared threats breed new forms of solidarity. People who once viewed each other's concerns as unrelated discover common ground. Anxiety thus becomes a unifying force, bringing together communities once estranged by differing priorities. This collaboration yields more holistic solutions. Instead of fragmentary efforts, integrated approaches emerge—policy packages that address both environmental integrity and social equity. Seeing diverse coalitions emerge from environmental distress reassures that anxiety can catalyze inclusive moral action, weaving ecological care into the broader fabric of social justice and global human rights.

Communicating Constructively Within Groups
Working alongside others means navigating disagreements and conflicting strategies. Eco-anxiety can fuel impatience—why waste time debating incremental reforms versus radical transformations when the clock is ticking? Yet collective action thrives when members respect each other's perspectives. Apply empathy, active listening, and nonjudgmental dialogue within your chosen community. Validate differing experiences: some volunteers might focus on education, others on direct protest, still others on lobbying for policy change. Recognizing that multiple tactics can complement each other prevents internal tension from derailing efforts.

Anxiety managed constructively means seeing diversity of approaches as a strength. While one subgroup informs policymakers, another improves community composting systems, and another works on public awareness campaigns. Each thread contributes to a richer moral ecosystem. Instead of exhausting yourself trying to do everything alone, trust that others cover complementary fronts. This sense of teamwork eases anxiety: you know that while you focus on what you do best—organizing events, writing op-eds, leading workshops—others tackle aspects you find intimidating or less appealing. Sharing burdens and celebrating each role keeps the collective energized and resilient.

Transforming Public Spaces and Narratives
Channeled anxiety can also reshape cultural narratives and public spaces. Consider starting a local environmental newsletter, blog, or podcast that highlights small victories, interviews experts, and features community stories. Offering balanced coverage—acknowledging fear while emphasizing constructive steps—demonstrates moral maturity. By influencing public discourse, you help steer conversation from gloom to nuanced engagement. Artistic expressions—murals depicting thriving ecosystems, poetry slams on climate themes, or photo exhibitions documenting restoration successes—touch hearts, making the environmental cause feel less like abstract policy and more like a shared moral journey.

Such cultural initiatives inspire others to join. People hesitant to volunteer or advocate might find the courage after encountering your storytelling or creative work. Anxiety, previously paralyzing, transforms into a wellspring of inspiration. The public narrative shifts from fatalism to possibility. Schools incorporate environmental case studies into curricula; local media covers community-led sustainability fairs. Seeing these changes confirmed by cultural output reassures participants that their efforts matter. Anxiety thus becomes a catalyst for cultural enrichment, ensuring that environmental stewardship integrates into community identity over time.

Mentoring and Supporting Newcomers
As you gain experience, turn your once-isolating anxiety into mentorship. Newcomers to environmental action may arrive timid, overwhelmed by the scale of problems. Your empathy and guidance can smooth their integration. Share your journey: how initial fear inspired you to volunteer at a local wetland restoration project, how small successes built confidence, how joining an advocacy group illuminated policy levers you never knew existed. Offering tips on coping with setbacks—like difficult policy battles or limited funding—helps newcomers persist. By nurturing a supportive environment where novices feel welcomed, you multiply the community's capacity to address environmental issues. Each newcomer's anxiety is converted into fresh energy and ideas, replenishing the collective's moral resources.

This mentorship cycle passes on moral resilience. Just as someone once helped you find your footing, now you assist others in harnessing their anxieties. Over time, a generational transfer of knowledge and encouragement emerges. Younger activists learn from older ones how to pace themselves, maintain hope amid challenges, and find effective niches for their talents. This continuity stabilizes movements so they withstand leadership changes, policy setbacks, or shifting public attention. Anxiety ceases to be a recurring crisis and becomes an enduring impetus for moral evolution and adaptability.

Learning and Skill-Building
Transforming anxiety into action also involves building practical skills. If civic participation intrigues you, learn how local governance works—who makes decisions, when public consultations occur, how to draft persuasive testimony. If advocacy interests you, develop communication skills to craft clear messages that resonate with policymakers and the public. If volunteering appeals, hone technical skills—tree identification, composting techniques, event

organization—that amplify your usefulness in community projects. As capabilities grow, so does confidence. What once felt like anxious flailing becomes directed effort backed by competence. Skill-building shows that moral intentions gain traction when coupled with tangible abilities. Anxiety transforms into motivation to become more informed, agile, and effective in navigating environmental landscapes.

Celebrating Incremental Wins and Persistence
Given the enormity of environmental challenges, it's crucial to celebrate incremental wins. Anxiety often stems from feeling that no effort suffices. But acknowledging small victories—like a successful neighborhood tree-planting drive, a local store banning plastic bags, a policy committee agreeing to review sustainable transport options—reinforces that change is possible. Each milestone is a morale booster, a reminder that anxiety-inspired action makes a difference. Celebrations, whether a community potluck after a clean-up or a social media post thanking volunteers, validate everyone's contributions. Public recognition encourages newcomers to join and reassures veterans that their anxiety, channeled into meaningful engagement, yields results.

Progress may come unevenly; some initiatives fail or yield partial outcomes. Anxiety might flare again at setbacks. Yet resilience grows as the group learns from these experiences. Perhaps a policy proposal didn't pass this time, but advocates identified shortcomings and will return stronger. Such reframing teaches that setbacks aren't terminal defeats but learning opportunities. Anxiety becomes a prompt not only for action but for continuous improvement and strategic refinement. With each cycle of effort, reflection, and adaptation, participants internalize that moral engagement is an ongoing practice, not a one-time fix.

Engaging in Broader Networks and Movements
Beyond local organizations, consider connecting with national or international coalitions. Online platforms link environmental defenders worldwide. Professional networks unite scientists, lawyers, or educators working toward sustainable goals. Social justice alliances integrate environmental issues with racial equity, gender justice, or indigenous rights, emphasizing that environmental care intersects with all forms of fairness. By participating in broader movements, you widen your perspective and access richer resources—policy briefings, expert guidance, global solidarity. This expansion reduces isolation. Instead of feeling that your corner of the world struggles alone, you see a global community collectively facing eco-anxiety and translating it into moral courage. Strengthening ties with larger movements also ensures that local successes or learnings are shared widely, amplifying influence and moral momentum.

Integrating Action with Personal Well-Being
Channeling anxiety into action doesn't mean neglecting self-care. Emotional well-being remains essential for sustained involvement. Without tending to your own mental and physical health, anxiety might resurface as burnout or despair. Consider personal practices that restore balance—mindfulness, nature walks, creative hobbies, or downtime with supportive friends outside the environmental sphere. Healthy boundaries prevent over-commitment. While engagement is fulfilling, you cannot shoulder every ecological crisis alone. By pacing your involvement, you ensure longevity in the struggle, maintaining moral clarity and resilience over years rather than months.

Sometimes, stepping back to recharge can yield fresh insights. During a break, reading about different ecological success stories worldwide might spark new approaches for your local group. Personal well-being and collective action feed each other: feeling stable allows more effective contribution, while meaningful engagement offers purpose and community support that sustains emotional health. Balancing these elements acknowledges that eco-anxiety and moral responsibility co-exist in a complex ecosystem of motives, actions, and rest. Wisdom lies in navigating this ecosystem gently and persistently.

Cultivating Moral Courage and Leadership
As you grow more comfortable translating anxiety into action, moral courage emerges. Courage means acting ethically despite uncertainty, acknowledging risks and remaining steady in your commitment. Civic engagement may confront corporate lobbyists, public apathy, or political gridlock. Advocacy might mean facing critics, skeptical relatives, or unforeseen obstacles. Volunteering can entail weather challenges, funding shortfalls, or slow progress. Moral courage doesn't guarantee easy triumphs; it guarantees perseverance grounded in conscience.

This courage can inspire others. Observing someone who once felt paralyzed by fear now leading a workshop, organizing a rally, or calmly articulating policy demands shows that transformation is possible. Your journey from anxious observer to active participant demonstrates that fear need not be a prison. Instead, it can be a signpost directing you toward constructive deeds. In turn, you become a role model for others seeking to channel their anxiety similarly. They see that moral courage isn't innate; it's cultivated through repeated choices to engage, learn, adapt, and collaborate.

As your influence grows within communities and movements, leadership may naturally follow. Leadership here doesn't mean hierarchical authority but moral guidance—encouraging respectful discourse, mentoring newcomers, and suggesting innovative collaborations. By exemplifying how anxiety transforms into meaningful action, you help maintain group coherence. When internal disputes arise or external opposition intensifies, your calm presence and experience reassure others that solutions are within reach if everyone persists ethically. Such moral leadership can shape a culture of trust, empathy, and steady progress, inoculating the collective against despair or fragmentation.

Linking Local Actions to Global Impact
While local efforts and civic engagements are essential, it's crucial to remember their global dimension. Anxiety often stems from understanding that challenges transcend borders. Embrace the knowledge that your local project contributes to a global mosaic of environmental stewardship. The trees you plant help stabilize a small habitat, but similar efforts worldwide form a green wave that cools the planet. Your advocacy for fair climate policies in one city reflects moral demands echoed in thousands of other communities, each adding moral weight to international negotiations.

Recognizing this interconnectedness reduces the sense of insignificance. Anxiety diminishes when you know that you're part of a vast network of moral agents pushing for sustainable transitions. The solidarity of countless unknown allies—activists in distant countries, policy wonks in think tanks, indigenous guardians of ancestral lands—gives meaning to your actions.

Anxiety becomes less about personal helplessness and more about a moral calling answered by people everywhere. This global moral fellowship lifts you beyond fear, imbuing your actions with a sense of universal purpose.

Sustaining Momentum Through Storytelling and Reflection
Ongoing reflection ensures that anxiety doesn't resurface as doubt or disillusionment. Periodically review why you care, what progress you've seen, and how your understanding of solutions evolved. Journaling, group reflections, or storytelling sessions within your network help integrate experiences. Sharing narratives of how you overcame initial fears or learned new advocacy skills reaffirms that action transforms not just the external world but your inner landscape. Instead of seeing environmental struggles as endless, you recognize them as moral journeys marked by growth.

Storytelling about successes, failures, lessons, and emotional breakthroughs allows others to learn from your path. These narratives become moral resources for future generations of advocates. Over time, a collective memory forms—an archive of how anxiety led to constructive responses, how persistence overcame obstacles, and how moral courage consistently renewed engagement. This historical continuity reassures that anxiety need not be cyclical despair; it can fuel cyclical renewal, each phase deepening moral insight and reinforcing communal bonds.

Embracing the Complexity of Moral Agency
Channeling fear into engagement also means embracing complexity. Not every volunteer project yields immediate results, not every advocacy campaign passes ideal policies, and not every civic endeavor convinces skeptics. Recognize these outcomes not as moral failures but as inevitable facets of navigating large-scale moral issues. Complexity means there are no quick fixes, yet moral agency remains essential. Anxiety signals that urgent questions loom; action confirms that despite uncertainty, you choose responsibility over resignation.

This acceptance prevents idealistic disappointments. Instead of expecting linear progress, you anticipate nonlinear developments—small steps forward, occasional setbacks, periodic shifts in strategies. Understanding that complexity is normal fosters patience and maturity. Moral steadiness grows. Anxiety, once crippling, becomes a familiar companion reminding you to stay alert, adaptive, and compassionate. Knowing that many others also embrace this complexity encourages collective perseverance.

A Lifelong Commitment to Evolving Engagement
Finally, channeling fear into meaningful engagement is not a one-time event but a lifelong commitment. As environmental conditions and knowledge evolve, so will your methods. Early volunteering might lead to more focused advocacy later. Initial civic actions might inspire future policy leadership. Over decades, anxieties morph as new challenges emerge—climate adaptation, climate migration, regenerative agriculture, circular economies—and so do the responses. Lifelong engagement draws moral energy not just from anxiety but from love of life, intellectual curiosity, moral principles, and faith in human creativity.

This long arc of engagement ensures that anxiety never reverts to paralysis. Instead, each new fear about emerging crises spurs another phase of moral exploration. While uncertainty never disappears entirely, your capacity to respond grows stronger. In this ongoing dance between fear and action, moral resilience flourishes. You develop trust in your moral instincts, your communities, and the collective wisdom that arises from countless individuals worldwide doing likewise—converting worry into will, dread into direction, and unease into unwavering ethical stewardship.

Working Together on Local Projects (Community Gardens, Beach Cleanups, Habitat Restoration)

As concern for the planet deepens and people seek meaningful ways to channel eco-anxiety into constructive engagement, local projects become an accessible avenue for collective action. Unlike distant policy battles or abstract debates, community gardens, beach cleanups, and habitat restoration efforts offer tangible opportunities to unite neighbors, friends, and new acquaintances around shared values and common goals. In these hands-on initiatives, participants transform worry into purposeful work. Each seed planted, each piece of litter removed, each invasive species uprooted affirms that care for the environment can yield immediate, visible improvements. By collaborating on local projects, communities not only nurture ecosystems but also cultivate moral courage, solidarity, and hope—ingredients vital for sustaining long-term environmental stewardship.

Bridging Ideals and Practical Action
Local projects bridge the gap between lofty environmental ideals and practical action. When anxiety hovers like an uninvited guest, threatening to paralyze moral ambition, stepping outdoors to engage in communal tasks provides relief. Instead of feeling helpless before global trends, participants concentrate on what they can achieve right here, right now. Community gardens turn vacant lots into vibrant, productive spaces providing fresh produce and biodiversity havens. Beach cleanups reclaim shores from plastic debris, protecting marine life and enhancing coastal enjoyment. Habitat restoration efforts heal damaged ecosystems, supporting native species and revitalizing ecological balance.

In these moments, eco-anxiety morphs from an internal stressor into a motivator spurring tangible results. It's one thing to mourn habitat loss abstractly; it's another to actively restore a plot of land, watching pollinators return and native flora flourish again. Each shovel of compost, each carefully planted seedling, each collected plastic bottle resonates as proof that while environmental crises are complex, human hands can contribute to solutions. This realization nourishes optimism: if a small group of dedicated individuals can improve a patch of earth, imagine the cumulative effect when thousands of communities do similarly worldwide.

Building Moral and Emotional Resilience
Engaging in local projects fosters emotional resilience. Physical labor, fresh air, and sensory contact with nature soothe frazzled nerves. Instead of scrolling through grim headlines, volunteers immerse themselves in the present moment—feeling soil between their fingers, admiring the symmetry of a leaf, listening to waves lapping the shoreline. These sensory experiences anchor participants in reality, reminding them that despite global threats, the world still contains beauty and healing potential.

Emotional resilience also grows through teamwork. Anxiety often isolates people, making them think they must bear worries alone. But local projects are inherently social. Chatting while removing invasive plants or sorting recyclables for a community recycling drive breaks down barriers. People share their environmental concerns casually: "I've been worried about these droughts; it's good to know we're doing something." Such small confessions find immediate validation. In response, another might say, "Me too. This cleanup helps me feel less helpless." Mutual understanding emerges organically, reducing stigma around eco-anxiety and reinforcing that fear and hope coexist. Working side by side, individuals reinforce each other's moral commitment, dispelling loneliness and forging solidarity that outlasts the day's tasks.

Diverse Participation and Cultural Exchange
Local environmental projects often attract diverse participants—parents with children, retirees, college students, professionals on weekends, newcomers to the area. This diversity enriches the experience. While environmental discourse can sometimes feel polarized or dominated by certain demographics, a community garden or beach cleanup welcomes everyone willing to help. Differences in age, background, or political leaning often recede in the face of shared purpose. Conversations become more about practical solutions—how to improve soil health or prevent stormwater runoff—than ideological clashes.

Such interactions build empathy and broaden moral perspectives. A longtime resident might teach younger volunteers traditional gardening techniques, passing on cultural knowledge. A newly arrived immigrant might share insights from their homeland's sustainable farming traditions. These exchanges transform local projects into spaces of intercultural learning. Participants realize that environmental stewardship transcends identity boundaries. Anxiety about global crises turns into curiosity about how different cultures approach sustainability, spurring cross-pollination of ideas. Moral imagination expands as people see that many roads lead to common goals—healthier ecosystems, stable communities, and a secure future.

Reconnecting with Place and Heritage
Environmental challenges can strip landscapes of their familiar patterns, unsettling a sense of home. Community gardens, cleanups, and restoration efforts help re-establish rootedness. Planting native species or restoring wetlands ties participants to the land's natural rhythms. They learn which plants thrive in local conditions, how pollinators rely on certain flowers, or how seasonal changes affect biodiversity. This knowledge fosters an intimate relationship with place, countering the disorientation caused by climate disruption. Instead of feeling that the environment is slipping through their fingers, participants witness regeneration and continuity.

In some regions, habitat restoration revives cultural heritages linked to particular ecosystems—reintroducing indigenous crops, reviving traditional land management practices, or protecting species tied to cultural stories. Engaging in these projects connects moral values with ancestral legacies. Anxiety transforms into reverence, and despair yields to determination to preserve what ancestors cherished. This intergenerational perspective fortifies resilience, as people understand their efforts continue a chain of care passing through time. Caring for local ecosystems, then, becomes not just a moral act but a cultural affirmation, bridging past wisdom and present moral responsibility.

Small Wins and Incremental Progress
In the face of vast environmental crises, small wins matter. Removing litter from a shoreline won't solve plastic pollution globally, but it removes a threat from local wildlife and improves the beach's beauty. Reviving a community garden won't halt global climate change, but it fosters local food security, reduces transportation emissions linked to grocery shipping, and educates residents about sustainable agriculture. These incremental gains restore faith in the power of collective action. Each visible improvement serves as an antidote to despair: "We did this together, and it made a difference."

Over time, a series of small wins accumulates into significant transformations. A network of community gardens might improve urban food deserts, providing affordable, nutritious produce and reducing reliance on resource-intensive supply chains. Beach cleanups could shift public attitudes about littering and consumption. Habitat restoration efforts might stabilize local species populations, inspire school curricula, and even influence municipal land-use policies. Seeing how modest, local actions ripple outward teaches that moral agency, far from useless, can shape broader narratives and outcomes.

Developing Civic Skills and Leadership
Local environmental projects often require coordination, planning, and problem-solving. Participants learn to organize volunteers, manage resources, delegate tasks, and handle unexpected challenges—like a shortage of gardening tools or sudden bad weather on cleanup day. These practical skills build civic competence. People realize they can navigate complexities, mediate conflicts, and sustain momentum. Such capabilities are transferable beyond the environmental realm. Once shy volunteers evolve into confident organizers who can apply these skills to other moral causes—educational initiatives, health campaigns, or social justice movements.

As individuals gain experience, some step into leadership roles—coordinating outreach, liaising with local officials, fundraising for supplies, or training new participants in habitat restoration techniques. Leadership grounded in moral purpose and collaborative effort strengthens communities. Leaders learn to communicate vision clearly, balance diverse interests, and maintain inclusive decision-making. Anxiety about helplessness dissipates as they see that moral courage combined with organizational prowess yields tangible achievements. In turn, community members respect and trust leaders who emerged naturally from within their ranks, demonstrating that anyone can become a moral steward if guided by sincerity, empathy, and accountability.

Attracting Partnerships and Resource Support
Working together on local projects doesn't just mobilize volunteers; it often attracts support from businesses, nonprofits, and government agencies. When stakeholders see consistent community engagement, they become more likely to invest resources—donating materials, providing grants, or offering expertise. Partnerships might form with environmental NGOs offering training in sustainable horticulture or with local eco-friendly businesses sponsoring tools or refreshments for volunteers. Government entities noticing the community's commitment may allocate funds or technical guidance. Media coverage may spotlight the project's successes, raising public awareness and potentially inspiring other neighborhoods to replicate the model.

These synergies amplify impact, ensuring that fear-inspired efforts extend beyond a single group of volunteers. Anxiety-driven action inspires trust and respect from institutional partners, proving that grassroots commitment is serious, resilient, and deserving of material backing. Over time, these alliances can steer policy changes. A cluster of successful community gardens, for instance, might influence city planners to integrate green spaces into development projects, acknowledging that local action prefigures viable long-term strategies for sustainability and climate adaptation.

Transforming Local Projects into Educational Hubs
Community gardens and restoration sites can double as outdoor classrooms. Schools organize field trips, scouts earn environmental badges, and neighbors attend workshops on composting or water conservation. Anxiety about future challenges finds a positive outlet in knowledge sharing. Participants become mentors and educators, explaining the importance of biodiversity, soil health, or coastal ecosystems. Children learn early that caring for the planet is not just an abstract principle but a lived practice, enjoyable and fulfilling. This educational dimension ensures that moral awareness endures across generations. Instead of passing anxiety forward, communities pass forward values, skills, and solutions-oriented mindsets.

These educational endeavors also reinforce community identity. The garden becomes a familiar landmark where lessons unfold, seasonal changes observed, and celebrations held. In restoration sites, schoolchildren return yearly to see how their planted saplings matured into small groves. Such continuity instills a narrative of renewal and moral responsibility. The fear of environmental instability recedes as new generations understand that though problems persist, collective action steadily nurtures resilience and adaptation.

Inspiring Others and Replicating Models
Local projects often become success stories that inspire other communities. Volunteers traveling or connecting online share their project experiences, offering templates for replication. A neighborhood's successful beach cleanup model—dividing tasks, involving local fisherfolk, partnering with a marine conservation group—might inspire a coastal town elsewhere to attempt a similar approach. Habitat restoration techniques developed through trial and error in one region—like using certain native plant species that thrive despite changing rainfall—can guide projects in areas facing parallel challenges.

As stories spread, anxiety transforms from a private burden into a shared moral motivator. Hearing that others across the globe overcame their fears to rejuvenate local ecologies encourages more people to follow suit. Replication builds momentum: what started as a small band of anxious volunteers restoring a single wetland can catalyze a chain reaction of restoration projects worldwide. Moral energy compounds when each success, however modest, gives others confidence that they too can make a difference.

Dealing with Setbacks and Learning from Failure
Not all efforts run smoothly. A community garden might suffer crop failures due to unexpected pests or droughts. A beach cleanup might encounter hazardous waste requiring professional intervention. A habitat restoration might face bureaucratic delays in accessing protected areas. Setbacks can trigger renewed anxiety. However, approaching these difficulties as learning

opportunities fosters growth. Reflecting together, volunteers identify what went wrong and adjust strategies. Maybe investing in better soil preparation, planning for irrigation, or consulting experts on waste disposal can improve outcomes next time.

Such iterative learning weaves moral resilience into the community's fabric. Anxiety no longer signals impending doom but prompts adaptability and perseverance. Rather than eroding motivation, setbacks become rungs on a ladder to greater competence and strategic clarity. Over time, these refinements deepen community capacity, ensuring long-term viability. A community that overcomes repeated challenges strengthens its internal trust and moral conviction, making members more willing to tackle even more ambitious projects.

Evolving Moral Horizons and Issue Awareness

Participating in local projects often broadens participants' understanding of interconnected issues. Garden volunteers might discover how soil health depends on pollinators and how local seed varieties resist pests better than imported ones. Beach cleanup crews learn about microplastics, ocean currents, and marine ecosystems, prompting interest in upstream waste management and corporate responsibilities. Habitat restoration teams become attuned to local species' needs and migration patterns, understanding that their efforts connect to global biodiversity treaties and conservation policies.

This expanded awareness encourages a holistic moral perspective. Anxiety shifts from a vague fear to an informed concern encompassing ecological complexity and social ramifications. Participants might become advocates for systemic changes they initially overlooked—pressing for agroforestry in their region, joining campaigns against single-use plastics at national levels, or supporting international wildlife corridors. In this way, local projects serve as gateways to deeper moral engagement, guiding participants from hands-on tasks to policy insights and broader alliances.

Ceremonies, Rituals, and Community Traditions

To sustain engagement and reinforce moral bonds, communities can integrate ceremonies or rituals into their projects. Perhaps each season's first harvest in a community garden is celebrated with a small festival where everyone shares dishes made from local produce, toasting not just the food but the shared commitment it represents. Beach cleanup teams might mark their annual event with a ritual of gratitude to the ocean, acknowledging that their efforts honor the marine world's value. Habitat restoration groups could celebrate the return of a particular bird species each spring, holding a short ceremony expressing joy, relief, and renewed dedication.

Such rituals deepen emotional connections. They remind participants that their labor is not transactional but spiritually or morally significant. Anxiety about the planet's future transforms into a sense of reverence, community pride, and continuity with natural cycles. These traditions persist through generations, ensuring that moral fervor outlasts any single cohort of volunteers. Over time, new members inherit these rituals, internalizing their moral meanings and contributing fresh interpretations that keep traditions alive and evolving.

Leveraging Technology and Media

Technology amplifies the impact of local projects. Social media platforms allow volunteers to share before-and-after photos of restored habitats, short videos demonstrating gardening techniques, or stories of overcoming challenges. Such content can attract new volunteers, inspire distant observers, and offer tangible proof that anxiety-driven action matters. Livestreaming a beach cleanup or posting updates on water quality improvements gives transparency and encourages virtual participation—people following from afar might donate funds or replicate similar cleanups in their hometowns.

Documenting successes digitally also holds authorities accountable. When local officials see that well-organized citizen initiatives produce positive outcomes, they may become more receptive to environmental proposals, land-use reforms, or integrating community-led projects into formal conservation plans. This feedback loop strengthens moral agency: communities learn that by publicly showcasing their achievements, they influence public opinion, media narratives, and political will.

Continual Adaptation and Innovation
Local projects thrive through adaptation. As environmental conditions evolve—shifting pollinator species due to climate changes, new pollution sources emerging—communities must remain flexible. Anxiety occasionally resurfaces as unforeseen problems arise. Yet past experience handling uncertainty builds confidence. Volunteers might experiment with more drought-tolerant crops, new composting methods, or innovative shoreline stabilization techniques. Learning from global best practices or consulting experts ensures the community never stagnates.

Such innovation transforms anxiety into a dynamic force prompting exploration. Instead of fearing change, communities anticipate it and prepare. They embrace trial and error, understanding that moral progress lies not in achieving static perfection but in ongoing responsiveness. Each adaptation reaffirms that moral engagement isn't a final destination but a continuous journey, fueled by creativity and collective intelligence. Over time, this culture of adaptation makes them morally resilient—able to face not only today's challenges but whatever tomorrow brings.

Integrating Local Projects with Larger Climate Strategies
While local projects offer immediate emotional relief and moral fulfillment, their significance grows when integrated into larger climate strategies. Communities can feed data from habitat restoration into scientific databases, contributing to research on species adaptability. They can advocate for local policy changes that reflect their project experiences—like pushing for greener zoning laws after restoring a wetland and seeing its flood mitigation benefits firsthand. They might join regional coalitions exchanging information and resources, aligning local priorities with provincial or national environmental goals. Through these alignments, anxiety-driven action scales up, influencing more substantial decision-making arenas.

When local successes inform higher-level policies, volunteers witness a gratifying loop: what began as fear-inspired collective effort now influences laws, resource allocation, and societal norms. This validation proves that personal anxiety, when channeled collectively, can shape institutions. Moral agency resonates through layers of governance, transforming despair into structural resilience. The lesson is powerful: never underestimate the moral force of grassroots engagement.

From Anxiety to Legacy
As years pass, local projects accumulate a legacy. They leave physical imprints—lush gardens, cleaner beaches, thriving habitats—and social imprints—cultivated relationships, shared traditions, accumulated know-how. They also leave moral imprints, demonstrating to future generations that when faced with environmental anxiety, communities can unite, act, learn, and create tangible good. Young people growing up in neighborhoods with community gardens or participating in annual beach cleanups internalize the message that caring for Earth is normal, enjoyable, and communal. Anxiety about climate change may still exist, but these children inherit examples of how worry turns into work, despair into determination, and fear into fellowship.

Reflecting on this transformation, participants realize that local projects did more than alleviate personal eco-anxiety. They anchored moral commitments in everyday life, ensuring that environmental stewardship permeates routines and relationships. By collaborating on local initiatives—whether planting seeds, removing debris, or restoring habitats—people learn that moral courage is not limited to grand gestures or distant negotiations. It manifests quietly, consistently, in the spaces they inhabit and the communities they cherish. In this ongoing process, eco-anxiety loses its grip as a paralyzing force and reemerges as a steady moral compass guiding hands and hearts toward a regenerative future.

Understanding That Collective Efforts Amplify Impact and Hope

As communities gather to transform eco-anxiety into tangible action, it becomes increasingly clear that working together does more than just ease individual stress—it amplifies both the scale of impact and the sense of hope. While solitary efforts can feel like drops in an immeasurable ocean, collective endeavors pool those drops into a meaningful tide. By uniting skills, resources, and moral convictions, groups magnify their power to effect change. More than that, witnessing the synergy of communal action reinforces the belief that solutions, though complex, are achievable. In unity, fears about the planet's future give way to confidence, and the narrative shifts from helpless individuals facing overwhelming odds to a chorus of voices and hands actively shaping outcomes.

From Individual Vulnerability to Collective Agency
Alone, anxiety often leads to feeling small and inadequate before global challenges. One might worry, "What can I, a single person, accomplish?" However, as soon as people collaborate, the dynamic shifts. A community garden that would be exhausting for one or two volunteers thrives under a dozen or more participants, each contributing an hour here or a skill there. Beach cleanups become faster and more thorough with coordinated teams. Habitat restoration succeeds when everyone brings different strengths—those knowledgeable about native species guide selection, skilled organizers ensure efficient logistics, storytellers raise community awareness. Each person's contribution, though modest, aggregates into a formidable collective output.

This collective agency reassures participants that the fate of the planet doesn't rest on their shoulders alone. Instead, everyone shoulders a piece, making the burden manageable and the journey less daunting. Anxiety stemming from perceived insignificance gives way to empowerment. Realizing that others share your concern, and are willing to act, confirms that your efforts matter. If once it felt futile to pick up a single plastic bottle from the beach, doing so as part of a team that clears entire shorelines illustrates exponential gains. Moral courage spreads; seeing others invest energy and time erodes excuses for inaction and generates mutual reinforcement. In this way, collective work transforms fear-induced paralysis into communal determination.

Exponential Growth Through Shared Networks
Collective efforts don't just add impact—they multiply it. A single community garden can inspire others, prompting neighborhoods across a city to replicate the model. Each new garden connects to a network of experienced gardeners, seed exchanges, educational workshops, and collaborative problem-solving. Similarly, beach cleanups evolve from sporadic events into recurring campaigns, each one improving techniques and outreach. Habitat restoration projects form alliances across towns, sharing best practices and resources. In this networked context, knowledge and innovation circulate rapidly, making once-difficult tasks easier over time.

The supportive network acts like a moral ecosystem where successes, stories, and solutions flow freely. The resulting synergy makes challenges less intimidating. If one community struggles with invasive plants, they can seek advice from another that overcame a similar issue. Lessons learned in one habitat restoration—about, say, planting certain shrubs to stabilize soil—pass to another region's efforts. The knowledge pool expands, decreasing dependency on trial-and-error approaches and allowing for more strategic, effective interventions. This cumulative wisdom fosters optimism: if communities around the world are honing their methods and sharing breakthroughs, progress feels continual and unstoppable.

Cultural Momentum and Shifts in Norms
As collective efforts become more visible, they influence cultural norms. When a neighborhood sees multiple families joining a local environmental project, sustainable behaviors gain social acceptance. Peer pressure shifts from "Why bother?" to "Why not join?" Over time, what was once considered extraordinary—volunteering weekends for habitat restoration or using reusable containers religiously—turns ordinary. This normalization reduces friction. People feel less like isolated moral pioneers and more like participants in a rising movement. Anxiety diminishes as environmental care becomes a widely embraced value, not a quirk of a concerned few.

This cultural momentum also affects language and narratives. Instead of describing environmental crises only in terms of threats and doom, conversations highlight progress and communal achievements. Media coverage may shift from grim warnings to balanced stories about communities reclaiming degraded lands, adopting green infrastructure, or protecting coastal habitats. Seeing success and agency represented in public discourse helps counter overwhelming despair. The cultural script evolves: where eco-anxiety once signaled helplessness, now it indicates a stepping-stone to collective empowerment. The feeling of "we're all in this together" transforms an isolating fear into a bonding moral principle.

Influencing Policy and Decision-Making
Collective action resonates beyond the local level and can influence policy landscapes. Elected officials, aware of community activities and their growing popularity, become more receptive to environmental proposals. Politicians who once hesitated to support ambitious climate policies might find encouragement in widespread civic support. Grassroots coalitions formed from multiple local groups gather strength and push for better regulations, funding for green initiatives, or more robust sustainability frameworks. Anxiety turns into advocacy: people anxious about the slow pace of change can now rally behind evidence of community-driven solutions, pressing decision-makers to scale up such endeavors.

Moreover, policymakers note that community efforts are not isolated outbursts but sustained commitments. This consistency signals that environmental care is not a passing trend but a moral baseline many citizens uphold. Officials face increasing pressure to align policies with these values, knowing that ignoring a well-organized, morally galvanized citizenry risks electoral backlash. As communities keep demonstrating tangible improvements—healthier ecosystems, stronger local economies tied to sustainable practices—policymakers see environmental action as politically safe, even advantageous. Collective efforts thus catalyze top-down changes, bridging the gap between grassroots initiative and institutional response.

Expanding Moral Imagination and Fostering Global Solidarity
Realizing that collective efforts amplify impact nurtures moral imagination. If a handful of volunteers can restore a small wetland, what about a coordinated network across entire watersheds or countries? Witnessing how pooling resources scales up impact encourages people to dream bigger. Maybe a city can become zero-waste with enough communal support, or a region can shift its agricultural patterns through community-driven innovations. Instead of confining themselves to incremental steps, communities aspire to larger transformations. Moral imagination flourishes when people understand that what once seemed impossible becomes plausible through shared commitment.

This expanded vision fosters global solidarity. Hearing about successful local initiatives abroad—restoration projects, citizen science campaigns, green education programs—inspires replication. International coalitions form, connecting coastal cleanups in one hemisphere to mangrove reforestation in another. Anxiety about global threats shifts from paralyzing fear to a determination to learn from and support allies worldwide. Individuals see their local efforts not as isolated gestures but as part of a global moral mosaic. This interconnectedness dissolves the notion that environmental stewardship is a solitary struggle; instead, it's a vast collective endeavor unfolding across diverse landscapes and cultures.

Reinforcing Trust and Reciprocity
Collective efforts amplify hope in part by reinforcing trust. Working side by side—sharing tools, dividing labor, cheering each other's progress—cultivates interpersonal bonds. Volunteers see that others are dependable, willing to endure challenges, and committed to outcomes. This trust spills into other community aspects: people become more likely to support neighbors in crises, cooperate on unrelated civic projects, and discuss moral dilemmas openly. Environmental work acts as a trust-building exercise, reweaving social fabrics frayed by polarization or isolation.

Reciprocity emerges naturally. When one group faces a shortage of supplies, others donate extras. If a community garden struggles with pest management, another garden's experienced volunteers offer guidance. This reciprocal ethos also extends to intangible resources like emotional support. If some members feel discouraged after a setback, others lift their spirits, reminding them of past triumphs and collective strengths. The knowledge that one is part of a morally committed network reduces stress and enhances psychological well-being. It assures individuals that their fears, far from burdensome, find a home in a community ready to share and address them together.

Empowering Marginalized Voices and Improving Equity
Collective efforts can also empower marginalized voices. Environmental problems often affect vulnerable populations first and hardest—low-income neighborhoods, indigenous communities, or regions historically sidelined from decision-making. Inviting and centering these voices in local projects ensures that solutions reflect broader moral values like justice and equity. Anxiety about injustice transforms into meaningful solidarity. Those previously excluded gain platforms to shape strategies, ensuring that collective actions benefit everyone.

Such inclusivity enriches moral discourse. Hearing firsthand accounts from communities living with contaminated water or shrinking coastlines personalizes distant abstractions. Empathy, already a cornerstone of collective engagement, grows deeper when people realize that moral responsibility includes supporting those unfairly burdened by environmental harm. This understanding fosters alliances that transcend typical divides, weaving moral fibers that strengthen not only environmental outcomes but also social cohesion and cultural sensitivity.

Inspiring Future Generations and Sustaining Long-Term Commitment
As these communal initiatives flourish, younger generations witness living examples of cooperation and moral efficacy. Children growing up in neighborhoods where collective environmental projects are routine incorporate these norms into their worldviews. They learn early that eco-anxiety need not freeze them in fear; it can propel them into teamwork, innovation, and meaningful action. Instead of inheriting cynicism, they inherit hope grounded in observable results. Over time, these generational shifts ensure that moral engagement doesn't fade but matures.

Sustaining long-term commitment also becomes easier. Projects that persist season after season create traditions—annual planting days, monthly shoreline surveys, yearly celebrations of restoration milestones. These rituals anchor moral energy across time. Even as conditions evolve, traditions remind participants of their shared journey. With each passing year, collective efforts accumulate moral capital—a reservoir of trust, experience, and pride fueling continued resilience. Anxiety once threatening burnout now serves as a reminder that vigilance and adaptation never cease, but neither does the community's capacity to respond.

Balancing Local, Regional, and Global Perspectives
Understanding that collective efforts amplify impact and hope encourages a balance of focus. Communities realize that while local projects provide immediate emotional and moral returns, they can also integrate regional and global thinking. Volunteers might coordinate with regional watershed councils, attend international workshops on sustainable agriculture, or support global

climate agreements through local advocacy letters. This multi-layered engagement satisfies the moral imperative to scale up solutions without neglecting the comfort and accessibility of local action. Anxiety, initially anchored in what seems too large to handle, disperses as participants recognize they can contribute meaningfully at multiple levels.

This interplay ensures that people don't become trapped in a parochial mindset or overwhelmed by global complexity. Instead, they navigate moral dimensions with agility—addressing immediate local problems, feeding insights upward, and absorbing global lessons downward. Over time, eco-anxiety transmutes into confident, strategic moral engagement that acknowledges both the intimacy of one's backyard and the magnitude of planetary challenges.

Celebrating Diversity of Approaches and Avoiding Dogmatism
Collective efforts show that not all solutions look the same. One community might excel at reforestation, another at composting programs, another at wildlife corridors. Anxiety about not doing enough fades as people witness the diversity of contributions. Instead of feeling pressured to adopt a single formula—such as only personal lifestyle changes or only political lobbying—collective engagement normalizes pluralism. Different groups bring distinct talents: some focus on science-driven monitoring, others on public outreach, others on cultural education.

This pluralism ensures moral flexibility. When confronted by critics who argue one approach is insufficient, communities can point to the ecosystem of actions unfolding simultaneously. Anxiety that emerges from questions like "Am I doing the right thing?" finds comfort in the understanding that many "right things" coexist. The collective mosaic of approaches reduces ideological rigidity, fostering a moral landscape where experimentation and adaptation thrive. Hope thrives as well, since failure in one avenue can be offset by progress in another, collectively weaving a safety net beneath moral aspirations.

Visibility and Accountability Through Storytelling
Another layer supporting the idea that collective efforts amplify impact and hope is the power of storytelling. Communities proud of their projects share stories—videos, blogs, photo essays—showcasing transformations achieved through cooperation. These narratives inspire others and hold the community accountable, reminding participants of their moral commitments. Transparency enhances credibility: outsiders see that claims of improvement aren't empty words but documented realities. This feedback loop strengthens morale—when people know their work inspires others and is recognized as valuable, they feel renewed determination to maintain momentum.

Moreover, storytelling encourages critical reflection. After each season's activities, recounting successes and setbacks fosters learning. What worked well can be repeated, what faltered can be adjusted. Anxiety about uncertainties morphs into confidence that continuous learning and iteration will refine strategies. Over time, these communal narratives build a legacy of moral wisdom, passed down as a cultural resource enabling future generations to start from a stronger base rather than reinventing the wheel.

Adapting to Changing Conditions and Scaling Impact
Environmental conditions are not static. As local projects mature, communities might face new

challenges—emerging pests, shifting weather patterns, demographic changes. The strength of collective effort lies in its adaptability. Having established trust, leadership experience, and diverse skill sets, groups can pivot, experiment, and adopt fresh solutions more readily than individuals acting in isolation. Anxiety arising from unexpected hurdles transforms into a stimulus for creativity. The sense that multiple minds are brainstorming and multiple sets of hands are implementing ensures that no single setback spells failure.

Scaling impact also grows simpler as confidence builds. Once communities prove their capacity through local successes, they feel bolder proposing larger projects—expanding garden networks, restoring multiple wetlands in a region, collaborating with neighboring towns on joint recycling systems. Each expanded effort invites new allies—engineers, policymakers, donors—who see the communities' track records. Anxiety about tackling bigger aims recedes as groups recognize their cumulative competence. They understand that, as moral beings acting in concert, they can handle complexity and move beyond what once seemed impossible.

Building Bridges Between Generations and Disciplines
As collective efforts expand, they attract people from various generations and professional backgrounds. Older members share experience and perspective, while younger participants bring energy, social media savvy, and fresh insights. Scientists join hands with artists, policymakers with local farmers, educators with entrepreneurs. This interdisciplinarity mirrors ecological diversity—just as healthy ecosystems rely on multiple species and niches, healthy moral ecosystems thrive on varied contributions.

Each time different generations, cultures, or professions collaborate, anxiety rooted in siloed thinking dissolves. People realize that no single group holds all answers. Intergenerational wisdom blends with modern innovations, forging strategies resilient against rapid changes. Anxiety, once fed by lack of direction, now diminishes as participants see they can build moral roadmaps collectively, drawing on multiple intellectual and cultural resources. Hope flourishes as these alliances generate robust solutions that endure the test of time and shifting conditions.

Instilling Long-Term Commitment Through Moral Rituals
Over years, communities develop their own moral rituals—annual celebration of restoration milestones, seasonal replanting ceremonies, or commemorations honoring volunteers' contributions. These rituals underscore continuity and shared purpose. Anxiety about future instability doesn't vanish, but it becomes woven into traditions that highlight adaptability and resilience. Each ritual says, "We've made it this far together, and we'll keep going." Such symbolic acts strengthen moral identity. They signal to newcomers that they join a community not only of action but also of meaning. Hope arises from the knowledge that moral courage is not episodic but institutionalized in community culture.

Reflecting on the Moral Transformations
Over time, people involved in collective projects notice subtle changes in themselves. Once, environmental headlines induced panic or avoidance. Now, they trigger reflection and determination. Instead of seeking escape from distressing realities, participants ask, "How can our community help here?" Anxiety once trapped inside private minds circulates in public spheres, where it's processed and channeled into endeavors. This moral maturity enriches

personal character and community ethos. Individuals feel less isolated, more capable of confronting moral dilemmas with integrity and resourcefulness.

In retrospect, the journey from isolated fears to collective endeavors clarifies that moral progress hinges on unity. The cumulative actions of many hands, guided by common values, reshape environments and human relationships alike. Anxiety, instead of undermining confidence, pushes communities to think bigger, try harder, and cooperate more openly. The synergy between moral conviction and collective effort offers a blueprint for responding to other crises—future pandemics, economic struggles, social injustices—reinforcing the principle that common challenges demand communal responses.

Expanding the Radius of Hope

Understanding that collective efforts amplify impact and hope also influences how communities relate to broader environmental goals—curbing emissions, safeguarding biodiversity, restoring global ecosystems. While no single local project can solve planetary problems, witnessing cumulative progress reassures participants that small threads weave a grand tapestry. Anxiety's voice that "it's too big to handle" becomes softer as communities realize they're not alone in this moral quest. A global movement arises from countless communities acting in parallel, each learning from the other, each contributing a piece of the puzzle.

This global perspective widens the radius of hope. People learn that their local endeavors resonate with distant efforts. A restored wetland here echoes the principles behind forest protection elsewhere. Volunteers removing plastic from beaches mirror the determination of those in other continents tackling river pollution. Collective action transcends borders, making environmental stewardship a universal moral language. The resulting hope is not naive; it's grounded in evidence that when moral impulses unite, they can shift trajectories. The planet's future, though uncertain, appears less foreboding when so many moral engines churn simultaneously.

In looking back at these explorations—harnessing anxiety as a catalyst for engagement, collaborating on tangible local projects, and realizing the magnified impact of working in concert—what emerges is a renewed sense of moral agency and shared purpose. Actions inspired by fear are transformed from solitary, uncertain gestures into coordinated endeavors that not only alleviate eco-anxiety but also inspire hope. Each volunteer initiative, advocacy campaign, and civic movement becomes a thread in a global tapestry of environmental stewardship, each reinforcing the other's effectiveness. Instead of individuals feeling alone and overwhelmed, they become part of a vibrant, evolving community, one that melds diverse skills, values, and cultural perspectives into a collective moral force. In this unity, fear loses its paralyzing grip, replaced by a sense that meaningful change is not only possible but already happening. By committing to continuous learning, nurturing trust, and celebrating incremental progress, the anxiety that once weighed heavily on hearts now fuels perseverance and moral imagination. Through collective action, people discover that their shared concerns can be channeled into constructive efforts, building momentum that fosters resilience, nurtures interconnectedness, and ultimately kindles lasting hope for the planet's future.

Part IV: Sustaining Hope and Engagement Over Time

Chapter 12: Recognizing Progress and Celebrating Wins

Avoiding the Trap of All-or-Nothing Thinking About Solutions

As the scale and urgency of environmental challenges become ever more apparent, it's easy to feel that only grand, sweeping solutions can genuinely matter. Confronted with headlines about dwindling biodiversity, melting polar ice, or intensifying storms, some people slip into an all-or-nothing mindset—believing that partial measures and incremental improvements amount to negligible gestures. At the same time, others might downplay the need for large-scale changes, contenting themselves with the smallest tweaks and dismissing calls for deeper reforms as overly idealistic. Both extremes hinder moral growth and practical progress. In reality, social transformation often emerges not from a single perfect leap forward, but from a mosaic of partial wins, imperfect policies, refined strategies, and continuous adaptation. Understanding that these incremental steps have moral significance and cumulative power can help us celebrate progress, maintain hope, and move beyond the paralyzing trap of all-or-nothing thinking about solutions.

This trap often takes the shape of perfectionism on one side and complacency on the other. The perfectionist stance demands that every environmental measure must immediately address all aspects of a problem. If a policy fails to achieve dramatic emission cuts instantly, it's dismissed as useless. If a community garden doesn't feed the entire neighborhood, it's deemed insignificant. Such rigid expectations can breed despair when reality inevitably falls short. On the opposite end, complacency may arise from skepticism about ambitious targets or systemic reforms. Those inclined to minimalism might argue that small personal choices—recycling a few bottles, using a reusable cup—are enough, resisting more substantial transitions and ignoring the magnitude of global environmental threats. This stance risks stagnation, obscuring the urgent need for cumulative, far-reaching changes.

In contrast, a balanced perspective acknowledges that partial successes, while not final answers, still shape moral and ecological trajectories. Even incremental improvements matter because they shift norms, build momentum, and demonstrate feasibility. A policy that reduces emissions by a modest percentage is not worthless—it's a starting point that can be strengthened. A single coastal cleanup that removes litter from a beach does not restore the entire ocean's health, but it protects local wildlife and engages the community, potentially inspiring larger initiatives. The effort to introduce bike lanes in one city might not solve global warming, but it encourages more citizens to cycle, lowering local carbon footprints and normalizing sustainable urban transport. When we perceive such steps as meaningful rather than negligible, we reinforce moral resolve and open pathways for continuous learning.

Recognizing incremental progress also respects the complexity of environmental dilemmas. Climate change, for instance, involves an interplay of economics, technology, social behavior, historical inequities, and ecological thresholds. No one policy, invention, or cultural shift will solve it in a single stroke. Instead, layered strategies—some focusing on cleaner energy, others on reforestation, others on sustainable agriculture, and still others on behavioral changes—must interact. Each partial solution addresses some components, not all. Insisting that only a total, one-time fix counts ignores how social transformations generally emerge from iterative processes. Past revolutions in health, civil rights, or technological advancement also unfolded through stages, experiments, and partial victories. Embracing incrementalism situates environmental efforts in this broader historical context of moral progress, alleviating despair that arises from comparing reality to an unattainable ideal.

Furthermore, incremental gains teach adaptability. Suppose a city adopts a policy to reduce plastic waste by 20% within five years. This goal, though modest, sets a direction, allowing officials, businesses, and citizens to experiment with alternatives—reusable packaging, bulk shopping, education campaigns. Over time, they'll discover what works best and what doesn't. Lessons from these initial efforts can inform stronger policies later, maybe aiming for a 50% reduction or extending the approach to other materials. If perfectionists dismissed the initial modest target as pointless, the community might never start, missing out on essential learning opportunities. Incremental progress acknowledges that moral and practical knowledge evolves through practice and revision, not from static blueprints waiting for perfect conditions.

From a moral standpoint, small wins also nourish hope and maintain emotional energy. Environmental challenges often feel overwhelming: climate disruption threatens entire ecosystems and human livelihoods, while deep-rooted injustices complicate every aspect of resource distribution and policy-making. Without acknowledging incremental successes, people may succumb to the weight of their anxieties, convinced that no effort suffices. But when we validate partial improvements, we reaffirm that progress is possible. A nature reserve's reintroduction of one endangered species, while not solving the crisis of extinction, offers tangible evidence that interventions matter. This evidence stokes moral courage: if we've restored one species, perhaps we can restore more. If we've greened one neighborhood, perhaps we can green the entire city. Incremental progress feeds moral imagination, fueling the conviction that larger breakthroughs lie ahead.

This perspective also counteracts cynicism. All-or-nothing thinking can foster cynicism when reality fails to align perfectly with ideals. Constant disappointment breeds distrust in institutions, activists, leaders, and even fellow citizens. By appreciating partial achievements, we guard against disillusionment. Instead of seeing a shortfall as a betrayal, we interpret it as a prompt to refine strategies and push further. Over time, this interpretive shift builds resilience. Communities learn that setbacks do not define failure; they represent lessons. Just as ecosystems recover gradually from disruptions, so too can human societies respond to ecological problems incrementally. Seeing incremental steps as legitimate moral progress aligns environmental action with ecological principles of gradual adaptation, strengthening moral convictions.

Avoiding all-or-nothing thinking also makes room for moral pluralism. Different actors contribute differently to environmental solutions. Some focus on technological innovation, others on policy lobbying, others on grassroots education, and still others on personal lifestyle changes. If we demanded a single, comprehensive "perfect" solution, we might dismiss these varied contributions as insufficient. Recognizing partial wins, however, celebrates diverse efforts. Each approach tackles a piece of the puzzle—engineers improving solar panel efficiency, farmers integrating regenerative practices, educators raising eco-literacy. This diversity leads to a more robust moral ecosystem, where multiple methods coexist and learn from each other. In this ecosystem, each partial success nudges the whole system toward sustainability, proving that moral complexity and multiplicity can be assets rather than liabilities.

Historical examples reinforce this viewpoint. Consider the struggle against ozone depletion. Initially, global recognition of the ozone crisis led to the Montreal Protocol, which phased out many ozone-depleting substances over time. The initial commitments were modest compared to the scale of the problem. Critics might have scorned these partial restrictions as too timid. Yet these incremental steps proved instrumental. As technology improved and international cooperation solidified, stricter standards followed. Over decades, these cumulative measures allowed the ozone layer to begin recovering—a testament to how incremental policies can steer humanity away from catastrophic outcomes. This historical case shows that moral perseverance and steady progress can achieve what once seemed impossible, highlighting the folly of dismissing partial actions as meaningless.

Another historical insight comes from shifts in energy systems. The transition from coal to cleaner fuels didn't happen overnight. Early adopters of wind and solar technology faced skepticism and meager initial market shares. If perfectionists had dismissed these early ventures as too small to matter, the renewable energy sector might never have matured. Instead, gradual improvements in solar panel efficiency, scale economies, supportive policies, and public awareness combined to expand renewables rapidly. Today, renewables form a major pillar of many national energy strategies. Recognizing small gains as milestones avoids snuffing out fragile innovations before they flourish.

The moral implications extend to personal behavior. If someone attempts to reduce their carbon footprint but can't immediately achieve a zero-waste lifestyle or afford an electric car, they might feel inadequate. Such frustration can turn eco-anxiety into self-blame. Embracing incremental progress reframes personal efforts. Reducing meat consumption by half or taking public transport twice a week still lessens impact. Instead of feeling guilty for not attaining perfection, individuals can celebrate improvements and remain motivated to keep evolving. This approach stabilizes moral identity. Instead of cycles of idealism followed by guilt and withdrawal, people experience continuous moral development—steadily making better choices as circumstances permit and knowledge grows.

Incremental progress also strengthens alliances between individuals, communities, and institutions. When activists present nuanced proposals that push forward without demanding instant perfection, policymakers and businesses find it easier to engage. This cooperative spirit reduces polarization. If all-or-nothing demands cause political gridlock or corporate defensiveness, more gradual proposals might secure initial agreements, laying the groundwork for stronger measures later. Moral sophistication lies in understanding that pushing too hard too fast might backfire, while pushing not at all yields stagnation. Incremental steps help maintain dialogue, trust, and willingness to deepen commitments over time. Appreciating partial achievements as stepping-stones can transform adversarial standoffs into evolving partnerships.

This doesn't mean settling for weak measures indefinitely. Recognizing incremental progress includes recognizing when it's time to push harder. Celebrating partial wins shouldn't breed complacency. Instead, each step can raise ambition. Once a certain emission reduction target is met, moral logic encourages aiming higher. Once a community garden thrives, the next step might be building a full-scale local food system. Acknowledging improvements encourages continuous moral aspiration. It's not about staying in place; it's about moving forward with confidence that progress, however small, is worthwhile and can be built upon.

Communicating these ideas effectively matters. If environmental leaders publicly highlight incremental achievements, they encourage public optimism and involvement. When citizens see that a recent policy delivered some pollution reduction or a habitat restoration attracted certain wildlife species back, they understand that their participation and support contributed. Such public recognition reinforces moral agency on a larger scale. Instead of spreading defeatism, leaders can say, "We've done X, now let's aim for Y." This inclusive narrative acknowledges complexity, applauds past efforts, and invites everyone to keep pushing. By repeatedly affirming that incrementalism is part of the moral journey rather than a departure from it, leaders shape cultural mindsets that resist despair and cynicism.

The role of education here is pivotal. Teaching younger generations that environmental solutions often arise through cumulative progress fosters patience and intellectual humility. Students learn historical examples—from ozone recovery to improved vehicle emissions standards—to internalize that moral and ecological gains emerge over time. They also study current cases, like regenerative agriculture scaling up gradually, city-by-city bans on single-use plastics accumulating global momentum, or renewable energy markets evolving from niche to mainstream. Armed with these stories, future citizens approach complexity without expecting miracles. They see themselves as participants in a long moral endeavor and embrace incremental changes as essential rungs on the ladder of collective achievement.

Creativity thrives when freed from all-or-nothing constraints. When we acknowledge that partial measures count, innovators feel encouraged to propose prototypes, pilots, and experimental interventions. Suppose a group wants to tackle food waste. Instead of insisting on a flawless zero-waste system from day one—a daunting task—they can start by implementing a pilot program in one school or restaurant, learning what works and what fails. Over months, they refine practices, expand to other institutions, and scale up. Eventually, what began as a small trial morphs into a city-wide initiative. Without celebrating the incremental success of the pilot phase, the group might never reach full potential. Moral recognition of each stage fuels iterative refinement and ambition.

Cultural acceptance of incrementalism also aligns with principles of ecological resilience. Nature rarely recovers instantaneously from disturbances. Forests regenerate through successive stages, coral reefs recover with careful management and gradual improvements in water quality, species reintroductions often start with a few individuals and scale up. Emulating nature's pattern of gradual healing offers moral guidance: just as ecosystems adapt step by step, human societies can ethically adapt their behaviors, policies, and infrastructures. Rejecting all-or-nothing thinking attunes our moral sensibilities to ecological rhythms, balancing urgency with patience and complexity with clarity.

This balanced approach can calm emotional turbulence. Eco-anxiety arises partly from the feeling that if we don't achieve sweeping solutions now, all is lost. Appreciating incremental progress soothes these nerves, showing that positive changes are not hypothetical—they're happening. Witnessing how earlier restrictions on certain pollutants led to noticeable environmental recoveries provides psychological reassurance. It proves that humans have previously addressed environmental harms through cumulative efforts and can do so again. This historical and contemporary evidence helps maintain moral optimism, counteracting paralyzing fears that the situation is irrevocably dire.

Supporting each other in recognizing incremental wins is crucial. Within environmental communities, sharing small victories—an organization meeting a modest fundraising goal, a local school adopting a composting program—encourages a culture of encouragement rather than constant critique. Of course, this doesn't mean ignoring shortcomings or ceasing to advocate for more robust policies. On the contrary, it means building moral confidence by celebrating what has been done, then using that confidence to tackle tougher challenges. If internal dialogues remain excessively critical and dismissive of anything less than perfection, morale suffers, and people burn out. Affirming incremental progress prevents moral exhaustion and keeps the flame of motivation burning brightly.

External critics might accuse incrementalism of being too slow given environmental urgency. This tension is real. While incremental approaches are realistic and morale-boosting, they must also acknowledge that time is limited. The moral skill lies in combining appreciation for partial wins with sustained pressure for acceleration. Incremental progress can lay foundations, but we must never confuse this acceptance with complacency. As incremental successes accumulate, they set precedents for bolder moves. At key junctures, activists and policymakers can leverage these successes as proof that larger steps are possible, thus using incrementalism as a strategic stepping-stone rather than an endpoint. The moral artistry involves knowing when to celebrate small gains and when to use them as leverage to demand bigger leaps.

Over the long term, integrating a non-all-or-nothing perspective into public discourse and policymaking frameworks enhances resilience. Policymakers might structure plans that include phased targets, acknowledging realistic transition times while maintaining a vision for more ambitious future benchmarks. Scientists and economists advising governments can highlight how incremental policies can prime markets for larger shifts, gradually building public support and technological capacity. Civil society groups can communicate that initial measures, while not exhaustive, create political capital, public trust, and institutional experience needed to later enact more transformative changes.

This logic also extends to personal choices. If someone cuts their meat consumption by half, they might later find it easier to adopt more plant-based meals as their culinary skills and tastes adapt. If a household reduces energy use by insulating their home, they might later invest in solar panels once the cost drops further. Each incremental change paves the way for deeper shifts by removing barriers, learning new habits, and gaining confidence. Thus, moral and practical readiness accumulate until making another, more significant change feels natural rather than daunting.

Embracing partial wins also humanizes moral discourse. If we expect people to abandon cherished habits instantly or transform their livelihoods overnight, we risk eliciting defensive reactions. Recognizing incremental steps allows space for people's fears, cultural attachments, and personal constraints. This inclusive stance invites more people to participate. It says: "You don't have to be perfect to join this journey. Just take a step." Over time, these newcomers may surprise themselves by how much further they go, once the first step proves manageable. Gradually, entire communities build moral momentum, each small decision reinforcing the next until collectively they reach impressive milestones.

In addition, acknowledging incremental results encourages reflective practice. After achieving a small goal, communities can pause to evaluate what enabled success. Was it community outreach, policy support, a technical innovation, or a shift in public attitudes? Understanding these factors helps refine future strategies, increasing the efficiency and impact of subsequent efforts. Continuous learning ensures that incremental gains don't remain static; they become catalysts for informed growth. Without this stepwise refinement, we risk recycling the same flawed approaches. Celebrating partial wins thus includes analyzing them, extracting lessons, and applying those lessons to scale up or diversify the methods used. Incremental gains become cumulative not only in their direct impact but also in the moral and strategic intelligence they generate.

This mode of thinking resonates well with moral philosophies that emphasize virtues like patience, prudence, and humility. Instead of placing justice or environmental integrity as distant, absolute ideals attainable only through total revolution, virtue ethics suggests that moral progress emerges through consistent moral practice. Each partial success can be seen as an exercise in moral virtue: practicing compassion through local conservation efforts, exercising wisdom in policy negotiations, applying temperance by making balanced choices. Recognizing incremental achievements affirms that moral character is built incrementally, like craftsmanship honed over time. Just as you wouldn't dismiss an apprentice's initial work as worthless because it's not a masterpiece, you wouldn't dismiss a policy that achieves moderate improvements because it's not a total solution.

In sum, avoiding the trap of all-or-nothing thinking about solutions involves appreciating that moral and environmental progress unfolds along a continuum. There's no denying the urgency or complexity of the issues at hand, nor the necessity for significant change. But demanding everything at once or accepting trivial efforts forever are both flawed. True moral wisdom finds a middle path: celebrate what's achieved, learn from it, use it to build momentum, and keep pushing toward higher goals. By cherishing incremental wins, we maintain hope, sustain moral engagement, refine strategies, welcome diverse contributions, and ultimately guide human societies toward deeper, more sustainable transformations. This approach not only improves emotional resilience and communal solidarity but also anchors moral action in the realistic dynamics of social change, thereby turning eco-anxiety and moral concern into steady moral progress over time.

Tracking Incremental Improvements (Policy Changes, Technological Advances, Restoration Successes)

As communities, activists, researchers, and policymakers attempt to navigate the moral and practical complexities of environmental challenges, the notion of celebrating partial wins gains resonance. Yet recognition alone is insufficient if it remains vague and anecdotal. Tracking incremental improvements—whether through policy changes, technological advances, or ecological restoration successes—provides a structured means to measure progress, learn from outcomes, and build confidence in long-term viability. By carefully documenting each step forward, we not only affirm that moral efforts matter, but also gain vital feedback loops that inform strategy, inspire innovation, and maintain momentum. Monitoring these gradual achievements ensures that what once seemed unattainable now appears as a continuum of reachable milestones, each representing growth in human capacities, commitments, and moral imagination.

Why Tracking Matters
Without systematic tracking, incremental improvements risk disappearing into the background noise of daily life. Policy reforms might pass unnoticed if not recorded and analyzed, technological breakthroughs could fade into the haze of competing news, and restored habitats could blend back into the landscape without public recognition that conscious effort drove their rejuvenation. Tracking ensures that progress leaves discernible footprints—data points, reports, visual comparisons, timelines—so that individuals and societies can reflect on what actually works. Quantifying improvements, however modest, turns abstract hopes into concrete evidence. This evidence reduces skepticism and cynicism, reinforcing the notion that moral and ecological enhancements are real and replicable.

Tracking also helps counter the psychological tendency to remember failures more vividly than successes. Human cognition often skews toward negative events, a trait that can heighten eco-anxiety. By documenting positive changes, we balance this bias, ensuring that uplifting milestones are front and center. When discouraged environmental advocates review a timeline of cumulative policy improvements, expanding renewable energy capacity, or recovering wildlife populations, they realize that while problems persist, they are not static. Just as scientists track population growth in restored habitats to measure the impact of interventions, communities track improvements in waste reduction or emissions cuts to gauge collective effort's efficacy. This impartial record keeps moral energy from dissipating under the weight of endless challenges.

Measuring Policy Changes: Beyond Headlines and Promises
Policymaking can feel abstract and slow. Laws pass through complex channels, and even after enactment, their real-world impact may take years to manifest. Monitoring incremental policy improvements clarifies that policy change is not just rhetoric. Data can show that a newly implemented renewable energy mandate gradually increases the share of solar and wind in the energy mix, month after month. Tracking can confirm that stricter vehicle emissions standards reduce urban air pollutants over time, correlating improved public health outcomes with earlier legislative victories. Instead of judging policies solely at their inception—when they inevitably fall short of perfection—we observe their trajectory.

Policy tracking also reduces disillusionment when initial targets are modest. For example, a city might start by pledging to cut emissions by 10% over a few years. Critics may sneer that this barely scratches the surface of what's needed. But careful monitoring may reveal that after achieving this initial goal, public support and administrative experience pave the way for more ambitious targets. Over multiple cycles, incremental policy goals stack up to substantial reforms.

By examining the metrics—tons of carbon reduced, hectares of protected wetlands established, percentage of public transport electrified—citizens see a clear line of ascent, not just sporadic leaps. Knowing these data points exist and are regularly updated encourages ongoing engagement rather than cynicism, because people can verify that their advocacy and compliance bear fruit.

Tracking also holds institutions accountable. Without measurable results, a government or corporation could claim environmental progress without substantiation. Monitoring ensures that promises align with outcomes. It lets watchdog groups, journalists, and citizens press for transparency: "The policy promised a 5% emissions drop by this year; have we achieved it? If not, why?" Such scrutiny prevents greenwashing, maintaining moral integrity in the political and corporate spheres. Accountability built on data fosters trust, which in turn nurtures public willingness to support new initiatives. The public learns that not only do incremental policies matter, but they can also be enforced and refined, avoiding the pitfall where policymaking reduces to empty gestures.

Technological Advances: Charting the Path of Innovation
Similarly, technological advancements rarely emerge overnight as fully formed miracles. Breakthroughs often proceed through iterative research, pilot programs, incremental improvements in efficiency, cost reduction, and user-friendliness. Solar panels, once prohibitively expensive and less efficient, now compete with or undercut fossil fuels in many regions—an evolution that spanned decades and countless incremental refinements. Energy storage solutions, precision agriculture tools, carbon capture methods, sustainable materials, and countless eco-technologies follow similar arcs. Without tracking each incremental jump—like solar panel efficiency rising from 10% to 15% to 20%, or battery costs halving over a series of years—the narrative of technological progress remains invisible. This invisibility can feed skepticism and apathy.

However, when we record and celebrate each technical milestone—such as a new solar cell design improving energy conversion by a few percentage points, a pilot project demonstrating viable direct air capture of CO2 at slightly reduced costs, or an eco-friendly plastic substitute meeting durability standards—we reinforce the message that innovation is alive and well. Anxiety about environmental limits diminishes as people witness how cumulative technical improvements turn once-futuristic visions into mainstream solutions. Moreover, this documentation can spur further innovation: researchers see that incremental gains are recognized and valued, encouraging them to tackle the next constraint, improve the next component, or streamline the next process.

Tracking also allows for strategic decision-making. Policymakers and businesses can review data showing how quickly a particular technology matures, determining when to scale up deployment or invest in supportive infrastructure. Communities may learn that investing in a certain technology at a given stage could yield long-term dividends if incremental improvements are already on track. Investors gain confidence seeing steady progress rather than erratic leaps. This rational decision-making framework emerges only when incremental advancements are measured and communicated clearly.

Ecological Restoration Successes: Documenting Nature's Recovery

Ecological restoration exemplifies the power of tracking incremental improvements. Rewilding degraded landscapes, reviving coral reefs, reclaiming polluted rivers, and controlling invasive species often take years or decades. Without patient, regular monitoring, it might seem that efforts are futile—especially if immediate results are subtle. But careful measurement reveals progress: a once-polluted stream now supports certain fish species again, native vegetation returns after invasive removal, coral polyps spread across restored reef structures, and previously silent forests echo once more with bird calls. Each data point—a rising bird population, a slight increase in coral cover, improved water clarity—stands as evidence that diligent effort, guided by science and stewardship, nurtures life's resilience.

Ecological data also shape public perception. If volunteers or donors see that a wetland restoration gradually increases biodiversity indices, their faith in the project solidifies. They come to understand that healing nature isn't about quick fixes but about providing conditions for gradual recovery. This lesson extends moral lessons: just as ecosystems repair themselves step by step, humans can address climate and ecological crises incrementally, learning from nature's patient processes. Recording these ecological increments provides moral relief from anxiety—proving that nature responds positively to sustained care, and that humans, through consistent actions, can restore rather than just destroy.

Cumulative Gains and Synergistic Effects

One powerful aspect of tracking incremental improvements is identifying cumulative gains and synergistic effects. When policies, technologies, and restoration efforts interact, their combined impact surpasses what each could achieve alone. For example, improving soil health through regenerative agriculture raises yields slightly and reduces chemical inputs. Meanwhile, better-managed forests near farmland increase pollinators, enhancing crop resilience. As local renewable energy becomes cheaper, farms can power irrigation systems sustainably. Separately, these steps are minor wins. But tracked together, they reveal an interconnected tapestry of moral progress, where each improvement reinforces others, leading to stable, flourishing agro-ecosystems that reduce poverty, emissions, and ecological strain.

Recognizing these synergies dissolves all-or-nothing thinking by demonstrating how partial measures accumulate and interact. Observing multiple data sets in tandem—energy efficiency improvements lowering household bills, habitat restoration attracting ecotourism and thus economic benefits—reframes debates. Instead of fixating on a single ultimate solution, communities appreciate that layered approaches can yield more profound transformations. Anxiety subsides as people realize that complexity is an ally: small policies and technologies designed thoughtfully magnify each other's effects, accelerating social and ecological healing. This emergent property of cumulative improvements provides a moral blueprint: rather than seeking one heroic measure, we weave multiple threads into a robust moral fabric.

Guiding Long-Term Strategies

Tracking incremental improvements also informs long-term strategies. Organizations can chart how small advocacy efforts—like annual educational workshops or incremental fundraising targets—translate into expanded membership, higher public awareness, and eventually political clout. Policymakers can review how initial carbon pricing at a modest level gradually reshapes markets, incentivizing cleaner production and paving the way for stricter carbon targets. These data-driven insights guide when and where to scale up interventions. Recognizing partial successes not only fuels motivation but provides a rational basis for strategic planning.

This rational foundation reduces the emotional roller coaster of moral engagement. Instead of oscillating between doom and unrealistic optimism, participants anchor their judgments in evidence of steady progress. This stability fosters patience, open-mindedness, and willingness to adjust tactics as needed. Moral maturity arises from this iterative learning process: acknowledging partial gains, evaluating their significance, and planning the next steps accordingly. Anxiety loses its paralyzing force when people trust that adaptive management, informed by tracked improvements, continuously refines moral endeavors.

The Role of Communication and Public Engagement
To maximize the benefits of tracking incremental improvements, communication is key. Data must be accessible, transparent, and narratively engaging. Raw numbers alone may seem dry; placing them in context—explaining why a 2% increase in renewable share matters, or how a slight rise in a native fish population signals ecosystem resilience—conveys moral relevance. Visual storytelling—infographics showing emissions declining year over year, before-and-after photos of a restored wetland, timelines illustrating policy evolution—makes incremental achievements tangible. Public meetings where experts present these findings and community members ask questions build trust and inclusivity.

When communities celebrate these incremental wins publicly—perhaps holding small ceremonies every time a habitat restoration milestone is met—they reinforce collective pride. Integrating data presentations into moral rituals (like annual sustainability festivals) ensures that tracking isn't just technical analysis but part of the moral culture. Through repeated exposure, citizens internalize that modest improvements are stepping-stones, not mere footnotes. This cultural embedding of incrementalism fosters a long-term moral perspective resilient against defeatism.

Challenges in Interpreting Incremental Data
One must remain vigilant that data tracking can sometimes lead to complacency if not contextualized properly. For instance, a community might be proud of cutting local emissions by a few percentage points, but if global emissions keep rising, is that minor reduction sufficient? Acknowledging incremental improvements doesn't mean ignoring scale or urgency. Data should be interpreted within global benchmarks and targets aligned with scientific advice. If evidence shows that while local pollution drops slightly, the overall climate crisis intensifies, then celebrate the local win but double down on efforts to scale up solutions.

Balanced interpretation is essential. The goal is to avoid both extremes: not dismissing incremental gains as meaningless, but also not settling for far less than what science and justice demand. Incremental data must feed into bigger questions: Are these steps cumulatively steering us toward a safer climate trajectory? Are they creating conditions that enable future leaps forward? Are vulnerable communities truly benefiting from these improvements, or are increments concentrated in privileged areas? Asking tough questions ensures that incrementalism serves moral depth rather than becoming a veneer of progress.

Integrating Incremental Achievements into Moral Education
Moral education—at home, in schools, in community gatherings—can integrate the concept of incremental improvements into ethical frameworks. By teaching that environmental ethics involves continuous effort, adaptation, and cumulative success, educators prepare learners for a lifetime of moral engagement. Instead of imparting a vision of moral purity or total solutions, education can highlight historical and current cases where partial measures laid foundations for breakthroughs. Students might analyze how small protected marine areas, tested and refined, eventually inspired vast marine reserves that now safeguard crucial ocean habitats.

Such lessons equip future citizens to recognize that real change often emerges through a mosaic of improvements rather than a single heroic event. They learn to spot incremental wins in everyday life—rivers running slightly clearer, local farms reducing pesticide use, communities divesting modestly from fossil fuels—and understand that these shifts reflect moral traction. By normalizing incremental progress, moral education reduces eco-anxiety, as learners anticipate gradual achievements rather than holding out for all-or-nothing victories.

The Moral Imperative to Track and Celebrate
Given the importance of recognizing and communicating incremental improvements, there's a moral imperative to track and celebrate them diligently. Activists, researchers, policymakers, and community leaders bear responsibility for ensuring that these data points don't get lost. Allocating resources to monitoring, reporting, and verifying progress is not an administrative luxury but an ethical choice. Without data, arguments about moral momentum remain speculative. With robust tracking, moral claims gain groundedness. We can say, "Look, over the past five years, we reduced local emissions by 15%, we increased renewable installations by this margin, we restored that many hectares of wetlands," proving that collective will turns into measurable outcomes.

Moreover, celebrating progress does not diminish the moral seriousness of pending challenges. It's possible to cheer small improvements while acknowledging the long road ahead. This dual perspective—applauding what's achieved and demanding more—reflects moral maturity. It respects complexity and uncertainty without discouraging ongoing efforts. Communities can say, "We've come this far—great work! Now, let's leverage this success to push further," maintaining constructive tension. Anxiety about insufficient action transforms into determination fueled by the confidence gained from having tangible successes under our belts.

Empowering Individuals Through Transparent Tracking
Transparency in incremental improvements also empowers individuals who might otherwise doubt their contributions. A citizen who recycles diligently may wonder, "Does my action matter?" But if local data show that recycling rates rose from 30% to 50% citywide, reducing landfill waste and associated emissions, individuals see direct links to their behavior. This reduces the psychological distance between personal decisions and collective outcomes. Moral agency sharpens: individuals realize that aggregated small acts truly influence the bigger picture. Anxiety about irrelevance fades as people witness how their efforts join a larger current of change. Transparency creates a virtuous cycle: the more people see results, the more they engage, and the more results follow, continually reinforcing moral engagement.

Similarly, businesses and institutions observing data on incremental improvements may find it easier to justify investments in sustainable practices. If a company sees that gradually replacing its delivery fleet with electric vehicles cut operating costs and reduced carbon footprint, it gains confidence to expand these measures. Investors and stakeholders trust data-driven narratives, interpreting incremental improvements as indicators of long-term viability and risk mitigation. Thus, tracking benefits not only grassroots activists but also decision-makers who need empirical reassurance to implement progressive measures. Anxiety about uncertain returns or lack of impact dissipates when metrics demonstrate steady forward motion.

Inspiring New Moral Goals Through Reflective Analysis
As communities accumulate evidence of incremental improvements, they can periodically pause to reflect on their broader moral goals. Suppose a city set out initially to reduce plastics in local waterways. Over time, they track declining plastic waste counts and improved water quality. These successes might embolden them to tackle more systemic issues—restricting upstream plastic production or advocating for regional plastic bans. The data-driven confidence transforms initial modest goals into stepping-stones for bigger ambitions. Anxiety no longer revolves around whether action matters, but rather which new frontiers the community should explore next. Incremental progress thus morphs into a moral catalyst, urging continuous redefinition and expansion of objectives.

Simultaneously, reflection ensures that people don't become complacent, satisfied with partial wins forever. Analyzing the data might reveal that while water quality improved somewhat, certain pollutants remain stubbornly high or certain wildlife populations haven't rebounded fully. These insights provoke refined strategies: maybe stronger regulations, better habitat management, or international cooperation are needed. Incremental improvement becomes a moral mirror, showing both achievements and gaps, guiding future priorities.

Countering All-or-Nothing Thinking in Public Debate
In public debates, acknowledging incremental improvements can also serve as a counterweight to polarized extremes. Opponents of environmental policies sometimes dismiss small-scale solutions as trivial, while hardline critics of incremental steps may insist they are mere greenwashing. Presenting evidence of incremental progress effectively counters these arguments. It shows that what some ridicule as minor does, in fact, yield measurable gains, and that such gains can pave the way for more ambitious frameworks. Instead of reducing everything to a binary—either we solve all problems now or do nothing—communities demonstrate that there's a middle path built on accumulating moral capital through partial successes.

This balanced approach can soothe tensions in political arenas. Policymakers who fear backlash from demanding too much too soon can start with moderate measures, track results, and use that data to justify stronger actions later. Activists can support these phased strategies, understanding that while the ultimate goal remains large-scale transformation, incremental policies help shift the political terrain and lay foundations. By presenting quantifiable improvements, they persuade more stakeholders to join. Over time, repeated increments change what's considered politically feasible, ratcheting the baseline for acceptable environmental standards upward.

Sustaining Emotional Resilience

For individuals who struggle with eco-anxiety, the knowledge that incremental improvements are tracked and publicly acknowledged can significantly improve mental well-being. When faced with grim forecasts, they can consult the growing archives of partial wins—community emission reductions, reforestation successes, cleaner rivers after certain interventions—and find solace in these positive narratives. Data and case studies become emotional anchors. Instead of drifting into despair, anxious individuals recall that progress, though incomplete, is real and ongoing.

This emotional resilience, in turn, sustains engagement. People who trust that their efforts, combined with others', yield measurable outcomes are less likely to give up. They understand that while perfection may remain distant, each step contributes to moral evolution. Eco-anxiety no longer paralyzes; it motivates continued involvement, each cycle of improvement reinforcing moral commitment. The synergy of emotional resilience and measurable progress drives a long-term cycle of moral perseverance, ensuring that communities keep refining and amplifying their solutions despite inevitable hurdles.

Moral Pluralism and Continuous Improvement

Tracking incremental improvements also reinforces moral pluralism. Different ethical frameworks—utilitarian, deontological, virtue-based—can all appreciate evidence of incremental change. Utilitarians see the improved well-being or reduced harm validated by metrics. Deontologists value policies that progressively align with moral duties, even if gradually. Virtue ethicists cherish the character-building journey from small successes to grander responsibilities. Thus, incremental data can unify disparate moral approaches under a common appreciation of moral progress as a process. This unity reduces philosophical clashes and encourages diverse moral communities to collaborate, guided by shared evidence that effort matters.

Additionally, tracking incremental progress aligns with the concept of continuous improvement found in management and governance frameworks. Just as organizations adopt iterative processes—plan, do, check, act—communities addressing environmental issues can follow similar cycles. They propose an initiative, implement it, measure its outcomes, evaluate and learn, then refine or expand. Anxiety transforms into a driver for careful monitoring and adaptation, ensuring that moral engagement never stagnates. Instead, it cycles forward, each iteration polishing strategies and honing moral intuition.

Imagining a Future Steered by Incremental Gains

As cumulative incremental improvements reshape policies, technologies, and ecosystems, we can imagine a future where humanity navigates environmental crises not as doomed bystanders but as diligent caretakers. In this envisioned future, people understand that while absolute certainty or instant perfection is impossible, measured progress fosters hope and stability. Institutions become adept at analyzing incremental data, adjusting policies accordingly. Businesses thrive by innovating stepwise solutions that incrementally reduce footprints and improve resource efficiency. Civil society groups celebrate each minor milestone, reinforcing the cultural narrative that environmental care is an ongoing moral journey, not a one-time fix.

In this future, anxiety remains—no one claims that incremental progress eliminates all fears. But these anxieties operate within a moral framework that proves responsive to effort. Fear turns into a catalyst for recommitting to the process of incremental gains. Individuals and communities know how to read the data, draw courage from previous improvements, and chart new courses. They accept complexity and imperfection as natural conditions of moral life. Driven by the knowledge that each small success lays groundwork for the next, they approach environmental challenges with steady resolve, patience, and a deep-seated belief that incremental progress is both real and morally meaningful.

Empowering Future Generations Through Ethical Continuity
By tracking incremental improvements, today's generations pass a legacy of ethical continuity to those who follow. Future activists and policymakers will inherit not just crises, but records of resilience and creativity. These records become moral compasses, showing that when confronted with adversity, humanity responded with incremental steps that, over time, shifted paradigms. Young people analyzing historical data will recognize that small measures their ancestors took eventually influenced global transitions—much as the Montreal Protocol phased out harmful chemicals step-by-step, or renewable energy adoption rose gradually then surged exponentially.

This legacy empowers future generations to maintain moral agency under their unique conditions. They learn that eco-anxiety isn't a novel burden but a recurring prompt for steady moral engagement. Instead of despairing that previous efforts failed to achieve perfection, they honor that cumulative path. Their perspective roots in gratitude for incremental foundations laid, combined with a moral imperative to continue building. In this way, tracking incremental improvements creates a moral inheritance—an archive of steady progress to be cherished, critiqued, and extended.

Collective Moral Growth: The Final Picture
Ultimately, the act of tracking incremental improvements (be they policy changes, technological advances, or restoration successes) becomes a moral practice. It's an exercise in humility (accepting partial solutions), honesty (acknowledging results), gratitude (valuing each gain), and perseverance (viewing steps as part of a longer trajectory). This moral practice sculpts character and communal ethos, ensuring that no single success or failure distorts the entire moral horizon. Instead, people embrace complexity and incrementalism as defining features of ethical environmental stewardship.

This perspective liberates us from the crippling constraints of all-or-nothing thinking. Rather than waiting in vain for immaculate solutions or settling for token gestures, we travel along a continuum. Each improvement offers a platform for the next leap forward. Over time, these increments weave a moral narrative of collective learning and adaptation—a narrative that tempers anxiety with evidence, widens hope through cumulative achievements, and affirms that human moral agency, though imperfect, remains potent and essential for guiding the planet toward a more sustainable destiny.

Celebrating Small Steps to Maintain Momentum and Morale

As communities and individuals grapple with the enormity of environmental challenges, it's easy to lose sight of the incremental nature of moral and ecological progress. Achievements often unfold slowly, quietly, and in scattered increments rather than in one grand, decisive moment.

This gradual pace can frustrate those who anxiously long for swift resolutions. Yet learning to celebrate small steps offers a powerful antidote to despair, providing vital emotional sustenance and reinforcing moral commitment. Recognizing that each partial victory contributes to a cumulative trajectory toward sustainability transforms isolated successes into beacons of reassurance. In celebrating these milestones, communities maintain momentum, safeguard morale, and cultivate the patience necessary to persist amid uncertainty. Ultimately, this practice enriches the moral ecosystem of environmental engagement, ensuring that anxiety never fully eclipses hope.

Finding the Value in Partial Achievements
Many people feel reluctant to celebrate small steps because they worry it might signify complacency. They fear that by cheering modest accomplishments, the urgency of larger goals will be diluted. In reality, acknowledging incremental gains does not mean settling for less; it acknowledges that moral growth, policy shifts, cultural changes, and ecological recoveries seldom happen overnight. Just as a tree does not grow from seed to full height in a single season, moral and ecological progress also requires stages. Each new root, leaf, and branch—though small on its own—cumulatively creates a thriving canopy.

By honoring these incremental stages, individuals and communities affirm that their energies are not poured into a void. Instead, they see their hours of volunteering, careful negotiations, and creative problem-solving bearing fruit over time. This recognition counters the psychological toll of seeming stagnation. Without celebrating small steps, anxiety can accumulate as participants wonder whether any of their efforts matter. Each acknowledged milestone answers that doubt: yes, it matters, and here's the evidence. This affirmation steadies participants, motivating them to continue investing moral effort, refining strategies, and nurturing collaboration.

Defining Meaningful Milestones
The first challenge in celebrating small steps is knowing what to consider a "win." Without clarity, it's tempting to trivialize every minor improvement or, conversely, set the bar so high that no intermediate outcome ever seems worthy of recognition. Identifying meaningful milestones requires a combination of goal-setting, data analysis, and community consensus. A policy campaign might define milestones as passing a preliminary committee vote, gaining endorsements from respected stakeholders, or shifting public opinion by a measurable margin. A restoration project could celebrate each increase in native species population, each area cleared of invasive plants, or each metric of water quality improvement.

When communities co-create these definitions, the process itself fosters moral cohesion. Participants debate what counts as progress—Is a 1% emission reduction worth celebrating? If so, how do we mark it? Setting such criteria encourages transparency and realism. It invites reflection on what kinds of changes hold intrinsic value and what success looks like. This preemptive calibration mitigates future disappointment. Because everyone knows that, say, every 5% improvement in local recycling rates earns a public thank-you event, no one feels these milestones are arbitrary or insincere. Moral clarity emerges from agreeing on thresholds that reflect the community's ambitions and constraints.

Emotional and Psychological Benefits

Celebrating small steps holds emotional power. Facing grim news about melting ice sheets or acidifying oceans can erode morale. But acknowledging that a local wetland restoration has attracted its first nesting pair of rare birds, or that a policy campaign successfully included an environmental provision in a broader legislative package, injects a dose of optimism. This optimism counteracts despair, reminding participants that not all trends move downward and that human initiative can bend trajectories.

Emotional benefits extend to individuals who doubted their ability to contribute meaningfully. They see their volunteer hours helping improve tree canopy coverage or their research guiding a small but pivotal policy amendment. Feeling seen, valued, and effective reduces anxiety about personal insignificance. The act of celebration—whether a small neighborhood gathering, a social media shout-out, or a local newspaper article—publicly affirms that each person's involvement contributed to something tangible. This recognition nurtures self-confidence, making participants more resilient to setbacks. Instead of feeling crushed by a failed initiative, they remember past mini-successes, drawing strength from their cumulative moral achievements.

Strengthening Social Bonds and Community Identity

Public celebrations of incremental wins strengthen social cohesion. When volunteers gather to commemorate restoring a certain percentage of a riverbank's vegetation or reaching a fundraising target for a solar panel installation, they share pride and gratitude. These events foster emotional bonds that transcend individual contributions. Everyone, from the planner who coordinated logistics to the student who contributed a single Saturday's labor, basks in collective accomplishment. Such solidarity transforms anxiety—once an isolating emotion—into a shared moral experience binding people together.

Moreover, these celebrations become part of community identity. Over time, they form traditions. Maybe each spring, the group marks the reappearance of a certain migratory bird species or the anniversary of passing a crucial climate ordinance. These rituals tell stories of progress etched into communal memory. Anxiety about future challenges feels less menacing when viewed against a backdrop of consistent, recognized achievements. As new members join, they learn the community's moral heritage, discovering that while perfection remains elusive, moral action accumulates. This historical continuity encourages sustained engagement and loyalty.

Inspiring Participation and Broadening Involvement

Highlighting incremental wins can attract new participants who might have hesitated to join earlier. Outsiders observing that the community doesn't only warn of catastrophes but also celebrates forward steps may feel more inclined to get involved. Positive reinforcement alters the external narrative: instead of perceiving environmental activism as endless struggle, newcomers see it as a balanced journey that acknowledges both difficulty and triumph.

This approach also motivates existing members to stay committed. Environmental work often requires long-term dedication, and burnout is a common hazard. By regularly marking progress, communities feed members' emotional reserves. Volunteers learn that moral investment yields not only moral satisfaction but also recognition and communal appreciation. This positive feedback loop stabilizes membership, reduces turnover, and enhances long-term capacity.

Countering Perfectionism and All-or-Nothing Thinking

One of the main moral lessons in celebrating small steps is countering perfectionism and all-or-nothing thinking. When anxiety leads to a sense that nothing short of total environmental salvation counts, even significant partial improvements appear trivial. This mindset can paralyze efforts—why bother reducing emissions by 10% if we need 90% cuts? But acknowledging partial achievements recalibrates moral expectations. Instead of holding every initiative to an impossible standard, communities learn to value partial solutions as essential building blocks.

This shift does not dilute moral ambition. Acknowledging small achievements doesn't mean lowering the ultimate goal. Rather, it contextualizes progress. A 10% emissions cut may not solve climate change, but it gets us closer, proving that change is possible and methodologies can work. This fosters moral resilience, as participants realize that incremental successes generate the conditions for further improvements. Celebrating small wins redefines failure: it's no longer failing to solve everything at once; it's failing to take any steps at all. Moral growth lies in summoning the courage to keep pushing, confident that each milestone eases the journey.

Integrating Celebrations into Decision-Making Cycles

To institutionalize the practice of acknowledging incremental progress, communities can integrate celebrations into decision-making cycles. For instance, after implementing a new policy trial for six months, the group reviews data, notes improvements, and holds a small ceremony—even if it's just a moment of collective applause—before deciding on next steps. This ritual not only boosts morale but also informs strategy. By reflecting on what worked, what partially worked, and what needs refinement, participants align emotional closure with rational planning. Anxiety recedes as people see that their community operates proactively, always learning and iterating instead of stagnating in either despair or complacency.

These cycles underscore that moral endeavor is a dynamic process. When communities celebrate progress regularly, they become adept at gracefully handling setbacks. Instead of treating a failure as the end of the road, they frame it as a lesson before returning to the data, adjusting tactics, and pursuing the next incremental target. Each cycle of attempt-measure-celebrate-adjust conditions participants to view environmental stewardship as a continuous journey where successes and learning moments alternate, each essential to moral evolution.

Artistic and Cultural Dimensions of Celebration

Celebrations of incremental wins need not be dry, technical announcements. Incorporating art, music, storytelling, or cultural rituals infuses these moments with deeper meaning. A community might commission a local artist to create a mural symbolizing the journey from a polluted river to a cleaner ecosystem, adding a small detail to the mural each time a target is met. Or they might host a themed storytelling night where volunteers share personal anecdotes of how anxiety propelled them toward meaningful action. Such creative expressions reinforce moral identity: participants see themselves as part of a living narrative, evolving with each achievement.

Artistic celebrations transcend language and ideological barriers. While policy reports might bore or intimidate some people, visual or musical representations of incremental progress resonate emotionally. These sensory experiences evoke empathy and pride. They capture the moral core of environmental work—love for life, commitment to future generations, reverence for nature's complexity—and tie it to tangible progress. Anxiety, once a silent, gnawing concern, finds voice in celebratory expressions that frame worry and hope as intertwined forces driving moral courage.

Encouraging Positive Peer Influence and Norm Cascades
In social psychology, recognizing others' positive behaviors often encourages similar behaviors. When communities applaud small steps, they create positive peer pressure. Individuals witness neighbors or peers being thanked for reducing their plastic usage, supporting local green businesses, or advocating for cleaner public transportation. These visible rewards encourage imitation. Moral standards shift, raising baseline expectations. Those who might have dismissed certain actions as too small or symbolic now see them as valued contributions.

This positive feedback loop fuels "norm cascades," where incremental improvements become self-reinforcing. As more people join in these behaviors—composting at home, backing energy efficiency measures—these practices become standard. Anxiety stemming from uncertainty—"Will anyone follow through?"—fades as people realize these actions are not marginal but increasingly common. Normalizing incremental improvements reorients moral discussions from whether to act at all, to how to refine and multiply such actions, ensuring momentum even if overall conditions remain challenging.

Linking Incremental Progress to Collective Morale
Moral progress and community morale are mutually reinforcing. Celebrations of small steps raise collective morale, which in turn bolsters moral effort. High morale enables communities to tackle tougher challenges confidently. Even if some targets remain out of reach, a community high in morale tries again, adapts, and collaborates widely. Moral energy, fueled by recognition of incremental gains, forms a protective shield against cynicism and fatalism. Without such bulwarks, anxiety might corrode unity and undermine will.

As morale rises, participants become more flexible and open-minded. Instead of rigidly insisting on only certain solutions, they consider multiple approaches, trusting that incremental refinements guide them forward. Increased morale also encourages mutual support: members cheer each other's personal contributions, whether small donations, a well-written letter to a policymaker, or a volunteer shift. This warmth and acknowledgment diffuse anxiety: fear cannot dominate a space filled with gratitude, praise, and moral affirmation.

Educational and Intergenerational Benefits
Celebrating incremental wins also educates future generations in moral resilience. Children growing up in communities that habitually mark progress with modest ceremonies or public acknowledgments learn that tackling environmental issues is an ongoing process. They understand that while big goals matter, it's normal and honorable to reach them through sequences of manageable steps. They inherit a mindset that values patience, collaborative

learning, and perseverance. Anxiety about uncertain futures will still arise, but they'll know how to temper it with evidence of consistent forward movement.

In classrooms or youth groups, teachers can use examples of incremental successes to illustrate moral principles. Students might analyze how each local improvement—such as improved recycling rates—connects to bigger ecological benefits, reinforcing lessons about interdependency. They see moral agency as cumulative: by starting small, they can grow into roles that enact larger transformations. Anxiety that might stifle young imaginations transforms into curiosity and ambition. They envision building on their elders' partial successes, pushing moral frontiers even further.

Global Exchange and Mutual Encouragement
The practice of celebrating small steps spreads globally. Online platforms, conferences, and media coverage allow communities to share their incremental success stories. Seeing another region's approach to composting or tree planting encourages cross-pollination of ideas. Communities mirror each other's celebratory practices, affirming that these approaches are not local peculiarities but universal moral tools. Witnessing how multiple societies keep morale high by applauding incremental gains suggests that this tactic transcends cultural differences.

This global exchange forms a supportive network of encouragement. Anxiety about slow global progress diminishes when people learn that similar incremental patterns unfold elsewhere: a coastal village achieving slight coral recovery, a metropolis incrementally reducing smog, a desert community restoring groundwater tables step by step. Recognizing the widespread adoption of incremental celebration underscores that humanity's moral response to ecological stress is not uniform despair but adaptive, hopeful, and collaborative. Small steps align humans across borders, forging solidarity in the shared struggle for planetary health.

Adapting Celebrations as Challenges Evolve
As environmental contexts shift—some ecosystems recovering, others still threatened—communities must adapt their celebration criteria. Over time, what once qualified as a significant incremental gain may become routine, prompting communities to raise the bar. This adaptive approach prevents stagnation. The community that once celebrated a 5% emissions reduction now aims for 15% and celebrates intermediate increments differently. By adjusting standards upward as experience and capacity grow, communities maintain motivational tension, always stretching but never dismissing smaller wins.

Conversely, if conditions worsen unexpectedly—like a natural disaster reversing some progress—communities can still highlight that previous incremental improvements built resilience. Even acknowledging partial rebound after a setback can restore faith in moral capacity. Thus, celebrating small steps supports dynamic goal-setting, flexible adaptation, and emotional recalibration. In this cyclical process, anxiety ceases to be a permanent emotional climate and becomes just another factor communities manage by highlighting achievable progress.

Moral Reflection in the Face of Complex Trade-Offs

Sometimes incremental improvements come with trade-offs, and celebrating them must reflect moral complexity. For example, transitioning a city's bus fleet to partially electric may still involve resource-intensive battery production. Recognizing this complexity—while still applauding reduced local air pollution—prevents naive triumphalism. By integrating nuanced moral reflection into celebratory practices, communities acknowledge that incremental steps aren't pure good. They come with costs, partial downsides, or insufficient scope. However, this does not negate their worth; rather, it encourages informed decision-making and continuous ethical inquiry.

This nuanced approach teaches that moral life seldom yields perfect solutions. A community that learns to appreciate incremental benefits while critiquing their limitations cultivates moral maturity. Anxiety about possible unintended consequences can be channeled into responsible monitoring and ethical debates rather than stifled progress. Celebrating small steps does not become a propaganda tool but a reflective practice: rejoicing in what's gained while questioning what remains undone. This honesty keeps communities focused on future improvements and maintains moral credibility.

Building Confidence for Larger Transformations
The ultimate aspiration in celebrating small steps is to build confidence for larger transformations. As people accumulate proof of moral agency and adaptability, they become less intimidated by ambitious goals. Seeing how incremental changes gradually improved local ecosystems or influenced policy frameworks, communities grow bolder. Anxiety about attempting bolder initiatives subsides: if we succeeded in these stepping-stones, why not try a more comprehensive renewable energy program, or a city-wide zero-waste policy?

By the time the community attempts bigger leaps, it doesn't enter them as novices. The habits of celebrating partial wins forged resilience, patience, and trust. Instead of panic at the scale of new challenges, participants recall past breakthroughs, no matter how small, which all indicate they can handle bigger tests. Moral muscle memory kicks in—facing a major reform or widespread restoration, everyone knows how to break down the challenge into increments, tackle them methodically, and celebrate each advance. Anxiety does not vanish but becomes manageable, channeled into methodical perseverance.

Spreading the Philosophy of Incremental Celebration
As the practice of celebrating incremental wins proves effective, communities share their philosophy outward. Educational materials, training sessions, and workshops can teach other groups how to identify meaningful milestones, maintain data transparency, integrate celebrations into moral rhythms, and communicate progress effectively. This diffusion of best practices helps diverse actors replicate the moral architecture that supports incrementalism. Over time, celebrating small steps might become standard practice in environmental governance, advocacy training, and community organizing worldwide.

This global spread strengthens the entire moral network of environmental engagement. With more communities adopting these methods, fewer people fall prey to despair or perfectionist paralysis. Instead, incremental celebrations normalize perseverance and adaptability. Anxiety transforms into a universal prompt for careful tracking, modest acknowledgments, and renewed effort. Cultural narratives evolve: children reading about environmental stories encounter repeated themes of incremental success—ecosystems recovering bit by bit, emissions declining gradually, policies evolving—and learn to trust the moral journey, not just the final destination.

Guarding Against Misuse or Manipulation
Of course, communities must guard against misuse of incremental celebrations. Some entities might highlight trivial steps as major accomplishments to deflect criticism. Transparency, accurate data, and critical thinking prevent greenwashing or tokenism. Celebrating small steps is only legitimate when those steps are genuine progress consistent with long-term moral goals. If a corporation trumpets a minor emissions tweak while expanding overall carbon-intensive operations, communities should call them out, insisting on integrity. This vigilance ensures that the tool of incremental celebration remains ethically sound.

Communities can establish guidelines, or even external review committees, to confirm that incremental improvements align with broader principles and remain on track toward ultimate sustainability targets. This moral accountability deters complacency and protects celebrations from devolving into PR stunts. Authenticating achievements preserves trust in the incremental paradigm, ensuring that anxiety reduction and morale-building remain grounded in truth and sincerity.

Evolving Meaning Over Time
As more and more communities integrate the practice of celebrating incremental wins, the meaning of these celebrations may evolve. Early on, they might feel like desperate attempts to salvage hope. Over time, as cumulative improvements accumulate, these celebrations become confident affirmations of collective capacity. Eventually, when moral and ecological paradigms shift significantly—imagine a future where emissions are halved, biodiversity partially restored, and sustainability widely integrated—the celebrations may highlight how far we've come, acknowledging that the incremental path was crucial in achieving systemic transformations.

In this future context, anxiety about environmental collapse might recede somewhat, replaced by anxiety about maintaining achievements or tackling new frontiers (like rewilding larger areas or ensuring climate resilience for marginalized communities). Even then, the habit of celebrating small steps remains valuable, reminding everyone that moral vigilance is never done. There will always be room for refinement, greater fairness, deeper restoration. Incremental celebrations become a permanent moral fixture, ensuring societies never lose their capacity to adapt and improve further.

A Durable Moral Tradition
Through these reflections, it's clear that celebrating small steps is more than a tactic—it's a moral tradition conducive to long-term environmental stewardship. It stems from humility (accepting that big problems need incremental solutions), honesty (admitting partial gains), compassion (seeing that emotional well-being improves when we validate progress), and perseverance (continuing even when goals remain distant). This tradition transcends any single issue, time period, or cultural context. It can apply to climate adaptation, pollution reduction,

sustainable agriculture, water management, or any domain demanding complex, ongoing moral work.

By solidifying this tradition, communities safeguard against fatalism. Anxiety can no longer claim that nothing changes or that incremental work is futile. Every recorded step forward negates that narrative. In cumulative increments, moral progress reveals itself as a practical reality. This reality, in turn, reinvigorates the ethical imagination, encouraging creative problem-solving, alliance-building, and moral growth. Instead of drifting into cynicism, communities become moral laboratories, testing and fine-tuning solutions, recognizing each small success as a stepping-stone toward broader ethical horizons.

In facing environmental challenges that loom large and complex, shifting perspective away from all-or-nothing expectations can bring both clarity and energy. By recognizing that moral and ecological gains emerge incrementally rather than in a single decisive moment, efforts become more resilient and creative. This approach disarms despair, since even partial advances signify tangible progress rather than empty gestures. Instead of despairing at the gap between a flawed reality and an ideal vision, communities learn to value the steps that bring them closer to ethical and ecological integrity. Tracking policy improvements, technological refinements, and restoration achievements validates that each contribution holds meaning. Sharing these insights publicly builds confidence, motivates broader participation, and encourages the continuous refinement of strategies. Engaging with complexity, adjusting goals as knowledge evolves, and celebrating these modest victories all nourish emotional well-being and foster a culture of perseverance. By embracing the cumulative power of incremental change, human societies protect against fatalism, maintain moral momentum, and keep alive the steady hope that collectively, step by step, they can guide the planet toward a more sustainable and just future.

Chapter 13: Hope as a Practice, Not a Fantasy

Distinguishing Grounded Hope from Naïve Optimism

As environmental challenges intensify and public awareness grows, the search for hope often becomes paramount. People who grapple with eco-anxiety, despair, or moral fatigue yearn for something to temper their fears—a sense that efforts matter, that all is not lost, and that the future remains open to positive transformation. Yet "hope" can be slippery. Amid the cacophony of dire warnings and idealistic promises, not all expressions of hope carry equal weight. Some resemble hollow mantras, wishful illusions untethered from reality. Others stand firmly rooted in understanding complexity, acknowledging difficulties, and committing to sustained moral action. Distinguishing grounded hope from naïve optimism is essential for navigating moral life with integrity. Without this discernment, we risk clinging to fantasies that crumble under scrutiny, or rejecting hope altogether when the world refuses to comply with comforting illusions.

Naïve optimism, on the one hand, often arises from the desire to escape anxiety by ignoring complexity. It thrives on simplistic assurances—claims that technology will magically fix everything, or that a single policy will deliver salvation, or that nature will effortlessly heal itself if only people "love the Earth" enough. Such optimism tends to discount inconvenient data, overlook entrenched power structures, and misread the scale of ecological disruption. By glossing over nuances, it offers emotional relief but fails to prepare anyone for the arduous moral and practical work ahead. Naïve optimism tells people what they want to hear, not what they need to know. It breeds complacency and shallow engagement, leaving people vulnerable to disillusionment when reality intrudes.

Grounded hope, in contrast, never denies complexity. It arises not from wishful thinking but from sober assessment, moral reflection, and an understanding that positive change, though possible, demands effort, courage, and adaptability. Instead of promising easy solutions, grounded hope urges a balance: believe in human capacities, learn from incremental successes, trust that creativity and cooperation can overcome obstacles, yet never underestimate the magnitude of challenges. This form of hope does not fear confronting difficulties. It acknowledges uncertainty, recognizes that progress may be partial, and accepts setbacks as part of an ongoing moral process. Grounded hope does not require certainty that everything will turn out perfectly; it only requires faith that deliberate actions and collective resilience can steer outcomes in better directions.

Understanding the difference between these two approaches is crucial. Naïve optimism might say, "Don't worry, technology will solve climate change soon," ignoring that technology alone cannot fix systemic injustices or transform consumerist cultures. Grounded hope, by contrast, might say, "Technological innovations can help, but we must also reform policies, shift mindsets, and distribute resources fairly. Let's celebrate that renewable energy is becoming more affordable and that communities worldwide are learning from each other, even as we push for deeper structural changes." Where naïve optimism relies on platitudes, grounded hope thrives on honest appraisals combined with moral resolve. It encourages continuous learning and adaptation rather than expecting a single miraculous breakthrough.

The roots of naïve optimism often lie in a discomfort with moral and emotional complexity. Faced with disturbing projections about ecosystem collapse or climate refugees, people may latch onto comforting narratives that downplay the need for tough decisions. For example, a naïve optimist might cling to the notion that some new geoengineering scheme will fix the atmosphere soon, requiring no significant changes to lifestyles, economic structures, or political priorities. This stance protects them from grappling with painful truths—such as the necessity of reducing consumption, redefining prosperity, or confronting vested interests. Rather than engaging with moral dilemmas, they outsource salvation to speculative future miracles.

Grounded hope, however, prompts a different response to complexity. Instead of fleeing discomfort, it calls for leaning in. If climate adaptation involves challenging ethical questions—who pays for damages, how to support vulnerable communities—grounded hope does not dodge these issues. Instead, it says: "Yes, this is hard. We must debate trade-offs openly, listen to those most affected, and strive for equitable solutions. Each partial step toward fairness or lower emissions is meaningful. Although no single measure suffices, cumulatively these efforts can avert the worst outcomes." Such hope rests on moral maturity—accepting that righteous ends demand sustained moral reasoning and compromise. It also refrains from guaranteeing neat, storybook endings. Instead, it finds solace in the ongoing moral journey itself, the never-ending practice of striving toward better arrangements.

Another hallmark distinguishing grounded hope from naïve optimism is the attitude toward evidence and uncertainty. Naïve optimism often cherry-picks data to support easy assurances. If one promising innovation emerges, it's hailed as definitive proof that worry is pointless. If a single country reduces emissions slightly, naïve optimism might exaggerate its significance to claim the tide has definitively turned. Yet it ignores contrary indicators, fails to question whether that improvement is robust, or glosses over systemic inertia. Grounded hope, conversely, respects facts, complexity, and the provisional nature of knowledge. It recognizes that positive trends must be contextualized: Are they scalable globally? Are they resilient under changing conditions? Grounded hope does not deny good news; it celebrates it while keeping an eye on what still needs to be done, ensuring that hope remains tethered to reality rather than floating on wishful illusions.

At times, people gravitate toward naïve optimism as a coping mechanism. The psychological burden of eco-anxiety, fear, and guilt can be immense. Embracing a narrative that proclaims "everything will be fine" provides temporary emotional relief. However, such comfort is fragile. When confronted with grim reports of deforestation or heatwaves surpassing past records, naïve optimism shatters, leaving individuals more disillusioned. They might swing from blind hope to despair and back again, trapped in an emotional rollercoaster. In contrast, grounded hope provides a more stable emotional ground. Because it never promised simplicity or instant triumph, it can absorb bad news without collapsing. It interprets setbacks as signals to adjust, learn, and persist rather than as final defeats. Thus, grounded hope better equips people to handle the emotional ebb and flow of moral struggles, offering steadiness instead of brittle reassurance.

The cultural implications of these distinctions are profound. If public discourse settles for naïve optimism—believing that minor tweaks or passive reliance on technology will somehow guarantee environmental salvation—societies delay essential reforms and fail to foster the moral depth needed to navigate crises. Political leaders might exploit naïve optimism to avoid making unpopular but necessary decisions. Businesses might greenwash minimal efforts, claiming moral high ground. Citizens lulled by complacent narratives may not support ambitious climate policies or recognize the need for systemic transformations. Over time, reality intrudes—extreme weather intensifies, resources dwindle—and the lack of genuine preparedness leads to bigger shocks and deeper disillusionment.

On the other hand, if grounded hope permeates cultural narratives, it encourages honest dialogue about what true sustainability demands. Yes, we acknowledge that no single action or policy is a panacea. Yes, changes are incremental and fragile. But we frame these realities not as reasons to give up, but as invitations to work harder, cooperate more widely, and celebrate each breakthrough as a foothold on a steep moral climb. Grounded hope encourages moral pluralism—appreciating diverse strategies and partial measures—and nurtures moral learning. Instead of clinging to any one fantasy cure-all, communities embrace multi-pronged efforts: reducing emissions where possible, restoring habitats step by step, innovating in clean technologies, refining policies, and building cross-sector alliances. This iterative approach, supported by grounded hope, proves more flexible, adaptive, and ultimately effective.

Educational institutions play a key role in teaching the difference between grounded hope and naïve optimism. Many young people approach environmental issues already sensing the gravity of challenges. If educators oversell simple fixes or gloss over complexities, students smell inauthenticity and lose trust. Alternatively, if educators present grim facts without acknowledging partial successes or potential for adaptive solutions, students might succumb to despair. Grounded hope provides a balanced educational ethos. Teachers can say: "These problems are big and real, but people are working on solutions. We have partial successes here and there—let's examine them critically. We can learn from what works and expand it, while also recognizing what remains undone." Such honesty empowers students. They leave the classroom neither lulled into complacency nor crushed by fatalism, but ready to engage morally and intelligently.

Moral and spiritual traditions also intersect with these distinctions. Many religious or philosophical frameworks caution against despair while warning against complacency. They recognize the virtue of patience, perseverance, humility, and compassion, none of which align with naïve optimism's shortcuts. Instead, they resonate with a hope grounded in reality and moral accountability. This synergy can inspire faith communities and moral philosophers to emphasize realistic hope in their teachings, guiding people to face environmental imperatives without panic or denial. Such moral counsel encourages believers and seekers to embrace moral complexity and incremental improvements, mirroring the adaptive strategies found in grounded hope.

In activism and advocacy, embracing grounded hope can redirect energy more productively. Activists sometimes struggle with burnout, partly caused by the constant onslaught of bad news and the frustration of slow progress. If activists rely on naïve optimism—believing a single campaign or march will solve everything—they risk heartbreak when the outcome falls short. Grounded hope allows activists to integrate small wins into their narrative. They celebrate incremental policy changes, community initiatives that took root, shifts in public discourse, and emergent alliances. Instead of seeing these partial results as insufficient, they view them as moral resources that replenish motivation. Over time, this mindset reduces burnout by calibrating expectations and maintaining a steady, reliable source of moral energy.

Likewise, policy and decision-makers benefit from distinguishing grounded hope from naïve optimism. If a politician embraces naïve optimism, they might promise constituents unrealistic transformations with minimal effort, undermining public trust when results fail to materialize. On the other hand, a policymaker guided by grounded hope acknowledges complexity, sets incremental targets, and communicates them honestly. They track progress, report challenges openly, and highlight that while no single measure solves the crisis, each step lays groundwork for deeper reforms. Over time, this transparency fosters public trust, stable support, and more robust political mandates to pursue ambitious but realistic environmental agendas.

The media, too, can influence whether public perceptions veer toward naïve optimism or grounded hope. Journalists who highlight only doomsday scenarios feed despair, while those who report good news without context risk naive optimism. Balanced reporting—showing both the seriousness of issues and the incremental improvements achieved—cultivates grounded hope. Instead of simplistic narratives, media can present stories of communities reducing emissions gradually, inventors refining sustainable technologies after multiple iterations, or governments incrementally strengthening environmental regulations. This nuanced coverage helps audiences understand that progress is not a sprint to a magical solution, but a marathon requiring stamina, learning, and collective moral commitment.

In personal life, adopting grounded hope involves how we talk to ourselves and each other about environmental threats. It means acknowledging fears openly and responding, "Yes, this is scary and complicated. Still, people are responding. We see renewable energy costs falling, more protected areas established, and dialogues about climate justice gaining traction. These don't fix everything, but they show we can improve." Such self-talk prevents slipping into cynicism. A friend might share a small success—like convincing a local café to cut plastic straws—and rather than scoff, we can say, "That's excellent, it's a start. If every café did this, imagine the difference, and maybe we can push for more changes soon." This kind of supportive acknowledgment builds moral solidarity. Anxiety eases because each person sees that others appreciate incremental wins and remain committed to progressing further.

Crucially, grounded hope never implies satisfaction with the status quo. It's not about praising tiny steps to shield oneself from moral responsibility. Instead, it sets a constructive, continuous growth mindset. If we celebrate that a city reduced its waste by a small percentage, we simultaneously reaffirm that there's more work ahead—further policies to adopt, more citizens to engage. Grounded hope says: "Look how far we've come, and let's keep going." It treats hope as an active practice—a discipline that requires moral vigilance, strategic thinking, and empathy—rather than as a passive fantasy.

Over time, this practice transforms how people conceptualize moral action. Instead of waiting passively for a hero solution, communities realize they are already part of the solution, each incremental measure contributing to a tapestry of moral innovation. This perspective diminishes the temptation to fall for silver-bullet narratives that promise effortless salvation or to dismiss all efforts as futile drops in the ocean. Grounded hope teaches that the ocean of environmental solutions is indeed formed by countless drops, each representing a moral decision, a technology refined, a policy strengthened, a restored habitat acre by acre.

Distinguishing grounded hope from naïve optimism also helps navigate moral dilemmas and tough choices. If we admit that no solution is perfect, no policy flawless, we accept that moral compromise and incremental progress are often necessary. This acceptance does not mean moral relativism—some measures are clearly better than others—but rather recognition that moral purity is rare in complex scenarios. Grounded hope thrives in this gray area, understanding that making things less bad today paves the way to make them even better tomorrow. Naïve optimism, by refusing complexity, often leads to disillusionment when reality intrudes with trade-offs. Grounded hope equips us to face trade-offs with courage and integrity.

Another advantage is that grounded hope nurtures intellectual honesty. While naïve optimism encourages selective attention to facts, grounded hope thrives on full-spectrum awareness. You can celebrate a new wind farm's contribution to emissions reductions while acknowledging the need to address raw material extraction or bird migration impacts. Embracing this complexity clarifies that hope is not about ignoring problems but about recognizing the capacity to mitigate them through layered actions. This honesty fosters credibility, strengthening moral authority. People trust voices that refuse to sugarcoat reality yet refuse despair, striking the delicate balance that characterizes grounded hope.

Cultural shifts often start with moral language. As phrases like "grounded hope" enter public discourse, communities gain a conceptual tool to navigate their emotions and expectations. Educational materials, activist handbooks, and policy briefings could explicitly differentiate grounded hope from naïve optimism, guiding readers toward more stable emotional footing. Moral leaders—clergy, philosophers, educators—can incorporate these distinctions into their teachings. Public figures who champion grounded hope set a tone that encourages reflection rather than delusion, perseverance rather than fantasy. Over time, this conceptual clarity shapes a more mature, resilient civic culture.

In practice, grounded hope is sustained by continuous dialogue. When activists debate strategies, reminding each other that incremental wins count prevents bitter arguments about "not doing enough." When policymakers face backlash for not achieving bold targets instantly, they can explain the stepped approach, referencing grounded hope as the ethical stance behind their methodology. When citizens talk to friends and family about environmental news, they can frame narratives that simultaneously acknowledge partial improvements and demand future action, weaving moral complexity into everyday conversations.

Eventually, as environmental conditions evolve—some improving due to collective efforts, others still demanding urgent response—grounded hope proves adaptable. If a technology initially seen as limited grows more efficient, grounded hope celebrates the journey from doubt to reliability. If a policy fails or backfires, grounded hope analyzes the lessons learned, salvaging moral insight from disappointment. This adaptability ensures that hope is never static or dogmatic; it's an evolving moral practice reflecting the changing landscape of environmental possibilities.

Overall, the distinction between grounded hope and naïve optimism stands as a vital moral guidepost. Without this distinction, communities risk oscillating between hollow reassurance and paralyzing despair. With it, they find a stable platform from which to face moral and ecological challenges honestly, creatively, and persistently. Grounded hope does not promise easy paths or guaranteed outcomes. Instead, it offers the moral compass we need to navigate complex moral terrains, to recognize partial successes as real gains, and to keep forging ahead with courage and compassion no matter how difficult the journey.

Finding Inspiration in Stories of Resilience, from Species Recoveries to Community Adaptations

As people grapple with the moral complexity of ecological crises and the need for grounded hope, many seek evidence that perseverance and wise choices can indeed steer environmental outcomes toward recovery. Empirical data and incremental policy shifts matter, but they often feel impersonal. Tangible hope often emerges most powerfully from stories—accounts of species rebounding from the brink of extinction, of degraded landscapes gradually restored, of communities adapting to new climatic conditions with ingenuity and solidarity. These narratives of resilience remind us that the natural world, when given the chance, can regenerate in surprising ways, and that human societies, when guided by empathy and vision, can respond creatively and adaptively to seemingly insurmountable challenges.

Yet these stories are not fairy tales. They never deny the struggle, complexity, or uncertainty inherent in environmental healing. On the contrary, what makes them inspiring is their honesty about the hurdles faced: how conservationists fought political apathy or economic pressure to protect a critical habitat, how scientists spent decades refining breeding programs to restore a threatened species, how communities invested time and learning into restructuring their agricultural practices, or how cultural values evolved to embrace stewardship in the face of disruptive changes. Stories of resilience chart moral, ecological, and social transformations through periods of doubt and incremental progress, culminating in partial but significant successes. Far from naïve optimism, these accounts reflect grounded hope in action, providing tangible reference points for what moral effort can achieve over time.

Consider the recovery of certain species once declared critically endangered. The story of the bald eagle in North America, for example, shows how banning harmful pesticides like DDT and protecting nesting sites enabled a gradual population rebound. At the outset, few dared to hope these iconic birds would flourish again, and skepticism abounded. Yet decades of vigilance, scientific monitoring, improved habitat conditions, and cultural reverence for the species allowed their numbers to climb. Today, eagle sightings are common in many regions where they'd once vanished. This narrative of patient work—adjusting policies, changing agricultural practices, enforcing protections—conveys that moral and environmental resilience stems from consistent moral commitments rather than instant cures. Witnessing how a species reclaims its ecological niche despite past harm reassures that human actions can indeed repair damage, given enough care and patience.

Similar stories unfold worldwide. The return of the Iberian lynx in Spain and Portugal, once considered nearly doomed, involved habitat restoration, prey management, reducing roadkill risks, captive breeding programs, and painstaking reintroduction efforts. None of these steps alone sufficed; each incremental improvement contributed to stabilizing and then expanding lynx populations. Observing these orchestrated efforts highlights that resilience emerges from synergistic approaches—biologists, policymakers, local communities, farmers, and NGOs all playing roles. Anxiety that humanity's clumsy attempts always fail yields to evidence that complexity can be managed. If the intricate puzzle of saving the Iberian lynx could be solved piece by piece, perhaps other ecological riddles also have solutions waiting to be discovered.

In marine environments, certain coral reefs, once bleached and barren, have shown signs of gradual regrowth when protected from overfishing, pollution, and direct physical damage. While coral ecosystems face profound threats from warming oceans, localized successes—such as carefully managed marine reserves—demonstrate that reducing local stressors can enhance reefs' capacity to recover. This doesn't solve the global climate crisis instantly, but it affirms that local moral interventions—establishing no-take zones, improving wastewater treatment, promoting sustainable tourism—yield tangible improvements. Such stories challenge the notion that complex problems are beyond remedy. Instead, they illustrate how moral will and strategic interventions can strengthen natural resilience. Anxiety about inevitable coral demise gives way to hope that, while far from assured, coral ecosystems need not be passively mourned; they can be actively supported through informed stewardship.

Beyond species and habitats, human communities adapting to environmental changes also offer powerful narratives of resilience. Consider coastal towns facing sea-level rise. Initially, many approached the threat with denial, hoping shorelines would remain stable. But as storms intensified and erosion advanced, some communities embraced adaptation measures. They built natural buffers like restored wetlands or living shorelines instead of relying solely on concrete seawalls. They shifted urban planning to retreat from the most vulnerable areas, diversified their local economies, and engaged in public dialogue about fair relocation or rebuilding strategies. Over time, incremental successes emerged: fewer homes lost in each successive storm, healthier coastal ecosystems that buffer storm surges, and stronger communal bonds forged through collective decision-making. These stories highlight that even in the face of relentless climatic shifts, human societies are not helpless. They can adjust, learn, and distribute burdens more equitably. Anxiety that climate impacts will tear communities apart diminishes as evidence grows that, while painful, adaptation can foster solidarity and innovation.

In regions where agricultural practices faced drought and soil degradation, stories of regenerative farming provide another source of inspiration. Farmers transitioning from monocultures reliant on heavy chemical inputs to diversified agro-ecological methods often undergo a learning curve. Early yields might falter, critics might scoff, and initial costs might strain budgets. Yet persistent experimentation—using cover crops, compost, rotational grazing, and water harvesting techniques—gradually improves soil health. Over years, biodiversity returns, soil fertility stabilizes, and yields become more resilient against erratic weather. Farmers share their experiences through case studies, workshops, and farmer-to-farmer exchanges, spreading the ethos that healing land and securing food supplies doesn't demand unattainable miracles, but consistent moral effort and knowledge exchange. Anxiety about feeding populations under changing climates transforms into a more hopeful outlook, grounded in real examples of farmland regeneration.

In these stories of resilience, a common pattern emerges: moral courage meeting persistence, guided by evidence and adaptation, leading to incremental progress that, over time, transforms once-dire situations. Such narratives reinforce that hope is not passive or naive. It's not about sitting back and expecting nature or technology to fix everything automatically. Instead, hope here is an active moral stance. It compels conservationists, farmers, local leaders, and citizens to confront difficulties head-on, to try different approaches, to fail occasionally but learn, to refine and persist. Each story is a moral lesson illustrating that while the scale of environmental problems is enormous, humans can improve conditions through steady, engaged intervention.

Cultural contexts enrich these narratives. In indigenous communities, traditions and ecological knowledge help restore degraded landscapes. Their stories of resilience integrate spiritual and moral dimensions—caring for ancestors' lands, respecting non-human kin, and living in harmony with natural cycles. When indigenous-led reforestation projects regenerate forests or restore watersheds, they demonstrate that ethical principles, cultural continuity, and environmental stewardship can reinforce each other. Anxiety rooted in cultural loss and environmental harm recedes as these communities reclaim autonomy and ecological integrity simultaneously. Their successes encourage others to respect indigenous wisdom and collaborate with these communities as equal partners in moral action.

Similarly, stories of community adaptations to environmental stress often highlight social justice. When neighborhoods vulnerable to heatwaves install green roofs and community cooling centers, ensuring that the poorest residents also benefit, the result is not just ecological improvement but moral progress in equity. Seeing that resilience narratives can address not only environmental stability but also moral fairness reassures us that environmental problems need not exacerbate inequality. Instead, they can prompt more inclusive solutions. Anxiety about the future often stems from the fear that only privileged groups will escape harm while marginalized communities suffer. But adaptation stories that center justice and participation show that different moral outcomes are possible when values guide implementation. This recognition amplifies hope: not only can we cope with environmental threats, but we can also become more ethical societies in the process.

Even at larger scales—such as multinational efforts to protect migratory bird routes or transboundary river management—stories highlight incremental diplomatic accords, data-sharing initiatives, and joint monitoring programs that gradually improve conditions. While no single treaty may solve all ecological threats in a region, these ongoing collaborations illustrate how patient diplomacy, moral trust-building, and reciprocal support can emerge amid tension and competition. Anxiety that international relations doom environmental solutions softens as evidence accumulates that states can cooperate incrementally, building frameworks and norms that, over decades, bolster environmental governance and resource equity.

For these stories of resilience to truly inspire, they must be communicated thoughtfully. Educators, journalists, artists, and community leaders can showcase them in ways that balance honesty and hope. Emphasizing the complexity and timeframes involved prevents slipping into naïve optimism; the audience understands these wins took sustained effort and faced setbacks. Including personal testimonies—from a farmer who once doubted new methods but now swears by them, or a wildlife biologist who spent years perfecting a breeding protocol—humanizes the narrative. Such authenticity resonates emotionally, letting listeners or readers see themselves as potential agents in similar stories. Anxiety transforms into motivation, as people realize that if others overcame obstacles, they might too.

Moreover, presenting multiple narratives together reveals patterns. While each case differs—some tackle marine life, others soil health, still others renewable energy—common threads emerge: moral commitment, iterative learning, community engagement, respect for local knowledge, and strategic use of scientific input. Understanding these shared elements helps individuals and communities replicate success. Anxiety about reinventing the wheel or facing a unique, unsolvable problem eases as they recognize underlying moral and practical principles that travel well across contexts. If one can adapt these principles, even if conditions differ, similar positive trajectories become attainable.

These stories of resilience also challenge the misconception that all environmental news is grim. This is not to deny the gravity of ongoing crises—rising emissions, accelerating extinction rates, widespread pollution remain dire. But acknowledging resilience stories ensures that complexity does not devolve into a single storyline of relentless doom. Complex reality includes pockets of improvement, moral breakthroughs, and renewed ecosystems. Balancing these narratives encourages a more accurate assessment of the world. Such complexity, ironically, can calm anxiety by refuting the fatalistic assumption that destruction is the only trend. Knowing some landscapes recover, some species rebound, and some policies improve clarifies that moral agency and ecological responsiveness persist amid adversity.

Additionally, these stories can bridge generational gaps. Older generations, who recall certain species declines or pollution horrors, can appreciate how younger conservationists and technologists reverse or mitigate some of those losses. Younger people, alarmed by current projections, gain perspective by seeing historical successes that their elders witnessed or helped achieve. This intergenerational dialogue reduces anxiety by showing moral progress across time. The old can share how incremental steps worked in the past; the young bring fresh energy, urging bolder reforms inspired by proven strategies. Together, they weave a moral continuum that neither idolizes the past nor despairs about the future, recognizing that resilience is cumulative and collaborative.

When communities incorporate these stories into their moral rituals and educational curricula, they anchor hope as a practiced virtue. Just as athletes study successful techniques to refine their skills, environmental stewards study resilience stories to refine moral action. Discussing a successful species recovery after dinner, reflecting on a community adaptation case study at a school assembly, or screening a documentary about a restored forest at a neighborhood event makes these narratives part of collective memory. This communal engagement transforms anxiety into a shared problem that the community knows can be addressed, not alone through fantasy or denial, but by learning from proven instances of moral and ecological resilience.

Some worry that focusing on success stories might downplay the continuing severity of crises. But grounded hope can handle nuance. Celebrating resilience narratives doesn't imply ignoring harsh realities; on the contrary, these stories gain their power from being exceptions to dire trends. They prove that different outcomes are possible if humans commit to better choices. By contrasting these uplifting examples with ongoing struggles, communities gain a clearer moral picture: progress is hard-fought but real, partial but meaningful, contingent on moral effort rather than guaranteed. Anxiety remains a motivator, prodding people to apply lessons from resilience stories more broadly and intensively.

This thoughtful engagement with stories of resilience can also influence policymaking and investment. Decision-makers who might be risk-averse or hesitant to support untested approaches gain confidence from documented cases of success. If they see that certain policies revived a local fishery or improved urban air quality elsewhere, they become more open to implementing similar measures at home. Investors funding green technologies or conservation projects find reassurance in precedents showing that patience, innovation, and collaboration can yield returns—economic, social, and moral. Anxiety about wasting resources on uncertain solutions diminishes when historical examples confirm that careful stewardship pays dividends over time.

In activism, resilience stories guard against burnout. Campaigners who repeatedly encounter setbacks can recall how others overcame even greater obstacles. This memory reduces the temptation to quit, turning anxiety into a signal that more strategic adjustments might be needed rather than a reason to give up. Seasoned activists might share these accounts with newcomers to inoculate them against despair, providing emotional training to withstand disappointment. The moral support drawn from these narratives transforms isolated struggles into chapters of a collective epic, where moral perseverance outlasts each setback.

At the global scale, storytelling about resilience encourages cross-cultural empathy. People in one region can appreciate that others, facing different environmental contexts, managed to restore mangroves, reintroduce wolves, or stabilize a water table. This empathy broadens moral horizons: not every solution applies universally, but learning that others overcame cultural, political, or technical hurdles suggests that we, too, can adapt solutions to our local realities. Anxiety that our local challenges are uniquely doomed fades as we see the universal potential of moral ingenuity. We understand that resilience transcends language, geography, and political systems. The human capacity for moral learning and ecological restoration is widespread and culturally adaptive.

As conditions evolve, new stories of resilience will emerge. Perhaps breakthroughs in carbon sequestration, large-scale reforestation, or sustainable materials lead to narrative arcs we haven't yet imagined. Our grandchildren may celebrate how communities curbed microplastic pollution in oceans through coordinated cleanup technology and responsible consumption shifts. By anticipating future narratives, we prepare ourselves to look for and amplify positive developments, encouraging each generation to keep updating the lexicon of environmental hope. In doing so, we establish resilience stories as living moral resources—constantly renewed, shared, and contextualized against fresh challenges.

Ultimately, these resilience narratives teach that hope, like moral effort, is an ongoing practice. Just as grounded hope rejects naive optimism, it also recognizes that hope itself must be fed—by analyzing outcomes, refining strategies, and inspiring each other through success stories. Instead of assuming hope is innate or constant, communities actively cultivate it by seeking out, sharing, and learning from examples where moral intention meets ecological response. Anxiety still occurs, but it now stands side by side with testimony that good things can happen. This duality creates a healthy moral tension, balancing concern with confirmation that not all battles are lost, that moral agency wields real power, and that human endeavors can indeed foster environmental resilience.

Embracing Uncertainty as a Space of Possibility

As individuals and societies navigate the moral complexities of environmental change, one persistent source of anxiety lies in the uncertain nature of future outcomes. Climate models project wide ranges of temperature increases, some species may adapt while others decline unpredictably, and new policy initiatives may either spark larger transformations or falter quietly. Faced with this spectrum of possibilities, it's tempting to seek clear-cut assurances—guarantees that certain actions will yield definitive results. Yet the environmental sphere inherently resists absolute certainty. Instead of seeing uncertainty as an intimidating void or proof of moral futility, embracing it as a space of possibility can reframe anxiety and invite creative, adaptive moral engagement. When understood as fertile ground for innovation, dialogue, and experimentation, uncertainty encourages humility, curiosity, and a willingness to explore diverse paths toward sustainability.

Uncertainty arises from multiple sources: scientific complexity, evolving social values, variable political landscapes, and changing ecological baselines. Attempting to control or eliminate all uncertainty can lead to frustration and rigidity. A community that demands to know exactly how fast an ecosystem will recover or precisely how effective a new technology will be in cutting emissions sets itself up for disappointment. Instead, by recognizing that uncertainty reflects our evolving understanding and the complexity of natural and social systems, moral agents learn to be agile. They accept that moral life does not unfold on a predetermined script; rather, it involves continuous learning and revision. This mindset transforms anxiety from a paralyzing force into a motivator prompting openness to new evidence, alternative strategies, and flexible thinking.

In moral terms, embracing uncertainty as a space of possibility fosters virtues like patience, courage, and intellectual humility. Patience arises because no immediate, definitive answers exist, encouraging steady effort rather than instant gratification. Courage emerges from acting ethically even without complete assurances—taking leaps of faith informed by best knowledge available, but without waiting for perfect certainty. Intellectual humility acknowledges that we may never fully grasp all ecological interactions or anticipate all policy outcomes. Instead of viewing this partial knowledge as weakness, humility treats it as a prompt to remain vigilant, continue gathering information, and welcome differing perspectives that might shed new light.

Embracing uncertainty also breaks the monotony of all-or-nothing thinking. If people believed only absolute solutions counted, they might discard partial successes or emerging ideas too soon. By welcoming uncertainty, communities appreciate that each attempt—whether a pilot project for urban agriculture, a policy tweak, or a technological prototype—offers valuable insights. Even if results differ from initial predictions, these outcomes expand the knowledge pool. Failures or unexpected results need not be moral defeats; they become lessons indicating which adjustments could improve future endeavors. Freed from the expectation that every action must yield guaranteed outcomes, moral actors can treat environmental challenges as ongoing inquiries, each step refining moral strategies over time.

Uncertainty also fosters collaborations that transcend disciplinary boundaries and cultural differences. If the future were perfectly known, there might be less incentive for dialogue, as everyone would agree on a single path. But uncertainty ensures no single actor or perspective holds a monopoly on truth. Scientists, local communities, indigenous knowledge holders, policymakers, entrepreneurs—all bring partial insights to the table. Accepting uncertainty motivates seeking guidance from diverse sources. For example, scientists might quantify probabilities around species adaptation under warming scenarios, but indigenous communities offer long-term ecological wisdom and historical adaptation strategies. Policymakers bring pragmatic understanding of political feasibility, while entrepreneurs spot niches for new solutions. Each lens clarifies a piece of the puzzle. As a result, uncertainty encourages moral pluralism: recognizing that complex problems benefit from multiple vantage points rather than a single absolute narrative.

This pluralism challenges anxiety-laden perceptions that complexity equates to chaos or impotence. Instead, complexity and uncertainty become allies in discovering novel pathways. Suppose a community wants to protect a coastal ecosystem vulnerable to sea-level rise. No one can guarantee that a chosen adaptation measure—like restoring mangroves or erecting a living shoreline—will fully prevent flooding decades from now. Yet uncertainty invites experimentation: implement living shorelines on a small scale, monitor results, involve local fishers and ecologists, and remain ready to pivot if evidence suggests another approach would be more effective. In this process, uncertainty is less a barrier than a condition that makes careful observation, iterative design, and moral deliberation indispensable. The anxiety stemming from the unknown is managed through adaptive action, moral imagination, and continuous feedback loops.

Moreover, embracing uncertainty can reduce polarization. When people cling to dogmatic certainty—whether in denying environmental problems or insisting on a single perfect solution—dialogue shuts down. Uncertainty, openly acknowledged, defuses rigid stances. It frames environmental decision-making as a shared learning journey. Opponents no longer have to fear conceding ground by admitting partial ignorance or acknowledging valid concerns from the other side. Instead, both sides can meet in the space of uncertainty, acknowledging that no one holds all answers and that each perspective might contribute clues. This approach transforms anxiety-driven stalemates into moral negotiations, where the absence of guaranteed outcomes encourages listening, respect, and compromise.

Embracing uncertainty also reshapes how communities approach time. Anxiety often fixates on unknown futures, imagining worst-case scenarios without considering that our actions today can influence which scenario becomes reality. Accepting uncertainty means understanding that the future is not a predetermined endpoint but a range of possibilities we help sculpt through present choices. Instead of paralyzing us, uncertainty should galvanize effort: if multiple outcomes are possible, then moral agency has real leverage. By reducing emissions, restoring habitats, or developing resilient infrastructure, we tilt probabilities toward better outcomes. In this sense, uncertainty bestows moral responsibility—knowing that the future's shape partly depends on what we do now. Anxiety about not knowing exactly what will happen evolves into a sense of moral empowerment to create conditions favorable to more hopeful trajectories.

Uncertainty as a space of possibility also complements the practice of hope as not just a feeling but a moral habit. Hope, as discussed, involves persistent moral action despite unknown results. Embracing uncertainty aligns perfectly with this stance. It acknowledges that we can strive for good outcomes, guided by values and informed by partial knowledge, without demanding guarantees. Just as hope encourages resilience under uncertainty, viewing uncertainty as an open field of potential outcomes affirms that hope and moral endeavor matter precisely because the future isn't fixed. If outcomes were predetermined—either wholly doomed or miraculously saved—moral choices would lose significance. Uncertainty, conversely, makes moral agency meaningful. Each choice and action, rooted in ethical reflection, alters probabilities and opens new avenues.

Looking to historical precedents helps contextualize this approach. Consider major societal shifts: the end of slavery, expansion of human rights, technological revolutions in sanitation and medicine. None were accomplished in a climate of perfect foresight. Reformers and innovators ventured into unknowns, unsure if their efforts would spark lasting change. Their moral courage found footing in the possibility that things could be different. They leveraged uncertainty—no one could prove their cause would fail or succeed definitively—and pushed forward, learning from incremental wins, forging alliances, adjusting tactics. We celebrate these achievements now, recognizing that moral progress often required navigating uncharted territory. Today's environmental struggles follow a similar pattern, demanding that we accept uncertainty as a natural milieu for moral creation.

In ecological contexts, uncertainty about how species and ecosystems will respond to interventions compels careful monitoring and adaptive management. Instead of expecting certain results, conservationists set goals, implement measures, and then observe nature's response. If a restoration plan doesn't yield the expected outcome—maybe a certain plant species doesn't re-establish as anticipated—the team revisits their strategy. This adaptive approach, made necessary by uncertainty, leads to more refined, context-specific solutions over time. Anxiety about unpredictable ecological responses is mitigated by a moral methodology: start with best knowledge, act, observe, learn, adjust. In this cycle, uncertainty fosters a culture of moral attentiveness and humility rather than resignation.

Technological innovation thrives in uncertainty as well. Inventors and engineers rarely know if their new design will revolutionize an industry or fail quietly. They test prototypes, analyze performance data, and iterate. Environmental technology—from solar panels to electric buses or water filtration systems—improves because developers accept they must learn from trial and error. Anxiety about unknown scalability or cost-effectiveness pushes innovators to refine their inventions. Rather than giving up because they lack crystal-clear predictions, they embrace the uncertain development process. Ultimately, this yields more robust and accessible technologies. Moral engagement means supporting research, pilot programs, and demonstrations, trusting that even if not all attempts succeed, some will prosper, forging paths to cleaner energy or more resilient food systems.

Embracing uncertainty as possibility also encourages moral imagination. Without fixed outcomes, there's room to imagine multiple futures and choose which one we try to realize. This imagination goes beyond wishful thinking; it involves scenario planning, envisioning different trajectories, and crafting strategies to influence which trajectory unfolds. Anxiety about unknown futures inspires creativity. If coastal communities cannot know exactly how fast seas will rise, they can design flexible adaptation measures—floating infrastructure, modular breakwaters, flexible urban plans—and remain open to adjusting as data accumulates. Instead of rigid solutions vulnerable to being proven wrong, they adopt flexible solutions that can be scaled up or down. Thus, uncertainty trains moral actors to think in terms of resilience rather than stability, complexity rather than simplicity, and continuous moral negotiation rather than definitive end states.

Culturally, embracing uncertainty might conflict with societies accustomed to seeking security and control. Indeed, many social institutions and economic models are built on assumptions of predictability. Challenging these assumptions morally reorients priorities. Instead of fixating on growth or consumption patterns premised on stable environmental conditions, communities accept variability. They develop moral codes that value foresight, adaptability, risk distribution, and solidarity. Anxiety over not knowing if agriculture will face drought next season transforms into moral preparedness—diversifying crops, conserving water, building social safety nets. Embracing uncertainty, therefore, can shift cultural norms towards cherishing foresight, precautionary principles, and moral investments in resilience rather than short-term certainty.

In education, teaching younger generations that uncertainty is inherent in environmental stewardship fosters moral fortitude. Students learn that no one can promise them a stable climate or abundant natural resources forever, but that does not justify apathy or fatalism. Instead, they understand that their decisions, innovations, and alliances contribute to shaping unknown futures. Anxiety that might overwhelm them turns into motivation to learn critical thinking, collaborate with diverse peers, and cultivate ethical sensitivity. Teachers can present case studies where communities confronted unknown ecological responses and still achieved partial improvements—embedding a lesson that action under uncertainty is not foolish but necessary and often rewarding. This educational approach produces morally agile citizens prepared to navigate complexity with a steady moral compass.

In terms of communication strategies, embracing uncertainty requires honest messaging. Environmental advocates must resist the temptation to claim absolute certainties. Over-promising outcomes sets the stage for disillusionment if results differ. Instead, acknowledging uncertainty underscores transparency and builds trust. When leaders say, "We don't know exactly how quickly this policy will reduce emissions, but we estimate within a range, and we have plans to adjust," audiences appreciate the honesty. This forthright approach resonates morally, as it treats citizens as capable moral agents who can handle complexity. Anxiety over not being given definitive assurances fades as people recognize that moral engagement involves shared responsibility, ongoing evaluation, and democratic participation in decision-making under uncertain conditions.

Moreover, uncertainty clarifies that moral responsibility does not depend solely on guaranteed success. Acting ethically remains meaningful even if final outcomes remain unclear. Moral effort becomes an expression of values rather than a transactional bet on assured returns. Anxiety that moral actions are futile if not assured of success recedes as communities internalize that uncertainty makes moral agency profound. Engaging ethically amid unknowns demonstrates loyalty to principles, compassion for future generations, and reverence for life itself, not merely adherence to a cost-benefit calculation. Thus, moral identity strengthens when people accept uncertainty as the context in which their commitments unfold.

In activism, this approach liberates campaigns from rigid narratives. Instead of demanding that a single initiative solve everything, activists present a menu of plausible pathways. If one approach falters, another can be tried. Anxiety about wasted effort diminishes as groups openly acknowledge trial and error as part of moral work. Instead of being shamed for not achieving definitive results fast enough, they celebrate the courage to attempt solutions, learn from partial outcomes, and pivot. Over time, this dynamic approach can energize environmental movements, attracting supporters who appreciate sincerity and flexibility over grandiose promises. Anxiety transforms into solidarity that respects uncertainties while forging ahead.

On the policymaking front, uncertainty as possibility encourages flexible governance frameworks. Policies can be designed with built-in review periods, triggers for stronger measures if initial targets aren't met, and sunset clauses that require reauthorization based on performance data. Such adaptive governance reflects moral wisdom: acknowledging we cannot predict all variables, we structure decision-making to evolve with new insights. Anxiety about making a "wrong" policy choice lessens when policies are not rigid bets but evolving moral compacts that can be recalibrated as evidence accumulates. In this scenario, uncertainty supports moral accountability, as decision-makers must continuously justify adjustments rather than hiding behind static rules.

In ecological restoration and wildlife management, uncertainty-based strategies lead to more robust outcomes. Instead of relying on a single management plan presumed infallible, practitioners experiment with various interventions at small scales, track indicators, and scale up what works best. Anxiety about failing an ecosystem by committing to a flawed method diminishes because no single irreversible gamble is made. Instead, multiple smaller efforts, carefully monitored, produce a portfolio of experiments. Some may fail, but others succeed, cumulatively guiding moral practice toward sustainable paths. Adopting this approach acknowledges that uncertainty prompts moral humility, making ecological restoration a learning partnership with nature rather than a one-sided imposition of human will.

Embracing uncertainty also affects media narratives. Instead of pushing for definitive predictions that often feed anxiety (either over disaster or tech-utopian cures), media can highlight stories of ongoing moral experimentation. Journalists can portray environmental action as a series of reasoned bets under uncertainty, showing both the stakes and the evolving understanding that emerges from trials. Such narratives encourage public audiences to see complexity not as a failure of communication but as an honest depiction of the real moral terrain. Anxiety does not vanish, but media can portray it as a natural emotion that inspires vigilance and moral reflection, rather than a sign of hopelessness.

At a spiritual or philosophical level, embracing uncertainty resonates with traditions that view life's unpredictability as inherent and instructive. Many spiritual teachings, across cultures, emphasize humility before the unknown. They counsel courage in pursuing good deeds even when outcomes remain hidden. Integrating environmental care into these traditions enriches moral motivations. Anxiety about uncertain futures can then find solace in spiritual frameworks that celebrate moral effort for its intrinsic value, not just for guaranteed results. This alignment of ecological moral practice with spiritual or philosophical wisdom about uncertainty can strengthen people's resilience and sense of meaning.

Ultimately, embracing uncertainty as a space of possibility reframes environmental stewardship as an ongoing moral conversation, rather than a fixed contest with pre-declared winners and losers. In this conversation, anxiety signals moral seriousness: we care enough to worry about outcomes. Instead of letting worry paralyze us, we use it as fuel to keep asking questions, testing ideas, forging alliances, adjusting tactics, and learning from each iteration. Over time, these collective efforts guide humanity along adaptive pathways that preserve options, nurture biodiversity, secure livelihoods, and uphold justice. While we cannot know the final shape of ecological or social landscapes, our commitment, creativity, and collaboration can open pathways toward more hopeful prospects.

In this moral landscape, uncertainty does not represent failure or a lack of will; it reflects the authentic complexity of environmental systems and ethical challenges. Acknowledging that we cannot fully predict or control what lies ahead, we find moral liberation in flexibility, moral depth in humility, and moral strength in collective exploration. Anxiety, instead of a dead end, becomes a sign we are fully engaged in the moral work of shaping the future—one uncertain possibility at a time.

As the effort to address environmental challenges grows more urgent, it becomes increasingly clear that how we hold hope matters as much as whether we hold it. Approaching hope not as a static feeling or comforting illusion but as an active discipline prevents us from sinking into naïve optimism or retreating into despair. This disciplined form of hope acknowledges difficulties, respects complexity, and insists on honest moral engagement. Rather than promising an effortless future, it energizes people to try, learn, collaborate, and adapt even without guarantees.

Choosing grounded hope means recognizing that progress often emerges through partial wins, incremental policies, and slow cultural shifts. Such hope encourages resilience and openness to uncertainty. It does not rely on easy assurances or dismiss the scale of problems. Instead, it highlights that even amid uncertainty, new strategies can be tested, alliances forged, and incremental achievements celebrated. In these conditions, anxiety ceases to be an excuse for withdrawal; it becomes a prompt to stay alert, refine approaches, and draw lessons from resilience stories—where species recover, landscapes revive, and communities adapt through persistence and creativity.

By embracing uncertainty as an opportunity for moral growth rather than a reason for fatalism, we discover that paths forward multiply as understanding deepens. Hope then no longer serves as a fragile shield against grim realities, but as a guiding principle that sustains moral commitment. It propels us into a continuous cycle of learning, building alliances, and sharing knowledge. Under these circumstances, even the most daunting environmental challenges need not paralyze moral action. Instead, they inspire a more flexible, resourceful, and ethically mature engagement with the world's complexity—one in which hope is earned daily through practice, clarity, and the courage to navigate uncertain territories together.

Chapter 14: Learning from Nature's Resilience

Looking to Ecosystems for Lessons in Adaptation and Renewal

As societies grapple with profound environmental changes, many look inward—examining technologies, policies, and moral codes—for guidance. While human ingenuity certainly has its place, there is another, often overlooked teacher with millennia of experience in managing uncertainty and complexity: nature itself. Ecosystems, shaped by evolutionary forces and ecological interactions, offer living examples of adaptation and renewal against a backdrop of shifting climates, periodic disturbances, and resource limitations. By studying how forests regenerate after fires, how coral reefs respond to changing water conditions, or how wetlands evolve to handle floods and droughts, we can glean insights into resilience strategies that transcend human inventions. Far from romanticizing nature as a utopian model, this approach treats ecosystems as dynamic, learning systems—complex adaptive networks that survive and even thrive amid countless challenges. In observing and interpreting these patterns, we find moral and practical lessons to guide our responses to environmental crises.

This process of learning from nature's resilience rests on the understanding that ecosystems do not pursue goals or ideologies; they adjust, reorganize, and persist through a non-linear dance of species interactions, resource flows, and feedback loops. Their resilience does not come from top-down planning but from distributed, emergent properties—multiple niches, genetic diversity, symbiotic relationships, and slow accumulation of adaptations that proved successful over evolutionary time. A forest doesn't rely on a single species to handle every condition but spreads risk across countless organisms, each with its particular strengths and weaknesses. When disaster strikes—be it a storm, fire, pest outbreak—no single entity "decides" how to recover. Instead, a complex interplay of seeds, soil microbes, surviving trees, and opportunistic plants gradually reassemble a functioning community. The moral lesson is that solutions to environmental problems need not be perfectly orchestrated by a central authority or hinge on a single silver-bullet technology. Instead, they can emerge from pluralistic, layered strategies, each contributing a piece of the puzzle.

One of the key takeaways from ecosystem resilience is redundancy. Many ecological communities contain multiple species occupying similar niches. If one pollinator species declines, another can take on some of its role. If a particular predator fails to control a pest population, another predator might step in, preventing catastrophic imbalance. This redundancy is not mere duplication; it's a buffer against unpredictability. In human moral and strategic terms, this suggests that relying on a single "perfect" policy, technology, or cultural practice to solve environmental problems is risky. Instead, a variety of approaches—renewable energy sources, diverse agricultural methods, different conservation strategies, multiple avenues of civic engagement—create a moral tapestry resilient to unforeseen failures. Anxiety that arises from the fear of one solution failing recedes when people acknowledge that a portfolio of options can compensate, ensuring that moral and ecological functions persist even under stress.

Ecosystems also teach about dynamic equilibrium rather than static balance. Nature does not stabilize at a fixed point. Temperate forests evolve through successional stages after disturbances; coral reefs shift species composition as conditions change; savannas oscillate between woody and grassy dominance depending on fire and rainfall cycles. Rather than clinging to a stable "ideal" state, ecosystems continually adjust. Similarly, human efforts to address environmental issues can't aim for a permanent, flawless solution. Instead, moral and social resilience emerges from embracing change, accepting that conditions—economic, climatic, technological—will keep evolving. Policies must be revisited, infrastructures retrofitted, values debated, and education updated. Anxiety over not achieving a permanent fix gives way to a realistic understanding that success involves the capacity to adapt gracefully, not to freeze progress at a single "optimal" moment.

Observing nature's resilience also emphasizes the importance of diversity—biological, cultural, and methodological. In ecology, diversity enhances resilience. A more diverse ecosystem can better absorb shocks, because different species respond differently to stressors. Some flourish in heat, others in cooler conditions; some handle floods, others drought. This variety ensures that no single disturbance can eliminate the entire community. Translating this to human societies, moral resilience benefits from cultural diversity, multiple knowledge systems, and a spectrum of solutions. Indigenous environmental knowledge, scientific research, traditional farming, modern agroforestry—all contribute unique perspectives. Instead of anxiety stemming from conflicting opinions, communities can learn to appreciate moral plurality as a strength. Just as ecosystems thrive with many species, so do human problem-solving processes thrive when multiple voices and skill sets collaborate.

Another ecosystem principle is the role of keystone species—organisms whose influence on the ecosystem's structure and function is disproportionately large relative to their abundance. Wolves regulating prey populations in Yellowstone, sea otters maintaining kelp forest health by controlling sea urchins, or fig trees providing year-round fruit for tropical birds—these species offer stability and foster diversity. In human contexts, keystone "moral actors" or "moral institutions" might not be the most numerous, but they have a pivotal influence. For example, certain environmental NGOs, community groups, or policy frameworks can act like keystones, anchoring broader efforts. Protecting and supporting these keystone entities—even if they are few—can have outsized positive effects. Recognizing the existence and necessity of such moral keystones can guide strategic investments. Anxiety about complexity softens when we identify key leverage points that stabilize and support a wide range of sustainability endeavors.

Ecosystems also rely on feedback loops—processes where outputs affect inputs, creating cycles of reinforcement or stabilization. Negative feedback loops can maintain balance, such as when increasing prey populations boost predator numbers, which in turn reduce prey to sustainable levels. Positive feedback loops can drive change—like warming temperatures melting ice, which reduces reflectivity, causing further warming. Human environmental strategies also involve feedback loops: a well-designed policy that reduces emissions might lower climate risks, encouraging more robust climate action over time. A restored habitat that improves local livelihoods might foster stronger community support for further restoration. Understanding feedback loops teaches that actions can set in motion processes that either mitigate or exacerbate challenges. Moral practice can aim to create positive reinforcing cycles: each success builds support for more significant steps. Anxiety over not seeing immediate large-scale impacts recedes as we appreciate that small initial changes can, through feedback loops, catalyze larger transformations down the line.

Examining how nature copes with disturbances can also inform how we approach crises. Ecosystems rarely prevent disturbances altogether; they survive and recover. After a forest fire, certain seeds germinate only in ashes, pioneering plants stabilize soil, and over time a mature forest returns, sometimes even more diverse. This pattern illustrates resilience as not just bouncing back to the old state, but sometimes achieving a renewed configuration. In human moral endeavors, strict adherence to past paradigms might hinder adaptation. Accepting that disturbances—be they climate shocks, economic upheavals, or social conflicts—are inevitable encourages moral frameworks that emphasize regeneration rather than restoration of an old status quo. Anxiety about losing what once was is replaced by hope in creating something better suited to current realities.

Learning from nature's resilience also involves understanding the importance of scale. Ecological processes operate at multiple levels: gene pools, populations, communities, ecosystems, landscapes, and biomes. Resilience at one level does not guarantee resilience at another. A species might adapt well but if the broader ecosystem collapses, that species may eventually fail. Similarly, moral and environmental strategies must consider scale: local reforms help, but without broader policy frameworks, their impact may be limited. Conversely, global agreements set targets but rely on local implementation. Emulating ecological complexity means bridging scales—combining grassroots initiatives, municipal actions, national policies, and international treaties. Anxiety that arises from feeling powerless at a certain scale (like individuals feeling their personal habits are trivial) can be eased by recognizing that actions at one scale often ripple up and down. Planting local community gardens can influence regional agricultural networks, inspiring policy changes and eventually reshaping global food systems, especially when multiplied by countless similar efforts.

Another lesson from ecosystems is the inevitability of change and the futility of resisting it entirely. Landscapes shift as climates vary over geologic timescales; species evolve or go extinct; new relationships form as old ones end. Viewing environmental and moral challenges as part of ongoing evolutionary processes can reduce anxiety driven by the desire for static perfection. Instead, we learn that moral action should aim to enhance capacity for adaptive change. Societies that facilitate transitions—like guiding workers from fossil fuel sectors into renewable energy jobs—mimic nature's adaptive responses, ensuring continued functioning under changing conditions. Anxiety over instability is tempered by understanding that nature thrives not by freezing conditions, but by continually reshaping them.

From an ethical standpoint, learning from nature is not about blind imitation. Ecosystems display competition, predation, and parasites as well as cooperation and symbiosis. We must selectively interpret which aspects align with human moral values. For instance, we might admire nature's recycling of nutrients, its intricate cooperation (as seen in pollination), and its balanced checks and balances, but not necessarily endorse predatory hierarchies or survival-of-the-fittest ethos as moral principles. Instead, we look for analogies: the way ecosystems recycle nutrients might inspire circular economies. The mutualism between fungi and tree roots suggests models of partnership and resource sharing. The complexity of food webs might inspire diverse energy portfolios and integrated resource management. Applying ecological insights ethically requires discernment, ensuring we draw from patterns that enhance dignity, fairness, and compassion in human systems.

Involving ecologists, indigenous knowledge holders, and interdisciplinary researchers in policy-making exemplifies how to operationalize these lessons. By inviting voices deeply familiar with ecosystem dynamics, decision-makers gain insights into what makes systems resilient—diversity, modularity, redundancy, connectivity, adaptability. Applying these concepts to socio-economic planning or urban design can guide policies that mimic nature's strengths. For example, designing cities with green corridors, diverse transportation options, mixed land use, and responsive governance structures increases urban resilience. Anxiety about rigid urban infrastructures failing under climate stress transforms into confidence that, like ecosystems, cities can evolve adaptively if planned with complexity and uncertainty in mind.

Nature's resilience also highlights the role of time. Ecosystems seldom recover instantly; their healing may span years, decades, or centuries. Understanding this long-term perspective encourages patience and generational thinking in environmental efforts. Instead of demanding overnight results, moral actors can set long horizons, ensuring that short-term incremental steps accumulate into enduring shifts. Anxiety tied to impatience can be relieved by acknowledging that true resilience emerges over extended temporal scales. By committing to sustained efforts—like reforesting over multiple planting seasons, or phasing out harmful substances gradually but steadily—communities align their timelines with ecological rhythms, supporting stable rather than fleeting solutions.

Moreover, studying ecosystems can reveal how disturbance regimes—like periodic fires in savannas—maintain diversity and prevent stagnation. These systems show that occasional shocks prevent dominance by a few species and keep opportunities open for newcomers. Analogously, moral and environmental interventions might find that some disturbances in human systems—like policy debates, protests, or disruptive innovations—prevent moral stagnation. Anxiety about conflicts and disagreements can give way to understanding them as catalysts that maintain flexibility and openness to new ideas. This perspective doesn't glorify conflict but acknowledges that moral life, like ecology, can require tension and reshuffling to stay dynamic and resistant to decay.

Cultural narratives can incorporate these ecological lessons. Traditional tales or modern documentaries can emphasize how certain landscapes recovered after setbacks, highlighting patience, mutual reliance, and adaptiveness. These stories can influence collective morality. Instead of perceiving environmental stewardship as a linear conquest toward a final stable state, people come to see it as an ongoing moral practice of nurturing conditions that allow life's resilience to flourish. Anxiety about final outcomes dissipates in the recognition that moral success often lies in perpetually enabling renewal and responsiveness.

For future generations, including ecosystem-based lessons in education fosters moral imagination grounded in natural dynamics. Children learning about forest succession, coral symbioses, or migratory patterns understand early that complexity and adaptation are natural. This framing can prepare them to approach environmental issues as continuous moral puzzles, requiring patience, creativity, and humility. Instead of seeking static balances or absolute solutions, they learn from nature's capacity to reconfigure itself. Such an education encourages resilience not just in the environment, but in moral thinking, problem-solving, and cultural evolution.

In business or economic planning, nature's resilience principles suggest diversifying supply chains, building modular infrastructure, and designing for flexibility. Anxiety that arises from unpredictable markets or resources can be mitigated by economic models that, like ecosystems, contain redundancy and distributed functions. A single source dependency is risky; multiple suppliers, diversified energy inputs, and robust social safety nets mimic ecological patterns, fostering moral responsibility and stability. Decision-makers can thus draw parallels between ecological and socio-economic resilience, internalizing ecological wisdom as a foundation for sustainable development.

Spiritual and philosophical traditions that emphasize harmony with nature can also integrate these insights. Many faiths already celebrate nature as a teacher or moral exemplar. Incorporating modern ecological understanding reinforces that spiritual respect for nature aligns with practical lessons in resilience. Moral liturgies can invoke the patience of forests, the adaptability of wetlands, or the cyclical renewal of savannas as metaphors for human virtues. Anxiety about moral insufficiency transforms into faith that sustained, humble engagement with complexity, just like nature's adaptive processes, leads to growth. Spiritual solace emerges not from denying struggles but from learning how nature survives and thrives amidst them.

Even in activism, acknowledging nature's resilience can shift campaigning narratives. Instead of communicating only fear and deadlines, activists can highlight that ecosystems have remarkable capacities if given space and supportive conditions. This adds nuance to calls for action. Demanding protective legislation and habitat restoration is not just about preventing doom; it's about enabling natural resilience to do its work. Anxiety is balanced by trust in the world's inherent regenerative potential. Witnessing that nature can rebound when given chances can motivate more people to join environmental causes—since they see positive potential as well as danger.

From a research standpoint, interdisciplinary studies that integrate ecology, social sciences, and ethics can help translate ecosystem resilience lessons into concrete moral frameworks. Policy advisors might propose "ecological governance" principles derived from studying wetlands or coral reefs, guiding human systems to embed redundancy, modularity, and adaptive feedback loops. Over time, these principles could form a body of moral and institutional knowledge known as "biomimetic ethics"—an ethical approach modeled on nature's resilience strategies. Anxiety about moral cluelessness or lack of adequate frameworks dwindles as we realize nature's patterns offer robust starting points for moral and institutional design.

Ultimately, drawing lessons from nature's resilience is not about idealizing the nonhuman world or abandoning human ingenuity. It's about broadening our moral and intellectual horizons to include nature as a patient mentor. Ecosystems have navigated eons of change without giving up on complexity, diversity, and adaptability. By approaching these systems humbly and seeking analogies for human problem-solving, we enlarge our repertoire of moral strategies. Anxiety transforms into curiosity, despair into determination, as we comprehend that the environment we strive to protect and restore is itself a source of guidance.

In this exploration, nature's resilience suggests that moral life, like ecological life, is a continuous unfolding—no static endpoints, but ongoing processes of renewal, adaptation, and interaction. Embracing this view helps us accept uncertainty, celebrate incremental successes, integrate multiple perspectives, and persist despite obstacles. We learn that moral flourishing, like ecological flourishing, arises from combining diversity, redundancy, dynamic equilibrium, and constant learning. Thus, nature's resilience, observed in forests, reefs, savannas, wetlands, and countless other ecosystems, becomes a moral teacher: showing that anxiety and complexity need not deter action, that uncertainty can spur innovation, and that sustained care can coax life to rebound even after setbacks.

Engaging with Regenerative Practices and Indigenous Wisdom

As communities, policymakers, educators, and activists strive to move beyond purely extractive models of environmental engagement, the search for authentic, restorative practices intensifies. The aim is not merely to slow down degradation but to reverse it, healing landscapes, renewing biodiversity, and restoring the conditions that allow future generations—human and non-human—to flourish. Central to this shift is the idea of regeneration: going beyond minimizing harm to actually improving ecological and social systems over time. While the concept of "sustainability" often focuses on maintaining current levels, regenerative approaches emphasize enhancing vitality, fertility, and complexity. They remind us that it's possible to rebuild soil health, re-establish wildlife corridors, regenerate forests where deserts have encroached, and reinvigorate cultural traditions aligned with respectful stewardship of the land.

Within this exploration, Indigenous wisdom holds a special place. Long before the modern environmental crisis became widely acknowledged, many Indigenous peoples maintained intricate systems of ecological knowledge. Far from romanticizing or freezing Indigenous traditions in time, acknowledging Indigenous contributions means recognizing the ongoing relevance of ethical frameworks and practical strategies that have sustained communities for millennia. These traditions often challenge prevailing assumptions—refusing the false dichotomy between humans and nature, rejecting the idea that resources exist merely to be exploited, and embracing a relational understanding of life's interconnectedness. Integrating Indigenous wisdom into regenerative practices invites moral humility, cultural respect, and intellectual curiosity. Instead of substituting one universal solution with another, it encourages us to learn from proven patterns of coexistence between people and place.

Regenerative approaches differ from conventional environmental management in several key ways. Traditional resource management often treats nature as a stock of inputs to be extracted efficiently, then attempts damage control when depletion looms. Even some well-intentioned conservation models focus on preserving "pristine" fragments, isolating them from the broader socio-ecological fabric, without fully considering ongoing human-nature relationships or the need for ecological dynamism. Regeneration, by contrast, sees humans as participants in evolving, living systems. It does not assume that nature can or should revert to a mythical untouched state; rather, it aims to restore functional integrity, resilience, and evolutionary potential. This perspective accommodates human livelihoods and cultural expressions within ecological patterns, encouraging a reciprocal exchange: humans enhance biodiversity and ecosystem health, while ecosystems provide sustenance, meaning, and spiritual nourishment.

One practical example of regenerative practice lies in agriculture, where industrial farming methods have eroded soils, diminished genetic diversity in crops, and contributed to climate instability. Regenerative agriculture reverses these trends by improving soil fertility, increasing carbon sequestration, and fostering more complex biological interactions. Instead of synthetic inputs and monocultures, farmers rotate crops, integrate livestock, use cover crops, compost, and reduce tillage. Over time, this enriches soils, promotes water retention, supports pollinators, and mitigates greenhouse gas emissions. While regenerative agriculture draws on scientific research, it also resonates with traditional farming practices employed by Indigenous communities for generations—techniques that favored diversity, soil stewardship, and harmonious relationships with local landscapes. By incorporating these older methods, regenerative farmers enrich their moral frameworks, seeing agriculture not as a purely economic pursuit but a moral act of nurturing the Earth.

Indigenous wisdom has long recognized the value of reciprocal relationships. Many Indigenous cultures conceptualize land not as property but as kin—an extension of community and family. Instead of asking, "How can we exploit these resources?" the question becomes, "How can we maintain respectful relations so that land, water, animals, and plants continue to thrive and sustain life?" This approach aligns naturally with regenerative ethics. Rather than externalizing environmental costs or blaming distant polluters, regenerative mindsets and Indigenous traditions both see local action, local learning, and local accountability as fundamental. Anxiety about global inaction is partly tempered by realizing that healing can start at home, guided by place-based knowledges passed down through oral histories, ceremonies, and land management practices refined over centuries.

For instance, consider fire management. In many Indigenous communities, controlled burns have historically been used to maintain healthy forest understories, promote certain food plants, and reduce the risk of catastrophic wildfires. Modern fire suppression policies often led to excessive fuel build-up, making wildfires more destructive. Recently, some regions have begun to relearn Indigenous fire stewardship—using prescribed burns informed by ancestral knowledge and ecological monitoring. This integration of Indigenous wisdom into contemporary policy is a regenerative step: it not only prevents large-scale disasters but also revitalizes cultural practices and ensures ecosystems maintain their natural fire regimes. The anxiety tied to unstoppable mega-fires lessens as we restore centuries-old management techniques that maintain ecosystem balance.

Another dimension of Indigenous wisdom relevant to regeneration is the emphasis on long time horizons. Western environmental policies and economic models often operate on short-term cycles—quarterly profits, election terms, or immediate returns on investment. Indigenous perspectives frequently adopt intergenerational frameworks, considering the effects of decisions seven generations ahead. This temporal stretching encourages deeper moral responsibility. Instead of viewing regeneration as a quick fix for today's problems, it becomes a commitment that will unfold over decades and centuries, ensuring that descendants inherit healthier conditions. Anxiety from short-sighted approaches transforms into cautious optimism, as we adopt practices acknowledging that true ecological recovery and cultural resilience require patience, consistency, and foresight.

Integrating Indigenous wisdom also involves confronting historical injustices. Many Indigenous communities have faced dispossession, cultural suppression, and resource extraction imposed by colonial systems. Embracing their knowledge without addressing these injustices risks appropriation and moral hypocrisy. To truly learn from Indigenous teachings, societies must engage respectfully—acknowledging sovereignty, returning decision-making power over lands and resources, and valuing Indigenous rights. This moral reckoning is part of regenerative ethics. It confronts anxiety rooted in past wrongdoing and ongoing inequalities, offering a chance to restore not only ecosystems but also social and cultural balance. By weaving moral restitution into ecological revival, we align regeneration with justice and healing, ensuring that the knowledge guiding us emerges from ethical relationships rather than exploitation.

The principle of reciprocity, so central in many Indigenous philosophies, aligns well with regeneration. Instead of a linear flow of resources from the Earth to humans, reciprocity suggests giving back—planting to replace what we harvest, restoring habitats after extracting resources, investing in regenerative solutions that benefit more than just human interests. Reciprocity frames the moral landscape as relational rather than transactional. Anxiety over scarcity or ecological collapse transforms into a sense of moral obligation: we cannot simply take; we must also restore, honor, and replenish. Adopting such an ethic can reshape policies, business models, and community projects. For example, a timber company implementing regenerative forestry might rotate harvest areas, preserve old-growth refuges, replant diverse native species, and involve local communities in stewardship, thereby upholding reciprocity rather than pure extraction.

Inspiring examples abound. In the Arctic, certain Indigenous communities manage caribou herds and fish stocks sustainably through complex rules and taboos that limit overuse. These unwritten laws, honed over generations, maintain resource availability despite environmental fluctuations. By studying these practices, contemporary resource managers gain insights into flexible, adaptive governance structures that can cope with uncertainty. Rather than panicking under changing conditions, they learn to establish buffer zones, rotate harvest areas, and respect indicators from nature—like animal migration patterns. Anxiety about unpredictable resource futures yields to confidence in a moral and ecological approach proven effective over centuries. This trust allows us to move beyond rigid blueprints toward evolving protocols that mirror ecological fluidity.

Another example is found in efforts to restore river ecosystems by removing dams and reintroducing natural flow regimes. Indigenous communities have often supported or led these initiatives, recalling that salmon runs and healthy riparian zones were once abundant, supporting cultural and spiritual practices. Reconnecting rivers enables fish populations to rebound, floods to nourish floodplains, and wetlands to filter water. While the outcome isn't guaranteed—some rivers recover slowly or face new challenges—this willingness to experiment and learn from pre-dam conditions is a regenerative act. It respects historical knowledge that rivers were once lifelines for entire cultures. Anxiety about water scarcity or habitat fragmentation transforms into constructive energy as engineers, ecologists, Indigenous leaders, and policy-makers collaborate to restore waterways that can adapt to climate variations.

Education and community learning also play roles in integrating regenerative practices and Indigenous wisdom. Curricula that include Indigenous ecological knowledge and regenerative agriculture principles teach students that environmental care extends beyond recycling or emissions reduction. Students might learn about polyculture farming from Indigenous traditions that maintained soil fertility for millennia without synthetic fertilizers. They explore how seasonal rituals guided harvesting times, ensuring long-term availability of certain plants or fish. This exposure counters environmental fatalism by showing that people in various historical contexts managed ecosystems sustainably without modern technology. Anxiety about needing high-tech miracles lessens, replaced by curiosity about low-tech, place-based solutions that emphasize relationships, observation, and respect.

Cultural narratives that celebrate regeneration and Indigenous insights can shape public discourse. Films, novels, and documentaries portraying communities who restore degraded landscapes or revive endangered species not through brute force, but by applying traditional wisdom and regenerative science, can captivate audiences. These stories highlight that moral engagement is about working with nature, not dominating it. Anxiety that modern societies have lost the path to sustainability diminishes as people realize that guidance exists, often preserved in Indigenous languages, stories, and land-based teachings. The renewed appreciation of local knowledge keepers and environmental defenders encourages more inclusive dialogues, bridging gaps between scientific experts, policymakers, and Indigenous elders, all united by a regenerative vision.

In policy-making, incorporating Indigenous consultation and co-management structures reflects regeneration's moral dimension. Environmental laws that mandate Indigenous participation in land-use decisions or marine resource management signal an ethic of shared responsibility. Rather than treating Indigenous knowledge as a supplementary resource, governments recognize it as a foundational element of ecosystem governance. Anxiety over making top-down decisions that might fail under changing conditions fades, replaced by confidence that local knowledge holders offer context-specific insights into resilience strategies. This institutional shift not only enriches policies but also acknowledges moral obligations to rectify historical exclusions and honor Indigenous sovereignty.

Regenerative thinking also embraces experimentation and trial. Just as ecosystems do not have predetermined plans, human societies must try different regenerative approaches, evaluate outcomes, and revise techniques. Sometimes Indigenous methods can guide initial steps, like controlled burns to reduce wildfire risk, then modern science can add monitoring technologies that track improvements in biodiversity or carbon sequestration. The synergy between tradition and innovation characterizes regenerative adaptation—old and new knowledge systems interact like species in a healthy ecosystem. Anxiety about balancing respect for tradition with the need for novel solutions can be allayed by emphasizing ongoing dialogue and shared decision-making, ensuring that each innovation or old practice is tested, discussed, and adjusted collaboratively.

In the energy sector, regenerative logic might favor distributed, community-owned renewable grids rather than massive centralized plants. This mirrors how ecosystems distribute functions and resilience across multiple species and habitats. Similarly, adopting Indigenous water governance practices—like seasonal rules for fishing or harvesting medicinal plants—could inform adaptive water sharing agreements in regions facing drought. By blending these local ethical frameworks with modern hydrological modeling, water managers arrive at robust, flexible allocations. Anxiety about water wars or ecological collapse turns into cautious hope as people see that layered, regenerative strategies can defuse tensions and promote long-term sustainability.

Spiritual and philosophical reflection can deepen the moral significance of regenerative practices and Indigenous wisdom. Spiritual traditions often revere nature's cycles—birth, growth, decay, rebirth—depicting them as metaphors for moral perseverance. Regeneration aligns with such cycles: damaged soils replenished by compost mirror moral renewal after moral failings; forests recovering from storms symbolize healing after societal trauma. Indigenous ceremonies—honoring the first salmon run, giving thanks for a successful harvest—inscribe the moral duty of gratitude and reciprocity into the human-nature relationship. By participating in these rituals, people internalize the idea that caring for land is not just practical, but sacred. Anxiety about losing cultural or spiritual meaning under environmental pressures fades as traditions adapt and reaffirm their relevance, guiding communities through transitions with dignity and reverence.

This moral integration can inform how we measure success. Instead of just tallying species counts or water quality improvements, regenerative metrics might include indicators of cultural vitality, equitable resource sharing, and spiritual well-being. Observing how Indigenous wisdom incorporates moral values into land management suggests that environmental data should not stand alone, detached from human ethics and cultural identities. Quantifying greenhouse gas reductions is crucial, but so is noting whether communities feel empowered, whether traditional livelihoods revive, and whether younger generations find meaning and guidance in these efforts. Anxiety that moral values might be lost in technocratic approaches recedes when social and cultural dimensions are woven into evaluation frameworks.

Critics might argue that blending Indigenous wisdom with regenerative approaches risks tokenism or oversimplification. Indeed, care must be taken. Indigenous knowledge is diverse, context-specific, and often guarded as intellectual property. Extracting bits of it without proper consultation or cultural sensitivity is unethical. Regenerative practitioners should approach collaboration with humility, seeking permission, investing in long-term relationships, and acknowledging that Indigenous communities choose what knowledge to share. Rather than treating Indigenous insights as a toolkit to be mined, respectful partnerships must recognize Indigenous peoples as moral agents with sovereignty and rights, shaping the narrative on their own terms. Anxiety about misappropriation and cultural harm diminishes when the process is anchored in consent, fairness, and mutual benefit.

Over time, the fusion of regenerative practices with Indigenous wisdom can inspire policy frameworks that embody pluralistic ethics, adaptive governance, and relational thinking. For instance, treaties co-developed between governments and Indigenous nations could ensure that land and resource management adheres to regenerative principles validated by both scientific research and ancestral knowledge. Anxiety over legal uncertainties or bureaucratic inertia transforms into constructive energy as stakeholders witness how integrating multiple knowledge systems produces robust, enduring arrangements.

As communities apply these integrated approaches on the ground—restoring wetlands guided by elders' instructions, implementing agroforestry systems blending old and new techniques—the stories of success accumulate. Future generations will then recount these narratives of moral evolution: how people overcame eco-anxiety, recognized the limits of extractive models, turned to nature's resilient patterns and Indigenous wisdom for guidance, and forged a path of healing. This historical perspective assures that anxiety about whether humanity can learn from its mistakes finds an affirmative answer in tangible legacies: revived soils, healthier waters, thriving cultural traditions, and cohesive moral frameworks reflecting nature's enduring lessons.

By engaging with regenerative practices and Indigenous wisdom, we expand the moral imagination. We learn that solutions need not be simplistic or uniform; they can be adaptive mosaics that resonate with local ecologies and cultural nuances. Instead of despairing over uncertainty, complexity, and past failures, we discover that moral agency flourishes where multiple sources of knowledge converge. Rather than dreading the future, we approach it as an open landscape, informed by nature's resilience and guided by ethical traditions proven resilient over countless generations.

Drawing Strength from the Earth's Capacity to Rebound and Heal

In the face of escalating environmental crises, it is easy to dwell on the fragility of ecosystems and the scale of damage inflicted by unsustainable practices. Headlines about collapsing fisheries, drought-stricken farmlands, and unprecedented wildfires feed a narrative of irreversible decline. This focus, while realistic in acknowledging the severity of threats, can obscure an equally important truth: the Earth possesses remarkable capacities for regeneration, resilience, and renewal. Though these capacities do not justify complacency or diminish the urgency for decisive action, they can inspire moral courage and endurance. Drawing strength from nature's inherent ability to heal reminds us that catastrophe is not inevitable, that ecosystems can rebound if given the chance, and that human agency can facilitate this recovery rather than stand in its way.

Nature's resilience is not a new discovery. Ecologists have long documented how forests regrow after storms and fires, how coral reefs can rebound from bleaching events if local stressors are reduced, how rivers purified by wetlands can recover water quality once pollutants cease. Yet, as anxieties mount over climate change and biodiversity loss, these examples take on deeper moral significance. Recognizing Earth's healing potential affirms that moral commitments to conservation, restoration, and equitable resource management matter. Without the planet's latent regenerative capacities, our efforts would be futile; no matter how careful, humans could not restore what is fundamentally incapable of renewal. The Earth's responsiveness is thus a moral ally, supporting the notion that conscientious interventions are not in vain.

This capacity for rebound and healing is not uniform or guaranteed. It depends on context, severity of damage, the stability of surrounding conditions, and the presence of keystone species and essential ecological functions. Some ecosystems bounce back relatively quickly—grasslands might recover after moderate grazing pressure is relaxed—while others require decades or centuries to regain complexity. This variability invites moral humility. We must not assume that nature will always self-correct regardless of what we do. Instead, we acknowledge that ecosystems thrive best under supportive human stewardship, free from incessant stressors. Anxiety that we have pushed Earth past a tipping point encounters a countervailing perspective: while some thresholds may indeed be crossed, other systems remain recoverable, and even those that have tipped might find new equilibria if humans act responsibly and restore conditions conducive to renewal.

For example, consider large-scale rewilding projects. In certain regions, giving land back to natural processes—removing drainage systems, halting overgrazing, ceasing pesticide use—can trigger remarkable recoveries. Wetlands might re-emerge, sequestering carbon and filtering water. Predators reintroduced into ecosystems can rebalance trophic dynamics, leading to cascading improvements in vegetation and associated wildlife. Forests left to regenerate can show that once-barren hillsides can again host rich canopies, diverse understory plants, and teeming animal life. Each story of rewilding emphasizes that life's inherent drive to proliferate and diversify can rebound if we stop suppressing it. This lesson counters the despair that humans must orchestrate every outcome. Instead, we see that merely backing off from destructive pressures can unleash potent natural forces. Anxiety transforms into hope as we recognize that Earth is not a passive victim but an active participant in regeneration.

This perspective extends to marine ecosystems as well. Overfished areas, when protected by no-take zones or managed sustainably, often yield dramatic increases in fish biomass, reef complexity, and species richness over time. Initially, skeptics might dismiss these measures as insufficient or slow. Yet patient observation reveals that marine life can recolonize empty spaces, pioneer species create habitat for others, and once-fragmented populations interconnect. Such changes underscore that moral patience, informed by ecological knowledge, can pay dividends. Anxiety about barren oceans recedes as communities witness how carefully implemented marine reserves become seedbeds of renewal, spilling benefits beyond their boundaries as fish stocks grow and migrate. Drawing strength from these outcomes encourages more communities and nations to invest in similar conservation strategies, leveraging the ocean's capacity to mend itself if given respite.

Even within agricultural landscapes, nature's healing potential is evident. Transitioning from industrial monocultures to agroforestry, permaculture, or regenerative farming often reveals that soils depleted over decades can regain fertility, structure, and microbial diversity. Earthworms return, nutrient cycles stabilize, and biodiversity flourishes in field margins. Although these transitions take time and require effort, their success proves that deserts of pesticide-laden fields can become mosaics of life. This fact reaffirms moral confidence that feeding human populations need not equate to perpetual exploitation. If we align farming practices with ecological principles, we harness nature's inclination to restore balance, improving yields, resilience, and long-term viability. Anxiety that global food security depends on destructive methods eases as we understand that well-chosen regenerative tactics reinforce soil and ecosystem integrity.

In urban contexts, nature's healing also surfaces. Vacant lots replanted with native species attract pollinators and birds, cooling and greening neighborhoods. Rivers, long confined by concrete channels, can be restored to meander through constructed wetlands, reducing flood risk and supporting amphibians and other wildlife. Even small pockets of green infrastructure—green roofs, bioswales, pocket parks—improve urban resilience against heatwaves and stormwater surges. While these measures won't single-handedly solve global warming, they illustrate that even in human-dominated environments, nature responds positively to invitations for regeneration. Anxiety about urban landscapes becoming uninhabitable under climate stress gives way to guarded optimism. By working with ecological processes rather than against them, cities can become patches of urban ecosystems where humans and other species co-exist more harmoniously.

This recognition of Earth's healing potential doesn't excuse humans from moral responsibility. Quite the opposite: it underscores our capacity to facilitate or hinder nature's resilience. Just as a doctor's treatment can support a patient's natural healing mechanisms, human policies, conservation efforts, and sustainable livelihoods can either nurture or stifle ecological resilience. We must note that nature's capacity to rebound is not infinite—relentless pressures like unmitigated climate warming, ocean acidification, or widespread habitat fragmentation can surpass thresholds where recovery becomes improbable. The moral implication is that acknowledging nature's resilience should propel us to work diligently, ensuring the conditions for healing persist. Anxiety about irreparable damage thus becomes motivation to act proactively before more thresholds are crossed, tapping into the Earth's regenerative abilities while they remain accessible.

Indigenous and local knowledge systems often anticipate this dynamic. Many traditions portray nature as cyclical, capable of renewal if humans uphold certain ethical standards. Rituals that mark seasonal transitions, manage harvesting levels, or maintain habitat corridors demonstrate cultural recognition that nature's health depends on respectful human conduct. Indigenous stories often highlight how certain species return each year, certain landscapes rebound after proper stewardship, reinforcing a moral relationship of reciprocity. By learning from these narratives, contemporary environmental movements gain moral strategies that connect hope to concrete actions. Anxiety recedes as communities see that well-established cultural practices have long aligned with nature's regenerative capacities, providing models for our present challenges.

Scientific research further affirms this view. Studies tracking post-disturbance recovery—like how forests regrow after logging stops, how mangroves recolonize coastal areas when pollution is curtailed—underscore that life's complexity and adaptability are not static relics but ongoing processes. These processes depend on genetic diversity, connectivity between habitats, and stable climate conditions. Moral actions that protect genetic reservoirs, maintain corridors, and mitigate climate change thus become catalysts enabling Earth's regenerative forces to operate unimpeded. Instead of feeling we must engineer every detail, we focus on removing barriers: ending unsustainable extraction, restoring natural flow regimes, enabling species mobility. Earth's regenerative response does the rest, turning human humility into an asset rather than a weakness.

In political discourse, recognizing nature's healing powers can shift narratives. Instead of framing environmental protection as a lost cause or a burden that yields no return, policymakers and communicators can highlight success stories of recovery. They can say: "We cleaned this river and now fish populations rebound annually, improving local fisheries and community livelihoods." Such messaging replaces fear with inspiration. Anxiety about endless environmental gloom gives way to a sense of moral possibility. This encourages more ambitious policies as public support grows, convinced that positive feedback loops exist—investing in restoration now leads to more stable ecosystems that, in turn, require fewer remedial measures later.

Educationally, embedding the concept of Earth's capacity to heal into curricula enables students to see the bigger picture. Instead of memorizing lists of environmental catastrophes, learners study case studies of recovery—forests regaining biodiversity after land abandonment, coral reefs rebounding under careful management, wildlife corridors reconnecting fragmented populations. Such examples teach moral resilience. Students learn that while they must acknowledge serious threats, they need not succumb to fatalism. They understand that moral agency makes a difference and that nature is not a static victim awaiting demise, but a dynamic force that can, under suitable conditions, restore complexity and stability. Anxiety about inheriting a ruined planet is tempered by the knowledge that they can participate in its healing.

Technological innovation can also draw inspiration from nature's regenerative patterns. Biomimicry, for instance, studies how ecosystems maintain cycles of energy and nutrients, then applies these insights to design cleaner production processes, waste recycling, and energy systems. Instead of producing linear waste streams, manufacturing could mimic circular nutrient flows, turning byproducts into inputs elsewhere. If ecosystems show that nothing is truly wasted—leaf litter nourishes soil, dead wood hosts new life—then human industries can adopt similar principles. Anxiety that modern societies must inevitably leave massive ecological footprints diminishes as we realize we can align our technologies with regenerative cycles. This moral direction encourages engineers, architects, and industrial designers to incorporate ecological thinking, not just patching problems but generating systems that self-improve over time.

The practice of ecological restoration itself, as a field, gains moral depth from understanding that nature's healing capacity is real. Restoration ecologists who reintroduce certain keystone species or restructure plant communities trust that once triggered, ecological processes will carry the system toward greater complexity and resilience. They do not attempt to micromanage every detail indefinitely. Instead, they set conditions that allow nature's self-organization to unfold—much like gardeners create conditions for plants to thrive rather than manually designing every leaf's position. Witnessing these self-organizing patterns intensifies moral appreciation for life's intrinsic drive toward renewal. Anxiety that the system might collapse under its own weight transforms into a sense of companionship: we are co-creators with nature, aiding her restorative efforts rather than controlling them.

Cultural traditions celebrating nature's regenerative cycles abound. Festivals marking the return of migrating birds or the blooming of certain flowers capture the emotional essence of nature's resilience. These cultural expressions have moral potency. They teach communities to anticipate renewal each year, finding moral reassurance in the rhythms of life. When droughts persist or storms ravage coasts, people recall these cultural memories of past renewals, understanding that hardships can give way to revival if conditions permit. Anxiety about perpetual decline coexists with a cultural memory of nature's comeback stories, encouraging moral patience and renewed collective action to restore lost conditions.

A critical aspect of drawing strength from Earth's regenerative power is avoiding complacency. Acknowledging nature's healing capacity does not mean we can leave all to nature's devices. Human activities have introduced unprecedented stressors—rapid climate shifts, synthetic chemicals, invasive species—that can overwhelm natural resilience. Just as a severely wounded body may need antibiotics, surgery, or extensive care to recover, ecosystems under severe stress often need targeted interventions. Understanding that nature can heal if given a chance inspires us to create those conditions, not to assume healing will happen regardless. Anxiety remains a productive alert that we must reduce pressures—cut emissions, halt deforestation, control pollutants—to ensure Earth's resilience can operate fully.

Policy frameworks that leverage nature's resilience might prioritize green infrastructure over concrete solutions. For example, restoring wetlands to mitigate floods instead of building higher levees harnesses natural processes that replenish ecosystems even as they provide human benefits. This approach acknowledges uncertainty—no one can guarantee how quickly wetlands fully restore—but trusts that given proper space and time, natural processes will enhance stability. Anxiety that a single flood-protection measure might fail is offset by the knowledge that wetlands can adapt to changing water levels, maintaining protective capacities across varied conditions.

Embracing Earth's capacity to heal also empowers marginalized voices. Indigenous peoples, local communities, and farmers often understand local ecosystem dynamics intimately. Their involvement in regenerative projects ensures that interventions align with actual conditions and cultural contexts. Recognizing that healing emerges from nature's patterns encourages decentralized decision-making and participatory governance. Instead of top-down mandates ignoring local realities, policies can incorporate on-the-ground observations and traditional knowledge, aligning management with natural rhythms. Anxiety that solutions remain elite-driven and disconnected from frontline experiences dissipates as communities co-create restoration efforts, ensuring they resonate ethically and ecologically.

Over time, as more communities apply these principles, evidence accumulates: polluted rivers cleansed, deserts partially reclaimed, overfished stocks replenished, degraded farmland reclaimed as fertile fields. Each success story reinforces that healing is not a myth. The more we see Earth responding positively to reduced pressures and supportive interventions, the more moral courage we gain to tackle bigger challenges—like halting climate change at safe thresholds or restoring continental-scale migration corridors. Anxiety about our collective future coexists with the resolve to keep pushing boundaries, knowing incremental restorations prove our capacity for beneficial change.

This mindset can influence global negotiations. Instead of direly pitting countries against each other over dwindling resources, diplomatic forums can highlight successful restoration and resilience projects that yield shared gains. Nations that restore coral reefs or protect forests contribute not only to their local well-being but to global ecological stability—carbon sequestration, biodiversity conservation, and maintenance of weather patterns that benefit distant regions. Anxiety about geopolitical tensions over climate refugees or resource conflicts softens as we see that cooperating to support Earth's healing fosters mutual security. Diplomatic language can evolve from zero-sum resource claims to joint commitments that recognize ecosystems as allies in achieving sustainable peace and development.

In educational and media narratives, weaving stories of Earth's healing capacity helps counter cynicism. While not sugarcoating crises, these narratives emphasize that Earth is not a passive stage for human drama; it's an active participant with regenerative strengths. Students and citizens learn that acting morally—reducing pollution, restoring habitats, supporting indigenous land management—is rewarded by nature's positive feedback. This moral reciprocity reduces anxiety, reminding us that we are not alone in our moral struggles. Nature, as a living partner, complements our actions, amplifying their effects when guided by empathy and respect.

In personal life, internalizing nature's resilience can shift our emotional responses. Instead of breaking down when confronted with grim predictions, we ask: "What can I do to support Earth's healing here and now?" The answer might be small: planting a native garden, participating in a local restoration day, supporting policies that protect wetlands or mangroves. Such acts, multiplied across society, help tilt probabilities towards regenerative outcomes. Anxiety becomes a driver of constructive moral behavior, encouraging each person to contribute their part, trusting that combined efforts matter because nature responds. A hopeful humility emerges: we cannot guarantee outcomes, but we know Earth's resilience aligns with moral principles that restore conditions conducive to life.

Over the long arc of environmental ethics, recognizing Earth's healing capacity marks a moral evolution. Initially, environmental discourse framed nature as fragile and doomed by human misdeeds. Now, a more nuanced understanding emerges: nature is indeed stressed and battered, yet it harbors immense regenerative capabilities that moral agents can unlock or obstruct. This shift encourages balanced narratives—urgent enough to spur action, yet hopeful enough to inspire perseverance. Anxiety still signals danger, but not hopelessness. Instead, anxiety serves as a spur to engage morally, channeling effort into removing obstacles to regeneration rather than surrendering to defeat.

As communities gain confidence in Earth's capacity to rebound, they find greater moral agency to shape an adaptive, regenerative future. Instead of fantasizing about pristine pasts or perfect futures, they embrace the evolving dance of restoration: partial successes, lessons learned, and incremental expansions of ecological integrity. Cultural traditions, Indigenous wisdom, scientific insights, and technological innovations coalesce into a multi-faceted moral toolkit. Each component supports nature's healing, ensuring that while some systems may remain compromised, many others will thrive anew. Anxiety transforms into a kind of vigilant hope—a recognition that the Earth's resilience, once nurtured, can outperform our fears, guiding us towards healthier ecologies and more just societies.

In a time when environmental uncertainty can feel overwhelming, turning to nature's own resilience reveals fresh pathways forward. Ecological systems, shaped by evolutionary processes and cultural stewardship, show that adaptability and healing are not distant ideals but active possibilities. By observing how forests recover after disturbances, how coral reefs respond to changing conditions, or how regenerative farming restores soil health, we gain insights into strategies that harmonize human needs with ecological complexity. These lessons emphasize that solutions need not be linear or singular; they can emerge through diverse, layered efforts supported by continuous learning and adaptation.

Incorporating Indigenous wisdom into regenerative practices further enriches this perspective. Ethical frameworks developed over centuries highlight reciprocal relationships with the land, multiple scales of stewardship, and long-term horizons. Such guidance helps balance the urgency of modern challenges with enduring principles of respect, equity, and cultural continuity. Rather than expecting a single innovation or policy to deliver salvation, this approach embraces complexity, moral pluralism, and incremental change. Anxiety about the future turns into patient determination, as communities understand that no matter how daunting the task, ecosystems and human societies share an inherent capacity for renewal—provided we align our efforts with the natural rhythms and interconnectedness of life.

This evolving understanding reframes both the environmental crisis and the moral landscape. Instead of viewing nature merely as a backdrop for human drama, we rediscover it as a mentor that demonstrates resilience in action. Recognizing that forests, oceans, and grasslands possess remarkable abilities to regenerate when given proper conditions instills hope grounded in tangible evidence. Moral endeavors gain purpose and direction as humans assume roles not as conquerors or passive observers, but as active participants supporting the Earth's continuous ability to adapt, flourish, and restore. The result is a deeper, more durable kind of hope, one defined by courage, humility, reciprocity, and the shared conviction that we can learn from nature's resilience to shape a more vibrant, inclusive, and enduring world.

Chapter 15: A Lifelong Journey of Caring

Staying Informed Without Burning Out: Setting Boundaries with Media Intake

In an age of ceaseless information flow, staying informed about environmental issues can feel like a double-edged sword. On one hand, knowledge fuels moral engagement, advocacy, and responsible decision-making. It illuminates the stakes, reveals emerging threats, and highlights breakthroughs worth celebrating. On the other hand, an unfiltered stream of grim headlines, dire predictions, and tragic stories can easily overwhelm the psyche. Too much exposure to bleak news can spark despair, apathy, or cynicism—the very emotions that erode the willingness to take action. Managing this tension becomes a moral skill in its own right: striking a balance where we remain sufficiently informed to act wisely, yet guarded enough to prevent emotional exhaustion.

Finding this equilibrium is not simply a matter of personal preference or emotional self-care; it's a strategic moral choice. If caring for the planet is a lifelong journey, as this chapter suggests, then learning how to engage with media sustainably is as important as understanding ecological principles or mastering advocacy techniques. Without careful boundaries, the intense sadness and outrage invoked by relentless negative coverage can lead to burnout, undermining long-term moral commitment. Conversely, cutting oneself off entirely from environmental news may lead to ignorance, detachment, and missed opportunities for meaningful involvement.

The challenge of media intake is compounded by the structure of modern news and social media platforms. Algorithms amplify sensationalist or emotionally charged content. Fearful or shocking headlines draw attention, generating more clicks, shares, and debates. While these dynamics may not be maliciously designed to cause despair, their effect can nonetheless trap individuals in cycles of anxiety, continuously reinforcing the belief that everything is getting worse. Additionally, the sheer volume of environmental crises—spanning climate change, mass extinctions, pollution, deforestation, and more—makes it nearly impossible to stay current without feeling inundated. With so many fronts of concern, it's easy to believe that keeping up with every scientific report, policy debate, or catastrophic event is a moral duty. But in reality, this feeling of obligation can turn into moral self-punishment, as though failing to read the latest distressing article constitutes a lack of caring.

Setting boundaries with media intake helps transform anxiety into productive moral energy rather than a corrosive force. Boundaries remind us that moral engagement does not hinge on constant exposure to negative stimuli. Instead, it emerges from thoughtful, deliberate actions guided by informed perspectives. Paradoxically, by controlling the flow of information, we can become more effective advocates. When we create intentional spaces to digest content at our own pace, prioritize quality over quantity, and interpret news in the light of moral principles, we emerge less stressed and more focused.

To begin, consider the different roles that information can play in moral life. We need enough data to understand the scale and urgency of environmental problems. Without facts, we might drift into denial or unrealistic optimism. We also need stories of success, adaptation, and resilience to preserve hope and model solutions. Beyond that, do we need to track every micro-change in atmospheric CO2 levels, or follow every environmental scandal unfolding across the globe? Probably not. Embracing the idea that we cannot—and need not—know everything is liberating. Being selectively informed is not a moral failing; it's a practical necessity. The world's complexity outstrips any single individual's capacity for omniscient awareness.

This recognition leads to the first principle of setting boundaries: identify what information truly serves your moral engagement. Are you an activist focusing on pollinator conservation? Then regularly checking local pesticide regulations and pollinator habitat programs might be essential, while daily scanning of global fisheries reports may not be. If your personal contribution lies in community-based reforestation, you can focus on climate policy updates relevant to forestry, or research on seedling survival rates, rather than drowning in every piece of climate news. By tailoring information intake to your moral domain, you avoid indiscriminate overexposure. Anxiety that comes from feeling obligated to track all environmental ills dissipates as you curate a manageable scope aligned with your priorities.

Another useful strategy involves creating temporal and spatial boundaries around media consumption. Instead of scrolling through environmental news during every free moment—on waking, during meals, and before sleep—set designated times for reading, listening, or watching updates. Perhaps allocate thirty minutes every other day to check a few reliable sources. Outside these times, give yourself permission to disconnect without guilt. This approach ensures that information remains a deliberate choice, not a reflex. It also provides emotional recovery periods, crucial for processing what you've learned and assimilating it into actionable insights. Without breaks, negative headlines blur into an unending cascade, dulling empathy or nurturing cynicism. With structured intervals, you reassert control, reducing the sense of drowning in information and allowing your emotional state to reset.

Selecting trusted sources matters too. In a media landscape fraught with misinformation, sensationalism, and partial truths, choosing where you get your environmental information is a moral act. Quality sources—reputable news outlets, peer-reviewed journals, credible environmental organizations—offer more balanced, accurate coverage. They present not only problems but also context, potential solutions, and measured assessments of progress. In these accounts, negativity is tempered by evidence-based hope. Reading a thoughtful report on biodiversity loss that also highlights conservation successes is less likely to induce despair than a random social media post proclaiming imminent ecological doom. By curating sources that respect complexity and highlight incremental improvements, you reduce anxiety and reinforce grounded hope.

It's also helpful to diversify the types of stories you consume. If you only read about catastrophic climate tipping points, you may adopt a distorted view that leaves no room for positive change. Complementing such stories with accounts of community adaptation, technological breakthroughs in clean energy, or policy reforms that have begun to curb emissions provides a fuller picture. Just as a balanced diet includes various nutrients, balanced media intake includes various types of information—data on challenges, narratives of resilience, policy analyses, cultural reflections, and success stories. This diversity prevents emotional exhaustion. Anxiety, if fueled solely by negative headlines, can be offset by concrete examples of improvements that affirm your moral investments are not futile.

In forming boundaries, consider the emotional signals your body and mind send. If reading yet another grim headline leaves you feeling paralyzed rather than inspired, take note. When you sense a creeping numbness or a desire to tune out completely, that's a sign to step back. Adjusting boundaries is an ongoing process, akin to calibrating the volume on a radio. Sometimes, when a major policy summit or environmental disaster dominates the news, you may allow more intake to stay current. Other times, when anxiety grows too strong, pull back. Remind yourself that moral engagement isn't measured by how many grim reports you endure, but by the quality of your understanding and the actions it inspires.

Another dimension of setting boundaries involves communal support. Discuss with friends, family, or activist peers how they handle environmental media. Sharing coping strategies transforms isolation into solidarity. Maybe a colleague learned to read environmental news through a weekly summary rather than daily feeds, or another found comfort in reading solutions-focused journalism first. By exchanging tips, you'll see that your struggle with overexposure is common. Anxiety lessens when you know others face it too and that together you can model healthier norms. Communities can develop guidelines—like social contracts—to discourage alarmist re-posting of dire news without context, or to regularly highlight local success stories. Such cultural shifts can reduce everyone's stress load, fostering moral resilience on a group scale.

Moreover, boundaries need not imply disengagement from moral responsibility. They are about quality over quantity, enabling you to remain informed enough to act effectively. If your goal is to support a local river restoration, staying updated on that project's milestones and related local ordinances matters. Scanning every global environmental crisis can be counterproductive if it drains your emotional reserves. By focusing on the river project's progress, you can contribute meaningfully—attending community meetings, volunteering for cleanup days, advocating for better management policies. In this way, boundaries direct your moral energy where it can have tangible impact, rather than scattering it across an overwhelming array of distant problems.

In professional spheres like environmental journalism, science communication, or advocacy leadership, setting boundaries can also prevent burnout. Journalists covering environmental beats face the burden of constantly reporting grim findings. Without self-imposed limits, they risk losing their capacity to tell nuanced stories. Scientists reviewing endless data on declining species might become jaded without balancing their reading with stories of successful recovery. Advocates engaged in policy battles must pace their engagement with relentless news cycles. Setting boundaries in these professions is not just personal self-care; it ensures that these moral agents remain clear-headed, compassionate, and able to convey balanced narratives. Anxiety does not have to become a professional hazard if professionals learn to filter content strategically.

Incorporating rituals can reinforce these boundaries. For example, after consuming environmental news, you might engage in a brief reflection practice—journaling about what you learned, identifying one positive takeaway, or noting a potential action you can take. By ritualizing the processing of information, you prevent passive absorption of negativity. Another ritual could be spending a few minutes outdoors after reading difficult reports, reconnecting with a local green space, a reminder that while global problems loom large, nature's presence and resilience can still be felt intimately. Such mindful interludes channel anxiety into thoughtful planning and reaffirm moral purpose.

Social media platforms require special attention. Algorithms often amplify contentious, emotionally charged content to maximize engagement. Curating your feed by following organizations or individuals known for balanced reporting, solutions-oriented journalism, or positive storytelling helps mitigate the barrage of despair. Unfollowing or muting accounts that consistently post alarmist content without nuance can restore mental equilibrium. Creating a separate social media account dedicated solely to following a curated set of environmental experts, researchers, and solution-driven projects can further shield you from doomscrolling. Here, anxiety emerges not from fear of missing out but from the recognition that selective exposure is an active moral choice to safeguard mental health and moral clarity.

As these boundaries prove effective, reflect on the changes in your emotional landscape and moral practice. You might find that you're less reactive, more patient, and better able to engage in constructive conversations. Instead of impulsively sharing every dire headline on social media, you might choose to pass along a well-researched article that explains both the problem and a set of response strategies. This shift from panic-driven communication to intentional sharing enhances the moral quality of discourse, inspiring hope and encouraging others to seek nuanced understanding rather than simplistic outrage. Anxiety no longer controls the narrative; it becomes a background alarm that spurs careful curation and thoughtful dialogue.

Boundaries with media intake also help differentiate between moral responsibility and self-imposed moral martyrdom. Consuming endless tragedies might feel like a tribute to caring, but it often backfires, reducing your capacity to care effectively. Recognizing that your moral agency depends on maintaining emotional resilience reframes boundaries as a responsible ethical practice. Just as a physician must rest to treat patients effectively, you must guard your mental energy to contribute meaningfully. Anxiety that taking breaks or limiting exposure equals moral negligence dissipates as you realize that maintaining long-term engagement requires sustainable emotional practices.

The wisdom of setting boundaries resonates with principles of moderation and self-knowledge found in various philosophical and spiritual traditions. Stoic thinkers advised focusing on what we can influence and releasing what we cannot. Buddhist teachings emphasize mindful awareness and balanced consumption of stimuli. Indigenous philosophies often stress harmony and inner balance as prerequisites for moral stewardship. Drawing on these traditions, environmental actors can reframe boundary-setting as a moral discipline: knowing one's limits, respecting emotional well-being, and cultivating an environment where informed action, not overwhelmed despair, is the norm.

In a broader moral context, setting media boundaries reminds us that knowledge alone cannot solve environmental crises. While awareness is crucial, action arises from a blend of understanding, willpower, strategic thinking, and emotional stability. Boundaries ensure that knowledge remains a tool for empowerment rather than a source of paralysis. When anxiety about the future flares, boundaries help maintain perspective: yes, the situation is complex, but not all developments are catastrophic. Yes, we must stay informed, but we can do so in moderated doses that support clarity rather than confusion.

As environmental discourse matures, social norms might shift so that boundaries become culturally sanctioned. Instead of glorifying those who read endless reports as more committed, communities might commend those who exhibit resilience and balanced engagement. Activist groups could advise newcomers to manage their media intake mindfully, thereby preventing burnout and ensuring the movement retains long-term moral momentum. In this scenario, anxiety about appearing uninformed or less dedicated fades as everyone recognizes that genuine commitment thrives in the careful management of emotional and informational resources.

While boundaries with media intake may seem like an individual coping mechanism, their moral implications extend outward. A well-informed, emotionally stable individual can contribute more effectively to collective efforts. When each participant avoids drowning in gloom, the group as a whole remains dynamic, solution-oriented, and capable of moral innovation. Anxiety no longer defines the group's identity; purpose does. Boundaries, therefore, serve as building blocks for communal resilience, helping communities steer environmental action with steady moral compasses.

In the end, making peace with the idea that you do not have to absorb every piece of grim news, every heated debate, or every apocalyptic prediction empowers you. It clarifies that moral worth isn't determined by how much suffering you witness, but by how you use the information you have to guide constructive, compassionate action. By setting boundaries, you forge a path of ethical engagement that remains energized, hopeful, and determined. In doing so, you transform eco-anxiety from a burden into a catalyst for a lifetime of caring, ensuring that moral vigilance endures without diminishing your capacity for empathy, inspiration, and meaningful contributions.

Continually Nurturing Emotional Well-Being and Updating Your Coping Toolkit

Acknowledging that caring for the planet is a lifelong moral engagement means accepting that emotional landscapes also evolve over time. Just as environmental issues shift, intensify, and present new complexities, your internal reactions—feelings of hope, anxiety, frustration, determination—also fluctuate. What worked yesterday to maintain emotional balance may not suffice tomorrow as challenges mount or personal circumstances change. The art of continually nurturing your emotional well-being and updating your coping toolkit is not an afterthought to moral action; it's integral to sustaining it. By viewing emotional resilience as a dynamic, ongoing practice rather than a one-time fix, you enhance your capacity to contribute meaningfully over decades rather than burning out after a few intense efforts.

Consider that moral engagement is not a linear journey toward an endpoint of steady calm. Sometimes, reading about a newly protected forest or a policy breakthrough invigorates your spirit. Other times, encountering dire climate projections or witnessing political setbacks may trigger despair. Instead of blaming yourself for these emotional swings, recognize them as natural responses to complex realities. Cultivating resilience means preparing for these fluctuations. You do this by assembling a coping toolkit—strategies, habits, mental frameworks, support networks—that you can adapt as conditions change. Just as ecosystems maintain resilience by diversifying species and ecological roles, you maintain moral resilience by diversifying your emotional support techniques, ensuring you never rely on a single coping method or assumption.

One aspect of updating your coping toolkit involves reassessing personal triggers and thresholds. Maybe you realize that certain topics—like coral reef collapse—hit you harder emotionally, leaving you feeling drained after reading related news. If that's the case, you might set boundaries around how often you expose yourself to that particular content, or pair reading about reefs with learning about reef restoration success stories to balance emotional impact. Over time, as you build emotional muscle, you might find you can handle more challenging content without spiraling into defeatism. Alternatively, you may decide to permanently limit exposure in certain areas, focusing your moral engagement elsewhere. This flexibility acknowledges that your emotional well-being isn't static; it's a resource you must steward wisely.

Continual nurturing also implies revisiting your goals and motivations. Early in your environmental journey, idealistic visions of quick victories may have energized you. But as you gain experience, you discover that progress is often slow and irregular. Adapting to this reality means recalibrating expectations. Instead of waiting for sweeping transformations to feel satisfied, you might celebrate partial policy shifts, small-scale restorations, or incremental lifestyle changes in your community. Doing so transforms the emotional calculus. Anxiety that once arose from unmet lofty goals is alleviated by acknowledging that moral value also lies in steady incremental improvements. Over months and years, you update your coping methods by incorporating more nuanced success criteria, learning to find inspiration in subtle but meaningful progress rather than only dramatic breakthroughs.

Another way to nurture emotional well-being is to connect with others who share similar values and struggles. Peer support, group discussions, and community gatherings provide emotional anchoring. While early in your journey, you might have relied solely on personal reflection, over time you realize that moral resilience is a collective undertaking. Engaging with a local environmental club, attending workshops, or joining online forums dedicated to solution-oriented environmental discourse can replenish your spirits. As you evolve emotionally, you might shift from attending occasional large protests—exciting but draining—to more frequent small meetings that offer sustained moral support and practical insights. By continually updating who you interact with and how, you ensure that your coping toolkit includes not just self-care techniques but also communal moral reinforcement.

In addition to social support, re-examining personal rituals can bolster emotional well-being. Early on, you might have found relief in nature walks or gardening, but as urbanization or life changes limit these options, you adapt—perhaps substituting indoor meditation with nature documentaries or volunteering at a local green rooftop garden. Similarly, as your understanding of environmental complexity grows, you may integrate mindfulness practices that focus on accepting uncertainty or rotate through different forms of stress relief—yoga, journaling, creative arts—that resonate best at various stages of your emotional journey. What begins as a small coping hobby might evolve into a more profound spiritual or philosophical practice, providing deeper emotional anchors. Continually revising your toolkit ensures that as your inner landscape changes, your methods of maintaining balance remain effective.

Consider the importance of learning new conceptual frameworks that reshape how you interpret challenging information. Initially, encountering negative news might have triggered a simplistic response—either panic or avoidance. Over time, you can adopt frameworks like complexity thinking, incremental progress paradigms, or resilience theory. By studying these models and integrating them into your worldview, you alter your emotional reactions to stimuli. When you read about an endangered species edging closer to extinction, complexity thinking reminds you that numerous conservation efforts are underway globally, and incremental progress paradigms reassure you that even small habitat restorations matter. Embracing these frameworks transforms raw anxiety into more measured emotional responses, allowing you to remain engaged without succumbing to despair.

Updating your coping toolkit also involves revisiting the moral narratives you tell yourself. Early in your journey, you might have constructed a narrative of impending doom if major reforms did not occur immediately. Over time, you refine this storyline to acknowledge that while urgency is real, moral solutions often emerge step by step. By rewriting your inner script, you free yourself from the all-or-nothing framing that intensifies anxiety. Now you might say: "Yes, the situation is serious, and I must do what I can, but I understand this is a long struggle filled with partial wins, learning opportunities, and evolving alliances." This reframed narrative reduces emotional pressure, allowing you to accept the complexity of moral challenges without feeling paralyzed by their enormity.

External changes can also prompt you to update your toolkit. Suppose a policy breakthrough you celebrated a year ago stalls due to political shifts. Initially, this regression might shatter your confidence. However, if you've incorporated flexibility into your moral thinking, you won't be entirely derailed. You adjust by seeking alternative advocacy routes, supporting different initiatives, or focusing on building local community resilience if national-level progress falters. Such adaptability prevents permanent emotional setbacks. Instead of viewing setbacks as evidence that hope was misplaced, you treat them as signals to modify strategies and emotional responses. This ongoing adaptation exemplifies how nurturing emotional well-being isn't about creating a bulletproof shield against disappointment, but maintaining moral agility that rebounds after each challenge.

Technological changes can influence your coping toolkit as well. Emerging communication platforms may offer new ways to access balanced, solution-focused environmental coverage. Mobile apps that compile aggregated environmental headlines accompanied by actionable suggestions can help channel anxiety into productive tasks, such as signing petitions, donating to reforestation campaigns, or participating in local cleanups. Over time, if one platform becomes too sensationalistic, you might switch to another that fosters calmer, more constructive dialogue. Just as ecosystems evolve under selective pressures, your media ecosystem evolves under moral and emotional considerations, selecting for outlets that stabilize rather than destabilize your moral engagement.

In professional contexts, such as environmental advocacy organizations or research institutions, leaders can model updating coping toolkits by offering training in emotional resilience, providing mental health resources, encouraging staff to rotate tasks to avoid burnout, and fostering environments where acknowledging emotional stress is accepted, not stigmatized. Over time, organizational cultures can shift from silently overloading staff with crisis narratives to equipping them with strategies to maintain emotional balance. This cultural adaptation, like ecological succession, produces more resilient organizations that can sustain moral effort over the long haul. Anxiety at the institutional level diminishes as staff learn that ongoing support, skill development, and reflection sessions are standard moral practices rather than emergency measures.

Intergenerational learning also plays a role. Veteran environmental activists who learned painful lessons about emotional burnout in previous decades can mentor younger ones, sharing how they adapted their coping techniques over time. Young activists, meanwhile, bring fresh ideas—like using certain apps for meditation or curating positive storytelling on social media—to enrich the older generations' toolkits. This exchange ensures that emotional well-being strategies remain current and context-sensitive. Anxiety about repeating past errors recedes as each generation inherits not just ecological knowledge but also moral lessons on emotional sustainability.

Another aspect involves integrating emotional resilience training into environmental education. Just as students learn about biodiversity or climate science, they can also learn how to handle eco-anxiety, how to navigate overwhelming information, and how to celebrate incremental improvements. Education becomes holistic, treating emotional resilience as integral to moral competence. Graduates enter adult life expecting that caring for Earth involves not only knowledge and ethical principles but also sustained emotional management. Anxiety about future activism burning them out is mitigated by proactive preparation and normalization of these moral skill sets.

Spiritual or philosophical traditions can further enrich your coping toolkit. If you resonate with certain spiritual teachings, you might incorporate rituals that ground your emotional energy—meditating on nature's cycles, reciting prayers of gratitude for small improvements, or reflecting on ancestral wisdom about perseverance. If philosophical perspectives inspire you, practicing stoicism, existentialism, or humanism can provide mental anchors against despair. For instance, stoicism encourages focusing on what you can control and accepting what you cannot, reducing anxiety fueled by unattainable certainties. Over time, you may weave these spiritual or philosophical threads into your emotional toolkit, enabling a richer, more stable moral identity.

Cultural expressions of resilience—music, art, poetry—can also become part of your evolving coping methods. At first, you might passively listen to a nature-inspired playlist to relax after reading tough news. Later, you might actively create art that channels your feelings, transforming anxiety into creative expression. By doing this, you enlarge your emotional repertoire, ensuring that when one method feels stale, another remains available. Just as ecosystems maintain resilience through diversity, your emotional toolkit thrives on variety, ensuring no single coping strategy becomes overtaxed or loses efficacy.

The practice of regularly evaluating and updating your toolkit cultivates self-awareness. Periodically ask: "Which coping methods still serve me well? Which feel outdated or ineffective? Have new challenges emerged that require different responses?" Through honest introspection, you avoid stagnation and prevent harmful patterns. If doomscrolling re-emerges as a habit, you respond by reinforcing media boundaries. If catastrophic thinking creeps back, you recall complexity frameworks that encourage incremental progress perspectives. This cycle of self-assessment mimics adaptive management in ecological restoration, where managers continually monitor conditions and adjust actions accordingly. Anxiety, in this analogy, is like ecological feedback—signaling that a shift in strategy may be needed.

On a communal scale, updating coping toolkits collectively strengthens moral cohesion. Groups can hold periodic check-ins to discuss emotional states, share new coping strategies, highlight fresh success stories, or retire outdated approaches. This communal reflective practice transforms anxiety from a solitary burden into a shared moral concern. Just as ecosystems rely on communication signals among species—warning calls, pollinator cues—communities rely on open conversations about emotional well-being to maintain collective moral health. Over time, communal resilience emerges: no one faces moral or emotional challenges alone, and a collective intelligence refines coping methods as external conditions shift.

In essence, continually nurturing emotional well-being and updating your coping toolkit accepts that moral engagement is not a static calling but a dynamic moral craft. Anxiety will never vanish; it remains a companion, alerting you to looming threats and moral imperatives. But by regularly fine-tuning how you handle information, interpret setbacks, celebrate successes, and integrate multiple sources of support, you transform anxiety into a guide rather than a tyrant. This approach, far from indicating weakness, displays moral maturity. It shows that you've learned to treat yourself as a long-term moral participant who must safeguard inner resources to contribute effectively.

As environmental challenges evolve—new science emerges, social movements shift emphasis, political climates change—your coping toolkit also evolves. The methods that once helped you cope with early fears may give way to more sophisticated techniques suitable for advanced moral challenges. Instead of lamenting that old comforts no longer suffice, welcome the chance to discover new strategies, refine old ones, or adopt hybrid approaches. This continuous learning mirrors the adaptive strategies you advocate for ecosystems and institutions. Internalizing that resilience is a process rather than a destination ensures that your moral journey remains energized, adaptable, and genuinely sustainable.

In the grand tapestry of environmental care, emotional well-being stands as a crucial thread. Without emotional resilience, moral insights and technical solutions risk unraveling under stress. By committing to ongoing nurturing of your emotional health and the regular updating of your coping toolkit, you ensure that your moral contributions endure. Your ability to remain informed without succumbing to despair, to persist through setbacks, and to celebrate incremental progress flows from this conscious commitment. Anxiety thus becomes a manageable signal rather than an insurmountable barrier, allowing you to approach environmental stewardship as a lifelong moral vocation buoyed by well-tended emotional resources.

Mentoring Others, Paying Forward Insights, and Embracing Your Role in a Generational Effort

As your environmental commitment deepens and evolves, one natural step in sustaining moral engagement is to mentor others. Though you may have begun as a novice grappling with eco-anxiety, over time you accumulate knowledge, hone coping strategies, and refine your ethical frameworks. These hard-won insights are not just personal achievements; they represent moral capital that can be passed on to newcomers. Mentoring involves more than offering factual information. It means guiding others through emotional turbulence, helping them distinguish grounded hope from naive optimism, and teaching them how to recognize incremental progress and celebrate small steps. In doing so, you pay forward the moral support once given to you by mentors, peers, or thought leaders who showed you that anxiety need not lead to paralysis and that moral perseverance is possible.

Mentoring might occur in formal or informal contexts. Perhaps you coach a student group at your old university, introducing them to the concepts and emotional skill sets you've gleaned from experience. You might facilitate workshops at a local environmental organization, blending pragmatic advice—like how to analyze policy changes or track incremental improvements—with moral reflections on resilience and complexity. Informal mentorship might involve supporting younger colleagues at your workplace who are overwhelmed by grim climate reports, offering them practical coping strategies to maintain their moral footing. The goal is to ensure that each new generation of advocates, policymakers, scientists, and engaged citizens does not have to reinvent the moral wheel. Instead, they inherit a tradition of ethical understanding, emotional resilience, and strategic flexibility that you have helped shape and refine.

Paying forward these insights fosters continuity and stability in moral movements. Without such generational transfer, hard-earned lessons risk vanishing, forcing new participants to stumble through avoidable mistakes. By contrast, transparent mentorship encourages collective learning. Just as ecosystems thrive when experienced individuals guide juveniles, teaching survival tactics and navigation routes, moral communities flourish when established voices nurture fresh talent. This process ensures that each cohort enters the fray better prepared, armed not only with knowledge about policies and technologies but also with emotional tools to handle distressing information, navigate uncertainty, and sustain hope. Anxiety diminishes when newcomers sense they are not alone—that they stand on moral ground enriched by predecessors who understand their struggles and have forged coping methods tailored to the complexity of environmental challenges.

Another dimension of mentoring involves emphasizing that moral engagement transcends any single generation. Environmental stewardship is not a sprint completed by one cohort; it unfolds as a relay race. Each generation takes the baton, advances a portion of the struggle—perhaps stabilizing certain greenhouse gas levels, restoring key habitats, or implementing groundbreaking laws—before passing it on. By mentoring others, you acknowledge this generational continuum, accepting that your role is to build on previous gains and prepare successors to push further. Instead of feeling anxious about not witnessing ultimate solutions in your lifetime, you recognize that moral value resides in securing a better starting point for those who follow. This perspective transforms anxiety about long-term outcomes into a sense of purposeful stewardship. You know you may not see all your dreams realized, but your mentorship ensures that future moral agents inherit robust intellectual, emotional, and ethical resources to continue the work.

In some instances, mentoring might involve bridging generational divides. Younger activists bring fresh energy, digital literacy, and new cultural expressions, while older participants have accumulated wisdom and patience. Through mentorship, these differences become complementary rather than sources of tension. Younger recruits learn how veterans overcame despair after policy setbacks, while experienced mentors learn from youth who propose creative outreach strategies or cutting-edge tools to track incremental improvements. Such cross-pollination strengthens moral ecosystems, encouraging diversity of approaches and mutual respect. Anxiety that divides in age, background, or approach could fragment moral efforts recedes when mentorship reveals that each generation can learn from the others' strengths and mitigate each others' weaknesses.

Mentoring also includes guiding others in their emotional journey. As newcomers encounter disheartening data or face their first significant policy defeat, mentors can share stories of resilience. They can describe how they once felt crushed by setbacks but discovered new strategies—focusing on solution-oriented news, joining a supportive community, or learning from nature's resilience patterns—to restore emotional equilibrium. These personal anecdotes humanize the struggle, showing that moral perseverance is not a matter of iron will alone but of adaptability, humility, and continuous moral growth. By normalizing emotional fluctuations, mentors reduce the stigma around eco-anxiety, inviting open discussions that lead to collective emotional support. Over time, a culture of honesty and empathy emerges, reinforcing that moral action does not require stoic detachment but can accommodate genuine feelings, doubts, and renewed determination.

In addition, paying forward insights involves helping others interpret complexity. Many arrive in environmental discourse expecting linear solutions—fix X problem by doing Y—and become disheartened when reality proves non-linear. A mentor can explain that incremental progress, policy cycles, and adaptive management define moral engagement. By presenting these complexities as features, not bugs, of moral life, mentors prevent novices from concluding that complexity equals failure. Instead, complexity becomes the playground of moral innovation. Anxiety over unpredictable outcomes can be reframed as a prompt to learn multiple strategies, build alliances, and remain open to adjustments. In this sense, mentorship weaves moral narratives that prepare newcomers to embrace uncertainty without losing sight of their agency.

Mentoring also extends beyond personal encounters. Writing articles, recording podcasts, producing educational videos, or maintaining accessible websites that compile case studies of successful incremental improvements or regenerative approaches can serve as mentoring at a larger scale. Such public mentorship broadens access to moral insights, ensuring that even those without direct contact with experienced environmentalists can learn from their accumulated wisdom. By democratizing the insights gleaned from decades of environmental activism and research, societies enhance collective resilience. Anxiety that only a select few have the "secret" to staying morally strong in the face of environmental challenges disappears as resources become widely available. Everyone willing to learn can find guidance, making moral engagement more inclusive and less elitist.

Acknowledging that moral engagement extends across generations also means recognizing that different historical contexts shape moral tasks. Past movements addressed pollution crises with partial successes, leaving a legacy of norms and institutions that today's activists build upon. Future generations will face new conditions—perhaps more advanced renewable technologies or more complex migration patterns due to climate change—and will in turn refine and adapt current strategies. Mentors can highlight this historical continuity, relieving newcomers of the pressure to solve everything now. Anxiety about leaving problems unresolved melts into an understanding that progress is cumulative, that their job is to improve conditions, not deliver final perfection. This perspective allows moral agents to celebrate partial victories and incremental shifts without feeling overshadowed by the still-enormous tasks awaiting future cohorts.

In corporate or political arenas, mentorship can guide moral reforms that outlast individual careers. Experienced policymakers can mentor rising staffers, imparting lessons about negotiating environmental clauses in trade agreements or building cross-party alliances. Similarly, in businesses moving toward sustainable supply chains, senior managers committed to environmental ethics can mentor younger colleagues who share these values. Over time, institutional cultures evolve as mentored individuals rise in ranks and maintain ethical priorities. Anxiety about changing internal cultures diminishes when mentorship ensures continuity of moral perspectives even as personnel rotates. This gradual embedding of moral insight into organizational DNA contributes to long-term shifts, reinforcing the idea that moral transformations are collective, slow, and enriched by guidance across ranks and generations.

From an educational standpoint, schools and universities can integrate mentorship programs connecting seasoned environmental educators, activists, scientists, or policymakers with students. Young people gain not only technical knowledge but also moral and emotional frameworks for confronting complex challenges. Mentors can share how they navigated disillusionment, learned to celebrate small steps, recognized nature's resilience, and integrated Indigenous wisdom. By receiving such guidance at an early stage, students develop moral resilience earlier, entering adulthood better prepared. Anxiety about future generations having to start from scratch is relieved. Instead, a legacy of moral strategies cascades downward through time, ensuring each generation starts closer to meaningful solutions than the last.

In spiritual or cultural communities, mentorship might take the form of elders guiding younger members in rituals that honor seasonal changes or acknowledge ecological connections. Transmitting these traditions ensures that moral sentiments—gratitude, humility, reverence—live on. Younger adherents learn that their role isn't just to inherit beliefs passively but to apply them in defending and restoring ecosystems. Anxiety that cultural shifts might erode environmental reverence diminishes as spiritual mentors show how old practices adapt to new ecological contexts. By blending ancestral ceremonies with contemporary knowledge, faith communities craft moral narratives that bridge temporal gaps, ensuring each generation understands its role as a link in a chain of care.

Mentoring also includes helping others recognize when to update their coping toolkit. Perhaps a mentee struggles with doomscrolling through negative environmental headlines. A mentor might suggest setting boundaries with media intake or pairing difficult news with hopeful solution stories. As the mentee grows more confident, they might later discover their own coping innovations—like organizing group meditation sessions before high-stakes activism events—and eventually mentor someone else. This continuous exchange of emotional management techniques ensures that moral resilience never stagnates but evolves as a shared cultural asset. Anxiety about coping alone or repeating old mistakes fades, replaced by trust in a network of moral support and intellectual exchange.

In the larger moral universe, mentoring others and paying forward insights validates the notion that moral engagement is not a solitary quest but a collective tapestry woven over decades and centuries. By embracing your role in this generational effort, you see that your time on Earth is an integral chapter in a longer story. Anxiety about your individual limits or the magnitude of global problems shrinks before the realization that others come after you, ready to continue the struggle armed with knowledge you helped refine. This understanding imbues your efforts with additional meaning: you're not just acting for your own generation's sake but ensuring future moral actors inherit a stronger foundation, better emotional tools, and clearer moral maps.

Mentoring can also alleviate burnout. Instead of keeping all your insights and burdens internalized, sharing them lightens the load. Knowing you're not alone in the moral trenches, that younger or newly engaged individuals absorb your lessons and benefit from them, infuses you with a sense of purpose. Anxiety that your personal contributions fade once you step back or slow down dissipates. You realize that by mentoring, you contribute to a legacy that persists after your direct involvement lessens. This legacy becomes a kind of moral immortality—your ideas, strategies, and emotional wisdom live on in those you've guided, shaping environmental action well into futures you will not personally witness.

Embracing your role in a generational effort also reframes time. Instead of fixating on immediate outcomes, you think in decades or longer. Just as slow ecological restoration teaches patience, the transfer of moral insights across generations promotes acceptance that meaningful change takes sustained effort. Anxiety about urgency doesn't vanish, but it's channeled into structured approaches to mentorship: setting up mentorship programs, writing guides for future activists, conducting workshops that integrate past lessons with emerging trends. Over time, this forward-looking perspective fosters moral equanimity. You do what you can now, knowing your careful documentation, storytelling, and mentorship will amplify the impact of your current work later.

At its core, mentoring and paying forward insights stand as acts of moral generosity and strategic foresight. They demonstrate that you understand your limitations, yet believe in the cumulative power of collective moral intelligence. Each time you help someone navigate eco-anxiety, understand incremental progress, or appreciate nature's resilience, you strengthen moral continuity. Anxiety over fragmentation or moral amnesia recedes as you witness fresh voices carrying forward principles and wisdom, ensuring that no moral lesson is lost to time. Instead, they evolve, resonate, and inspire new adaptations.

As environmental conditions remain uncertain, the importance of mentoring becomes ever clearer. The next generation will face different climate patterns, novel technologies, and altered political landscapes. By providing them with moral frameworks that embrace complexity, incrementalism, and relational ethics drawn from ecosystems and indigenous traditions, you equip them to handle scenarios you cannot predict. Instead of leaving them vulnerable to despair or naive optimism, you offer them a compass that navigates shifting moral terrains. Anxiety about the future's unknowability transforms into confidence that future agents will be well-equipped, flexible, and morally resilient, thanks to the groundwork you've helped lay.

In conclusion, mentoring others, sharing the lessons gleaned from your evolving experience, and embracing your generational role reinforce the moral continuity essential for long-term environmental care. Through mentorship, your insights transcend temporal boundaries, guiding an ongoing lineage of moral actors. Instead of isolating yourself within one era's challenges, you become part of a continuous moral dialogue that enriches past achievements and informs future endeavors. Your personal moral arc intertwines with that of humanity's long environmental journey, ensuring that anxiety and hardship, while ever-present, never fully overshadow hope, creativity, and moral progress across generations.

This long journey of moral engagement underscores that environmental stewardship is not a sprint to a static endpoint, but a dynamic process shaped by changing realities, evolving emotions, and shared responsibility across generations. By learning how to manage media intake, you protect your emotional stamina, ensuring that information becomes a tool for informed action rather than a source of paralysis. By continually refining your coping strategies, you stay flexible and resilient, better equipped to adapt as the world transforms and as your personal circumstances shift. And by mentoring newcomers, you pass forward insights that help others navigate complexity without succumbing to despair—reinforcing a chain of moral continuity that threads past wisdom into present efforts and future solutions. In doing so, you help forge a community that recognizes caring for the planet as an ongoing collective endeavor, one that grows stronger as it incorporates the lessons of each generation, nurtures emotional well-being, and transforms anxiety into steadfast, hopeful engagement.

Epilogue: Embracing Complexity, Choosing Courage

Reflecting on Your Personal Growth and Shifted Perspectives

As you find yourself at this juncture—having navigated the moral complexities, emotional stresses, and evolving strategies for environmental stewardship—take a moment to consider the transformation you've undergone. When you first confronted the tangled realities of climate change, biodiversity loss, and ecological injustices, your emotional landscape might have felt raw and unsteady. Anxiety may have overshadowed confidence, and the enormity of global challenges could have seemed paralyzing. Now, after reflecting on the incremental steps, resilience patterns from nature, the importance of grounded hope, and the enduring moral journey guided by wisdom from multiple traditions and voices, what emerges is a sense of having matured ethically, emotionally, and intellectually.

This shift does not imply that all doubts have vanished or that you've attained perfect moral clarity. On the contrary, it acknowledges that moral growth doesn't mean the absence of fear, uncertainty, or sadness; it means learning to work with those emotions constructively. Instead of seeking escapes or rigid certainties, you've learned to embrace complexity—the idea that solutions often arise from multifaceted efforts, incremental progress, cultural wisdom, adaptive policies, and evolving emotional coping strategies. Facing complexity no longer signals defeat but an invitation to continuous learning and moral innovation. The sense of being overwhelmed transforms into a humble recognition that your role is to contribute where you can, knowing that you are part of a larger, multigenerational endeavor.

Throughout this journey, you may have noticed your perspective shifting in subtle yet profound ways. Perhaps once you believed that only large-scale systemic changes mattered; now you understand that smaller, local efforts, too, hold moral and practical significance. You might have started expecting clear-cut good and bad actors in environmental narratives and ended up appreciating that ecosystems and societies thrive in nuanced relationships rather than binary categories. This layered understanding parallels the moral frameworks you've encountered—from incremental progress paradigms to relational ethics informed by indigenous insights—revealing that living ethically amid environmental challenges demands continuous adaptation rather than static rules.

Another sign of personal growth is your evolving relationship with eco-anxiety. Early on, anxiety might have felt like a moral verdict—that the world's troubles were too grave and your efforts too insignificant. Over time, you have come to see anxiety as neither an enemy nor a mandate for perfectionism, but as a signal guiding you toward prudent action and emotional resource management. You've learned to channel worry into setting media intake boundaries, seeking communities of support, celebrating small steps, and acknowledging nature's resilience. This recalibration of anxiety's role shows emotional intelligence at work: fear, instead of dictating inaction or despair, now informs balanced engagement and compassion. If at first you viewed your emotional upheavals as personal flaws, you now recognize them as natural responses that can be harnessed, tempered, and integrated into a committed moral life.

Experiences with incremental successes have also shaped your outlook. Observing how partial victories accumulate and reinforce each other teaches you that meaningful progress does not require immediate, grandiose changes. Recognizing incremental improvements—like a policy tweak that paves the way for stronger regulations later, or a restored wetland that slowly nurtures species back—provides moral nourishment. These successes affirm that your efforts, combined with countless others, create ripples that spread beyond initial expectations. This understanding disarms the all-or-nothing mentality, encouraging patience, perseverance, and trust in the iterative, collective process of environmental healing. Where once anxiety told you that small gestures were futile, now you accept them as building blocks of systemic shifts.

Moreover, your perspective on hope has evolved. Perhaps you once conflated hope with naive optimism—an expectation of effortless fixes or miraculous turnarounds. Through this journey, you've learned the difference between hope grounded in complexity and illusions that deny reality's weight. Hope, as you now practice it, is a moral stance requiring courage and continuous effort. It thrives on acknowledging the difficulties, the trade-offs, and the unknowns, without yielding to apathy. This hope does not promise easy endings; it pledges the resolve to keep moving forward despite uncertainty. Such a refined understanding of hope enriches your emotional toolkit, enabling you to remain morally present even when conditions appear grim.

Your views on collaboration and knowledge sources may have broadened as well. If initially you relied solely on scientific reports or mainstream media narratives, you might now appreciate the moral depth offered by indigenous wisdom, local traditions, artistic expressions, and spiritual teachings that emphasize reciprocity and relational thinking. Integrating these varied forms of knowledge helps break down ideological rigidities, opening moral discourse to pluralism and cultural humility. You realize that moral guidance can emerge from many places, and that environmental stewardship improves when diverse voices contribute their insights. Anxiety rooted in not having a single authoritative roadmap fades as you embrace that moral navigation improves through open dialogue, cross-cultural exchange, and the blending of multiple lenses.

At a personal level, you might feel more grounded in your moral identity. By learning to manage emotional well-being, you resist burnout, maintaining consistency and longevity in your engagement. Watching nature's resilience patterns, you appreciate that restoration and adaptation take time, and that moral work must be paced accordingly. Just as ecosystems benefit from diversity and redundancy, you benefit from a variety of emotional supports—communal rituals, personal reflection, hopeful storytelling—to stay committed. Instead of feeling guilt over not being able to handle every crisis report, you accept human limitations and direct your energies where they are most impactful. Anxiety about not doing enough transforms into clarity about where and how you can meaningfully contribute, reducing paralysis and encouraging proactive steps.

This journey has also likely softened certain judgments and absolutes. Where you once sought pure heroes or villains, you now recognize that the moral landscape is populated by actors responding to incentives, cultural histories, and conflicting values. While maintaining moral standards and accountability remains crucial, you understand that empathy and inclusive dialogue can foster cooperation. Anxiety about entrenched opposition or irreconcilable differences moderates as you see examples of policy negotiations, community dialogues, and blended approaches that transcend polarized positions. This maturing view encourages pursuing alliances, incremental policy shifts, and mutual learning, rather than demanding all-encompassing victories overnight.

Reflecting on these shifts, you can appreciate that embracing complexity and choosing courage isn't a one-time decision but a practice woven into your moral life. Each new environmental challenge—whether a surprise extinction event, an unexpected climate policy reversal, or an unforeseen technological innovation—tests and refines your perspectives. As conditions evolve, so does your moral understanding. You learn to greet future uncertainties not solely with anxiety, but with a readiness to adjust your coping strategies, moral frameworks, and collaborative efforts. Instead of seeing change as destabilizing, you accept it as inherent to the moral enterprise of stewardship. Your personal growth becomes visible in how you handle tomorrow's complexities with greater composure, insight, and willingness to adapt.

These reflections on your personal journey affirm that environmental moral engagement is not a static skill or a state you "arrive at," but an ongoing evolutionary process. If earlier you longed for neat formulas or absolute certainties, now you value resilience, adaptability, and relational thinking as guiding virtues. If fear once ruled your emotional landscape, now it coexists with hope, patience, and an understanding of incrementalism. If skepticism about collaborative solutions once dominated, now you see that alliances, mentorship, and intercultural dialogues enhance collective moral capacity. In essence, these changes in your perspective illustrate that moral growth emerges not from escaping complexity, but from embracing it, and not from eradicating anxiety, but from learning to dance with it.

In acknowledging these transformative insights, you pay homage to the moral journey you've undertaken. You recognize that your understanding of environmental challenges has deepened, your emotional resources broadened, and your sense of possibility expanded. While you may not have resolved all uncertainties, you have learned to coexist with them, turning them into spaces for ethical exploration. No single chapter, lesson, or technique delivered this transformation alone. Rather, it arose cumulatively, as you integrated ecological wisdom, Indigenous traditions, regenerative practices, rational policymaking frameworks, and strategic emotional coping methods.

The result of these integrations is a state of moral readiness. You stand better equipped to confront emerging crises, mentor newcomers, refine strategies, and evolve your emotional responses as needed. While the world you inhabit remains fraught with ecological instability and social tensions, your capacity to respond effectively, humanely, and creatively has grown. Anxiety does not vanish from this landscape, but it no longer defines your moral horizon. Instead, it becomes one voice among many in a rich moral conversation—prompting attention, caution, and ingenuity, but never drowning out the steady drumbeat of hope, determination, and ethical commitment.

Reflecting on personal growth and shifted perspectives reaffirms that moral life is, at its core, a continuously adaptive adventure. By looking back at how far you've come, you ensure that the lessons learned are not static memories but active guides shaping your future moral steps. Each challenge you overcame, each coping strategy you perfected, each new insight you integrated, prepares you for whatever lies ahead. In this sense, reflecting on your personal journey is not a farewell gesture; it's a moral checkpoint. You leave this chapter—not with all problems solved or doubts erased—but with a deeper understanding of complexity, a stronger moral compass, and the confidence to face uncertainty with courage and compassion.

Affirming the Human Capacity for Creativity, Compassion, and Sustained Engagement

As you reach the final segment of this moral exploration, consider one more vital recognition: humans possess a remarkable capacity not only to learn and adapt, but to persevere, innovate, and care over the long term. Environmental challenges, layered with ethical nuances and ever-shifting conditions, might seem tailor-made to wear down morale. Yet history and personal experience both show that people can rally their creativity to devise solutions, summon compassion to support each other, and sustain engagement even under daunting circumstances. Embracing complexity and choosing courage, as you have discovered, does not rest on isolated acts of heroism. Instead, it draws on a wellspring of human qualities: empathy, imagination, curiosity, and a refusal to give up on the future.

Creativity stands at the heart of moral resilience. Where environmental problems defy simple answers, innovative minds find new approaches. Consider how entrepreneurs develop green technologies, farmers experiment with regenerative methods, educators integrate environmental ethics into curriculums, and activists pioneer novel campaigns that bridge divides. None of these innovations emerge from static thinking. They result from daring to think differently, to challenge assumptions, and to blend insights from multiple fields. Creativity transforms anxiety over uncertainty into a stimulus for finding fresh paths. Instead of feeling trapped by complexity, creative thinkers see complexity as fertile ground for experimentation and breakthroughs. This capacity ensures that moral agents are not limited to old paradigms or hopeless scenarios; they can push the boundaries of what's possible, forging more sustainable models of living and cooperating.

Compassion weaves together the moral fabric necessary to support enduring engagement. Without compassion, the weight of environmental injustice, the suffering of marginalized communities, and the plight of endangered species might overwhelm us. Compassion, however, reminds us that caring is not a burden to be avoided but a core human strength. By empathizing with those most affected—be they frontline communities facing rising seas or future generations yet unborn—we find moral motivation to persist. Compassion bridges gaps, softening ideological conflicts as people recognize shared vulnerabilities and aspirations. This moral quality turns fear into solidarity, inviting allies to cooperate, share resources, and celebrate each other's efforts. Compassion makes moral struggles feel less like solitary trials and more like collective endeavors supported by a community of morally invested individuals.

Sustained engagement—the willingness to keep contributing, learning, and adapting over decades—depends on nurturing these qualities. If creativity supplies solutions and compassion fosters support, sustained engagement ensures that neither flashes of innovation nor moments of empathy vanish without impact. Through sustained engagement, ideas mature, initiatives scale up, and long-term policies replace short-term fixes. Anxiety about relapses into destructive practices recedes when we accept that moral persistence can outlast temporary setbacks.

Knowing that endurance matters as much as brilliance encourages everyone to pace themselves, to continually refine strategies, and to remain loyal to moral principles even when progress is uneven.

This collective moral capacity, extending through generations, offers hope that environmental care need not be the lonely domain of a few determined souls. Instead, it can be a broad cultural project that taps into human strengths widely distributed across societies. Each person who learns to manage eco-anxiety, celebrate incremental improvements, or embrace uncertainty as possibility contributes to a more resilient moral ecosystem. Each adaptation of emotional coping methods, each mentorship session guiding newcomers, each respectful integration of Indigenous wisdom into policy frameworks, builds this communal reservoir of moral potential. Over time, these accumulations pay off: cultural shifts in norms, heightened public awareness, improved governance structures, and more inclusive, empathetic dialogues about environmental futures.

Realizing that human beings can collectively rise to these challenges, despite persistent uncertainties and looming threats, is itself a powerful form of reassurance. It reminds us that moral engagements are not doomed to fail; they are actively shaped by the choices we make today. If creativity spurs solutions, compassion nurtures solidarity, and sustained engagement ensures continuity, then no matter how thick the fog of uncertainty, moral navigation remains feasible. Anxiety, while never completely absent, becomes an occasional companion rather than a tyrant. In its presence, people act anyway, drawing strength from the knowledge that they are part of a larger lineage of moral actors who faced similar doubts and pressed forward regardless.

In envisioning the future, recognizing human potential for creativity, compassion, and enduring commitment imbues moral landscapes with depth and promise. Creativity suggests that environmental solutions will not stagnate but evolve; compassion signals that nobody must shoulder burdens alone; sustained engagement ensures that temporary gains accumulate into structural transformations. While no single quality can guarantee success, the synergy of all three shapes a moral posture resilient enough to handle setbacks, flexible enough to incorporate new insights, and robust enough to confront unknown crises. Humanity's collective moral development unfolds as a story of learning from errors, building on partial victories, refining ethical convictions, and continuously reimagining how to live in harmony with a changing planet.

Reflecting on this capacity does not deny the gravity of environmental struggles. Rather, it reframes them in a perspective that honors human agency. Acknowledging the depth of creativity available means no challenge is beyond our resourcefulness. Trusting the abundance of compassion means we never have to face these challenges alone. Relying on sustained engagement means we don't need immediate perfection; we can commit to a path of steady moral improvement. Anxiety diminishes because the moral toolbox is full: no matter what the future holds, these qualities—cultivated, shared, and handed down through generations—equip us to respond ethically and effectively.

Just as ecosystems benefit from diverse species each playing unique roles, human moral ecosystems thrive when multiple strengths—intellectual, emotional, cultural—interact. Creativity, compassion, and sustained engagement ensure that no single setback defines us.

Instead, we interpret each difficulty as a call to reapply moral principles, adapt strategies, and persist. This pattern, repeated countless times across centuries, has allowed communities to overcome historical calamities and injustices. Today's environmental crises, though unprecedented in scale, meet a humanity that can draw upon expanded moral knowledge. Anxiety remains a motivator, reminding us not to be complacent, but it no longer paralyzes. Instead, it signals that we must activate the capabilities honed over these long arcs of learning.

In embracing complexity and choosing courage, you acknowledge these capacities not as distant ideals but as living potentials within yourself and your community. You've learned to navigate the moral journey of environmental care with greater subtlety—balancing awareness of dire headlines with a trust in nature's resilience, celebrating partial improvements rather than fixating on total solutions, and weaving together rational analysis with empathic understanding. These transformations testify that humans are not passive passengers on a doomed planet, but active participants shaping moral destinies. Creativity helps imagine possibilities, compassion ensures no one is left behind, and sustained engagement locks in moral progress over time.

Considering this moral landscape, you can find solace and determination. Instead of dwelling on everything that remains uncertain or undone, you recognize the tools and virtues that empower humanity to respond constructively. Anxiety may still whisper doubts in your ear, but it can't drown out the chorus of hope, curiosity, and solidarity you've come to trust. In the face of ongoing challenges, you rely on these strengths to orient your next steps, encourage others, and refine your methods. Your own transformation—from a person anxious about global crises to one who harnesses anxiety as a signal for growth—mirrors a collective moral evolution. As more people undergo similar changes, societies become better equipped to address environmental needs.

Looking forward, the path remains complex. But complexity itself becomes less intimidating when you know that creativity can generate a multitude of approaches, compassion can keep moral communities cohesive, and sustained engagement ensures that partial successes expand into systemic improvements. These human capacities form a moral scaffolding that can adapt to diverse conditions. Instead of hoping for an easy fix or a single generation's heroic efforts to solve everything, you trust in the slow, deliberate accumulation of moral capital. With each passing year, new voices join the dialogue, new insights emerge, and the human moral tapestry gains richer patterns and sturdier threads.

These reflections underline that the journey of caring never truly ends—it continually evolves as both external challenges and internal understandings shift. Instead of longing for a final closure, you embrace the open-endedness of moral life. There will always be more to learn, more to improve, more relationships to cultivate, and more challenges to meet. This open-endedness does not foster anxiety but encourages a vigilant, hopeful stance. Armed with creativity, compassion, and sustained engagement, you face the future not as a powerless spectator but as a moral agent capable of influencing ecological and cultural trajectories.

By affirming this human capacity to evolve and persist morally, you affirm your place within a grand moral narrative. You stand on shoulders of past generations who overcame struggles, benefiting from their legacies of wisdom and courage. You contribute your own refinements, making it easier for future generations to carry on. Just as ecosystems accumulate complexity and stability over evolutionary time, human moral ecology enriches over historical time. Anxiety, always present, is contained by the knowledge that you belong to a lineage of moral explorers continuously charting new ground, forging alliances, and crafting solutions.

Ultimately, by focusing on these human capacities—creativity, compassion, and sustained engagement—you secure a moral vantage point that resists despair and fatalism. As each new day brings fresh environmental challenges or altered political landscapes, you respond not by retreating or lashing out, but by engaging in collaborative problem-solving, empathic dialogue, and long-term moral planning. This approach eschews naive optimism or denial; it acknowledges difficulties while harnessing human potential to rise above them. Instead of anxiety ruling the horizon, the horizon stands open, a space where deliberate, courageous actions can reshape possibilities.

All these insights reinforce that environmental stewardship, far from being an impossible demand, is a moral opportunity to express the best human qualities. Understanding complexity and uncertainty doesn't drain energy; it recalibrates expectations, enabling you to focus on what you can achieve and how to achieve it ethically. By championing incremental steps, nurturing emotional well-being, and guiding others through mentorship, you leave the world slightly better than you found it and ensure that future generations discover an environment of moral guidance, hope, and resilience waiting for them.

Thus, by affirming the human capacity for creativity, compassion, and sustained engagement, you solidify your moral readiness to continue shaping a more just, healthy, and life-sustaining world. Each individual who embraces these capacities contributes a vital strand to a tapestry of moral growth that extends beyond any single era. It's a testament to human moral evolution—an ongoing collective achievement that transforms anxiety into initiative, despair into purpose, and fragmented efforts into coherent progress toward a more harmonious coexistence with the Earth.

Carrying Forward a Sense of Purpose, Rooted in Hope and Action

When confronting environmental challenges, one of the most transformative realizations is that purpose does not arise solely from achieving grand, immediate victories. Instead, purpose emerges gradually, nurtured by each incremental improvement, each relationship forged, each lesson learned. As you navigate complexity, embrace uncertainty, and find ways to cope with eco-anxiety, the moral path you follow begins to feel less like a series of isolated tasks and more like a cohesive narrative. This narrative, fueled by a sense of purpose, weaves together hope and action into a durable moral tapestry. No single heroic moment defines this fabric; it's the ongoing interplay of multiple threads—practical steps, compassionate choices, creative solutions, emotional resilience—that yields enduring meaning.

Purpose itself evolves as you progress through different stages of moral engagement. Early on, you might have felt driven primarily by alarm—fear for the planet's future and frustration at slow progress. Over time, as you integrate insights from complexity thinking, incremental progress models, and emotional coping strategies, your motivations deepen. You learn to find moral value in modest achievements, steady commitments, and the quiet persistence of communities striving to improve conditions. Purpose no longer hinges on grandiose transformations. Instead, it infuses the daily acts of planting a tree, advocating for a local wetland, speaking thoughtfully about policy reforms, or mentoring a newcomer. By embracing these incremental contributions, you anchor your sense of purpose in reality rather than unattainable ideals.

This shift reflects a more nuanced understanding of hope. Naïve optimism might have once promised that a single breakthrough would guarantee environmental salvation, but it left you vulnerable to disappointment. Grounded hope, by contrast, acknowledges that both setbacks and progress are part of the moral journey. Purpose guided by grounded hope recognizes that while you cannot guarantee perfect outcomes, your efforts can tilt probabilities toward better scenarios. In this framework, hope is not blind faith; it is a deliberate choice to remain engaged, to seek allies, to learn from each step. It's hope informed by evidence, humility, and empathy, creating a stable emotional foundation for sustained moral action.

Carrying forward a sense of purpose also involves recognizing that you are part of a generational continuum. Your moral endeavors build on past triumphs and failures, and they set the stage for future initiatives you cannot yet imagine. This awareness transforms anxiety about not achieving everything now into motivation to lay robust groundwork. Each partial improvement—like a cleaner river stretch, a restored patch of forest, a community more informed about climate adaptation—serves as a launchpad for those who will continue the work after you. You realize that your lifetime of caring is a meaningful chapter in a much larger story, and that purpose resides in contributing your best efforts, ensuring that future moral agents inherit richer resources, better tools, and a more enlightened moral discourse.

In this sense, carrying forward purpose involves letting go of the notion that you must personally resolve all environmental crises. Instead, you commit to doing your part responsibly and ethically, confident that countless others are doing theirs. Purpose flourishes when individuals grasp their role as co-authors of moral progress, each adding lines of complexity and hope. Instead of anxiety bred by isolation, you experience moral camaraderie—knowing that other hearts and minds share your values, strive for compatible goals, and refine their coping toolkits in parallel. This collective effort transforms caring from an isolating burden into a shared adventure, where purpose is reinforced by the presence of peers and allies dedicated to nurturing the planet.

Drawing on nature's resilience and Indigenous wisdom highlights that your purpose is not to dominate or "fix" nature once and for all. Rather, it's about participating respectfully in Earth's dynamic cycles, learning from its adaptive patterns, and supporting conditions that allow renewal. This relational understanding links moral purpose with humility. You accept that human cleverness alone cannot achieve stability; moral progress emerges from working with, not against, ecological processes. Purpose guided by this awareness acknowledges that while we wield great power, we also owe great responsibility to maintain life's diversity and integrities. Anxiety about failing to find neat solutions recedes, replaced by confidence that moral dedication, when harmonized with ecological principles and cultural wisdom, can steer societies toward healthier futures.

Hope and action reinforce each other. Purpose motivates you to act, and acting—even in small ways—generates evidence that positive change is possible. This evidence then fuels further hope, reinforcing purpose in a virtuous cycle. If you plant a garden that restores pollinators, you see bees returning, flowers thriving, and neighbors appreciating the green space. This tangible success updates your emotional memory, reminding you that your efforts matter. When eco-anxiety resurfaces, you recall these positive experiences, strengthening your resolve to continue. Over time, a portfolio of small successes forms, each adding weight to your belief that sustained action, rooted in hope, can shift trajectories. Anxiety becomes less menacing when you have concrete memories of moral effectiveness counterbalancing it.

Purpose also leads you to refine the stories you tell others about environmental issues. Instead of bombarding people with gloom, you balance honesty about difficulties with accounts of incremental improvements, cultural resilience, and creative problem-solving. These narratives inspire hope, not by lying about challenges, but by demonstrating that caring leads to results. Sharing such balanced stories with friends, family, or colleagues spreads purpose outward, amplifying the moral signal. In this way, your individual sense of purpose contributes to cultural transformations. Anxiety that conversation about environmental challenges always stirs panic gives way as more people adopt solution-oriented, morally grounded narratives, creating a supportive atmosphere for engagement.

Another aspect of carrying forward purpose is adapting as contexts shift. Over decades, your interests, responsibilities, and capacities might change. Early in your journey, you may focus on personal lifestyle adjustments; later, you might lead a regional coalition or mentor youth activists. Purpose does not require static roles. On the contrary, it thrives on evolving expressions. Perhaps you transition from front-line activism to policy analysis, or from direct fieldwork to educational outreach. Each transition can feel disorienting, sparking anxiety that you've lost your initial spark. But if purpose is rooted in hope and action, these shifts represent moral growth. You find new niches to apply your skills, ensuring continuous contribution while exploring fresh moral dimensions of environmental care.

Embracing cultural and intellectual diversity likewise sustains purpose. Moral engagement does not belong to any single tradition or worldview. Different philosophies, faiths, and cultural backgrounds offer distinct moral insights. By welcoming diverse perspectives—listening to Indigenous knowledge keepers, respecting different faith-based ecological ethics, or learning from secular humanist frameworks—you expand the moral vocabulary available. This expansion strengthens purpose because it prevents monotony or dogmatism from stifling moral creativity. Anxiety that competing moral visions create confusion diminishes when you understand that multiplicity of viewpoints is an asset. Different ethical lenses highlight distinct facets of environmental care, combining to form a more robust moral compass.

Furthermore, carrying forward purpose means periodically reassessing and updating your moral aims. Early on, maybe your focus was on reducing personal carbon footprints. As your perspective broadens, you may recognize the importance of challenging systemic injustices, advocating for indigenous sovereignty, or influencing climate finance mechanisms. Each new layer of awareness might initially trigger anxiety—can you handle these bigger moral arenas? But purpose, once internalized, gives you confidence that moral growth is natural. Expanding your concerns does not invalidate earlier efforts; it enriches them, placing personal actions within a larger context. This adaptability ensures that moral purpose never becomes stagnant or brittle. Instead, it's an evolving moral journey, always open to enrichment and recalibration.

In this extended moral adventure, hope acts as an anchor. Hope does not promise easy victories or absolute security. Rather, it assures that striving ethically makes a difference, that human potential and natural regenerative capacities intersect to yield meaningful outcomes. Hope encourages you to imagine futures where current struggles lead to new norms—zero-waste habits become mainstream, sustainable cities arise as standard, cultural respect for biodiversity replaces exploitative patterns. This imaginative capacity helps ward off anxiety, not by denying uncertainty but by reminding you that present actions can sculpt more favorable conditions.

Action, in turn, grounds hope in reality. Without action, hope risks drifting into abstraction. But as you consistently take steps—volunteering for restoration projects, supporting just policies, educating peers—hope materializes. Each action forms a bridge from present challenges to future possibilities. Instead of waiting passively for better times, you help create them. Anxiety loses its grip when you confirm through action that you are not powerless. Even if the impact seems small in isolation, collective actions aggregated over countless individuals and communities reshuffle moral landscapes. Purpose thrives in this interplay: hope guides you toward beneficial goals, and action validates that these goals are attainable, reinforcing hope again.

In communal contexts, when many people carry forward purpose aligned with hope and action, moral momentum builds. Societies can transform entrenched patterns—whether fossil fuel dependency or unsustainable consumption habits—into more ecologically harmonious systems. The narrative shifts from grim inevitability to a shared moral endeavor that acknowledges difficulties yet persistently works toward improvement. Anxiety becomes a background hum rather than a dominant soundtrack, as collective morale uplifts individuals. Everyone contributes differently, some focusing on policy, others on grassroots organizing, still others on cultural renewal or technological innovation. Diversity in roles and approaches, all inspired by a unifying sense of purpose, ensures moral synergy and resilience.

Intergenerational collaboration also strengthens purpose. Older generations, having endured initial waves of eco-anxiety and trial-and-error approaches, can mentor younger ones, transmitting moral lessons and effective coping methods. Younger generations bring fresh perspectives and creative tactics. Together, they refine the moral apparatus that sustains engagement. Anxiety about generational misunderstandings fades when both sides recognize their complementary contributions. By blending historical lessons with modern tools, the generational continuum forms a stable bridge over which moral purpose travels, ensuring future cohorts inherit not just problems but also moral and cultural assets to address them.

Reflecting on personal transformations, you might recall how once you felt immobilized by grim forecasts, uncertain whether your voice mattered. Today, you stand more confident—aware that small steps accumulate, incremental progress counts, and moral complexity is not a sign of defeat but of intellectual maturity. Your coping toolkit has expanded; you know how to manage media intake, celebrate partial wins, learn from nature's resilience, and integrate Indigenous wisdom. All these evolutions arose from actively engaging with complexity, choosing courage, and sustaining moral interest over time. The result is a moral identity more attuned to reality, more comfortable with uncertainty, and more eager to collaborate. Anxiety, while not vanquished, is now tamed into a motivator rather than a master.

As you move forward, you'll continue to encounter new obstacles, evolving scientific insights, political shifts, and cultural debates. Maintaining purpose and hope amid these fluxes will still demand moral attentiveness. The difference now is that you possess tested methods, emotional resilience, and a clarified moral vision. Instead of yielding to discouragement upon encountering resistance or slow policy changes, you interpret them through the lens of incremental progress and collective effort. You remind yourself that moral journeys are long and winding, and that your role is to contribute faithfully, knowing that no generation accomplishes everything but each can leave a better platform for the next.

By affirming the human capacity for creativity, compassion, and sustained engagement, you trust that our species is not irrevocably trapped by environmental crises. Though challenges loom large, human moral potential is equally formidable. Creativity ensures we never exhaust solutions. Compassion ensures we never stand alone. Sustained engagement ensures that improvements endure and accumulate. Thus, moral purpose thrives in this fertile intersection. Anxiety, rather than a herald of doom, signals areas needing attention. Guided by purpose, hope, and action, you can approach these signals without fear, confident that the human moral project—like nature itself—holds ample capacity for renewal and advancement.

Looking ahead, you might not know exactly which policies will catalyze major emissions cuts, which habitats will flourish through careful restoration, or how cultural attitudes toward consumption and waste might shift. But you do know that as long as humans tap into their moral capacities, possibilities abound. Your personal resolution to sustain caring for decades to come is part of a broader tapestry—one in which countless others invest their creativity, compassion, and perseverance. Collectively, these moral investments form a robust moral infrastructure capable of navigating complexity, encouraging continuous learning, and fostering alliances across differences.

In this collective moral endeavor, your purpose is no static pledge. It's a dynamic orientation—always refined by new insights, tested by setbacks, enriched by unexpected alliances. Seeing purpose as an evolving guide rather than a final objective releases the pressure to solve everything at once. You can feel fulfilled by each morally sound choice, each shared lesson, each moment when fear gives way to determined action. Instead of feeling rushed by anxiety, you respect the moral tempo of long-term stewardship.

This alignment of purpose with hope and action completes the moral arc you've been tracing. Having learned to cope with complexity, to embrace nature's resilience, to appreciate incremental progress, and to integrate diverse wisdom traditions, you now anchor these insights in a steadfast commitment. Your capacity to remain engaged, empathetic, resourceful, and forward-looking defines the moral legacy you contribute to. It's a legacy that can flow seamlessly into the hands of future generations, who will inherit not just environmental challenges but also moral foundations on which to build a more equitable, vibrant, and regenerative world.

In short, by carrying forward a sense of purpose rooted in hope and action, you affirm the best of human moral potentials. You declare that despite uncertainties, obstacles, and anxieties, humanity's creative, compassionate, and persistent nature can rise to meet environmental imperatives. Your journey, and that of countless others, transforms existential dread into a moral engine driving innovation, unity, and progress. By choosing this path, you invest in a future where caring matures into ever more sophisticated, heartfelt, and enduring forms, ensuring that the moral flame ignited in one generation continues to burn brightly for many to come.

 This entire exploration, from the first anxious questions to the final recognition of purpose, underscores that environmental engagement is never merely about technical fixes or isolated heroism. Instead, it emerges from embracing complexity, forging moral alliances, and cultivating emotional resilience. The insights shared across these chapters reflect a gradual unfolding of understanding: that uncertainty invites creativity rather than despair, that incremental successes matter as building blocks of deeper change, that diverse wisdom traditions enrich moral strategies, and that each generation inherits not only unresolved challenges but also a growing legacy of ethical principles and coping methods. By acknowledging fear without submitting to it, by balancing hope with action, and by continuously updating emotional and intellectual tools, we learn to navigate environmental care as an evolving moral journey rather than a static endpoint. In doing so, we affirm that humanity possesses the empathy, ingenuity, and steady resolve necessary to co-create a more harmonious relationship with the Earth. This enduring commitment does not promise a trouble-free future, but it does promise that despite uncertainties, moral agency can always find new ways to steer toward healing, justice, and the flourishing of life.

Appendices

Worksheets & Prompts: Tools for Reframing Anxious Thoughts, Tracking Emotional States, and Planning Next Steps

This section provides practical tools to help integrate the concepts discussed throughout the book into your daily life. These worksheets, prompts, and tracking methods are not prescriptions but flexible resources. You can customize them, adapt them over time, and select only what feels most relevant. The goal is to create a tangible framework that supports ongoing moral engagement, emotional resilience, and effective environmental action.

Worksheet A: Reframing Anxious Thoughts

Purpose: This worksheet helps you identify, examine, and reframe anxiety-driven thoughts about environmental problems. By acknowledging these thoughts and applying constructive reframing, you transform paralyzing fear into motivation or insight.

Step	Action	Example
1	Identify the anxious thought	"I feel overwhelmed because every day I read about more species going extinct."
2	Explore the origins	"I read multiple alarming headlines recently without any positive context."
3	Acknowledge emotions	"I feel sadness and fear. I worry that nothing I do can help."
4	Challenge the thought	"Is it really true that nothing helps? Have I seen any examples of recovery?"
5	Reframe constructively	"While extinction is a serious problem, I know some restoration projects succeed. Let me focus on local conservation groups and what I can support."
6	Plan a next step	"I'll look up a local habitat restoration effort and see how I can volunteer this month."

Instructions:

- Use this table when you catch yourself spiraling into anxious environmental thoughts.
- Fill in each column, starting with the troubling thought and working toward a reframed perspective and a concrete action.
- Over time, you'll internalize the skill of reframing and gain confidence in your capacity to respond constructively.

Worksheet B: Tracking Emotional States Over Time

Purpose: Monitoring your emotional responses to environmental information helps you identify patterns, triggers, and effective coping techniques. This chart allows you to note emotional states and what strategies worked on a weekly basis, guiding you toward sustainable engagement habits.

Weekly Emotional Tracking Chart

Day/Date	Emotional State (1-10 stress)	Trigger (event, news, conversation)	Coping Strategy Used (e.g., media boundary, meditation, community support)	Outcome (e.g., stress reduced, felt motivated)
Monday	7 (high stress)	Read grim IPCC report headline	Set media limit for the day, took a walk	Stress reduced to 5, felt calmer after walk
Tuesday	4 (mild unease)	Conversation about local river health	Re-read success stories of past restoration	Anxiety stable, ended day feeling hopeful
Wednesday	6 (moderate worry)	Social media doomscrolling	Unfollowed alarmist accounts, read solutions article	Stress lowered to 3, felt more balanced
Thursday	3 (low stress)	Listened to a positive podcast on regen. farming	Journaled gratitude for incremental wins	Maintained low stress, inspired for action
Friday	8 (high stress)	Extreme weather news broadcast	Reached out to local community group for dialogue	Stress to 5, support session helped a lot
Saturday	2 (very calm)	Guided nature walk	Just enjoyed nature's beauty, no interventions needed	Felt aligned, morally energized
Sunday	5 (moderate concern)	Deadline for advocacy letter	Broke task into steps, wrote letter, took breaks	Anxiety managed, felt accomplished after action

Instructions:

- Assign a number (1-10) to your emotional stress level each day.
- Identify major triggers for changes in mood.
- Record which coping strategies you attempted and their outcomes.
- Review weekly patterns to adjust strategies, set media boundaries, or find supportive communities as needed. Over time, this practice reveals what consistently helps you maintain emotional resilience.

Worksheet C: Planning Incremental Actions

Purpose: Often, anxiety arises from feeling helpless. This worksheet helps you outline small, attainable steps that contribute to environmental goals, reinforcing that moral value also resides in partial achievements.

Incremental Action Plan

Goal	Step 1 (Immediate)	Step 2 (1-2 months)	Step 3 (6 months)	Indicators of Progress
Support Pollinators	Plant native flowering plants in my backyard	Ask neighbors if they'll also plant pollinator-friendly species	Advocate for a local ordinance encouraging pollinator gardens	Count more bees, observe butterfly return, get local garden club involved
Reduce Plastic Waste	Switch to a reusable water bottle	Explore a local zero-waste store for bulk buying	Organize a neighborhood plastic-free challenge	Fewer plastic items in household waste, neighbors joining challenge, store sales of bulk items increasing
Influence Local Policy	Attend a city council meeting to understand climate measures	Speak during public comment to support emission reduction targets	Join a local policy advocacy group or campaign for a council candidate who prioritizes sustainability	Improved city emission targets, positive council feedback, increased membership in advocacy group

Instructions:

- Choose a moral objective (e.g., enhancing local biodiversity, lowering personal carbon footprint, influencing policy).
- Break it into three incremental steps spanning immediate actions to medium-term endeavors.
- Identify indicators to track progress, ensuring you can celebrate partial achievements.
- Review this plan periodically, adjusting steps as conditions evolve.

Worksheet D: Embracing Complexity and Uncertainty

Purpose: Environmental moral engagements often unfold in uncertain conditions. This worksheet encourages you to accept uncertainty as a space for moral creativity and adapt your mindset rather than seeking absolute certainties.

Complexity-Acceptance Framework

Uncertain Aspect	Old Response (Needing Certainty)	New Response (Embracing Complexity)	Potential Moral Benefit
Future Climate Policy	"We must have exact solutions now!"	"We have partial policies; let's refine and scale them up incrementally."	Reduces panic, encourages steady improvements
Species Adaptation	"We must know exactly which species will survive."	"We monitor populations, support corridors, and stay flexible in conservation priorities."	Fosters adaptive conservation, avoids paralysis by uncertainty
Energy Transition	"We need a silver-bullet energy source."	"We combine renewables, efficiency measures, storage solutions, and community ownership."	Encourages diversity in solutions, building resilience

Instructions:

- Identify areas where you feel compelled to seek absolute certainty.
- Reframe them by embracing uncertainty, focusing on adaptability, learning, and layered strategies.
- Note how this shift could mitigate anxiety and strengthen moral engagement.
- Revisit these pairs periodically to see if reframing complexity continues to support healthier emotional states and more strategic action.

Worksheet E: Mentoring and Intergenerational Wisdom Exchange

Purpose: This worksheet helps you plan how to share your hard-earned insights and coping methods with others, ensuring that moral lessons become durable cultural assets passed between generations and communities.

Mentorship Action Plan

Mentee/Audience	What I Can Offer	Format (One-on-One, Workshop, Online Content)	Initial Step	Desired Outcome
Local Youth Group	Emotional coping strategies, incremental success stories	Monthly in-person workshop	Contact youth coordinator to set a date	Youth learn sustainable engagement methods, reduce eco-anxiety
New Activists at Workplace	Policy advocacy tips, media boundary techniques	Lunch-and-learn sessions	Invite interested colleagues via email	Colleagues gain balanced approach, improving long-term morale
Online Community Subscribers	Solutions-focused story aggregation, moral reframing articles	Blog posts, newsletters	Draft first article outlining incremental wins in reforestation	Wider audience gains access to moral insights, engages more productively

Instructions:

- Identify specific groups or individuals who could benefit from your experience.
- Determine what insights you can share—emotional coping, understanding incremental progress, policy tactics, etc.
- Choose an appropriate format and take the first step toward making it happen.
- Reassess outcomes after initial attempts, refining your mentorship approach accordingly.

Worksheet F: Updating Your Coping Toolkit Regularly

Purpose: To maintain long-term moral engagement, periodically reassess and refresh your emotional coping methods, ensuring they match evolving circumstances and personal growth.

Coping Toolkit Review

Coping Method	Still Effective? (Y/N)	If No, Why Not?	Possible Replacement or Enhancement	Trial Period
Weekly Media Limit	Y	N/A	Maintain as is, maybe add a solutions-focused newsletter	Ongoing
Journaling Gratitudes	N	Feels repetitive, less comforting now	Try creating art inspired by nature for emotional release	Try for 1 month
Mentorship Calls with Peer	Y	N/A	Consider adding a monthly group reflection session	Test next quarter
Meditation on Nature's Cycles	N	No longer relieving stress as before	Incorporate short mindfulness walks outdoors instead	Try for 2 weeks

Instructions:

- List your current coping methods and assess their effectiveness.
- If a method no longer works, explore alternatives or variations.
- Set a trial period for new methods to see how they feel.
- Repeat this review every few months or after major emotional shifts.

Resources & Further Reading: Organizations, Books, Podcasts, and Online Communities for Ongoing Learning and Support

This section provides a curated list of resources to help you continue learning, engage with supportive communities, and remain informed without succumbing to emotional overload. These recommendations span organizations, literature, digital content, and interactive forums, allowing you to tailor your engagement to your interests and emotional capacities. Consider these resources as a menu rather than a required syllabus. Choose what resonates with your moral goals, and remember to apply the coping strategies and boundaries you've developed to maintain balanced, hope-informed engagement.

Organizations and Initiatives for Continued Engagement

1. Environmental Defense Fund (EDF)

- **Focus:** Policy advocacy, climate solutions, and corporate engagement.
- **Why It Matters:** EDF combines science, economics, and bipartisan advocacy to achieve measurable results. Engaging with their campaigns or newsletters helps you track policy progress and learn from solution-oriented frameworks.

2. Conservation International (CI)

- **Focus:** Biodiversity conservation, nature-based climate solutions, and support for local communities.
- **Why It Matters:** CI's projects often highlight regenerative approaches and indigenous-led conservation efforts. Their reports and case studies are rich sources of insight into incremental improvements on the ground.

3. The Climate Reality Project

- **Focus:** Grassroots organizing, climate education, and leadership training.
- **Why It Matters:** Founded by Al Gore, it empowers individuals to understand climate science and communicate effectively. Their leadership corps and workshops help translate anxiety into informed advocacy.

4. Indigenous Environmental Network (IEN)

- **Focus:** Indigenous-led activism, land rights, and climate justice.
- **Why It Matters:** The IEN's communications provide frontline perspectives, offering moral depth, cultural insight, and authentic examples of resilience and reciprocity with the land.

5. Global Alliance for the Rights of Nature

- **Focus:** Legal frameworks recognizing ecosystems as entities with rights.
- **Why It Matters:** Shifting legal structures toward granting nature rights can inspire new moral narratives and policy reforms that support Earth's regenerative capacity.

Books That Offer Moral Depth, Complexity, and Hope

1. "Braiding Sweetgrass" by Robin Wall Kimmerer

- **Focus:** Indigenous wisdom, scientific knowledge, and reciprocal relationships with nature.
- **Why It Matters:** Kimmerer's storytelling exemplifies blending ecological understanding with moral insight, showing how gratitude, ceremony, and reciprocity foster hope and resilience.

2. "The Future We Choose" by Christiana Figueres and Tom Rivett-Carnac

- **Focus:** Balanced outlook on climate action, highlighting both scientific urgency and plausible pathways forward.
- **Why It Matters:** Offers grounded hope, encouraging readers to choose courage and persistence rather than despair, aligning well with complexity-based moral engagement.

3. "Don't Even Think About It: Why Our Brains Are Wired to Ignore Climate Change" by George Marshall

- **Focus:** Psychological dimensions of climate communication and moral engagement.
- **Why It Matters:** Understanding cognitive barriers helps refine coping strategies and moral communication, reducing anxiety through empathy and better dialogue.

4. "A Paradise Built in Hell" by Rebecca Solnit

- **Focus:** How communities respond to disasters with unexpected cooperation and solidarity.
- **Why It Matters:** Reinforces that human capacities for compassion and creativity often surface during crises, inspiring long-term moral resilience.

5. "All We Can Save," edited by Ayana Elizabeth Johnson and Katharine K. Wilkinson

- **Focus:** Essays, poems, and reflections by diverse women in the climate movement.
- **Why It Matters:** Multiplicity of voices fosters moral pluralism and reveals the emotional, intellectual, and creative richness supporting sustained environmental care.

Podcasts for Regular Inspiration and Nuanced Updates

1. "How to Save a Planet"

   - **Focus:** Solutions-oriented climate stories, practical steps to make a difference.
   - **Why It Matters:** Promotes incremental progress, balanced emotional engagement, and grounded hope through real-life case studies and accessible storytelling.

2. "Outrage + Optimism"

   - **Focus:** Climate action debates, featuring activists, policymakers, and innovators.
   - **Why It Matters:** Combines emotional honesty (acknowledging outrage) with grounded optimism, embodying the complexity-embracing ethos.

3. "For the Wild"

   - **Focus:** Engaging conversations on ecological renewal, indigenous rights, and cultural shifts.
   - **Why It Matters:** Deep moral reflections and focus on regenerative practices and local wisdom support the integration of emotional resilience and ethical depth.

4. "The Energy Gang"

   - **Focus:** Clean energy transitions, industry trends, and policy developments.
   - **Why It Matters:** Keeps you informed about technological and policy increments without drowning in negative headlines, maintaining a solutions perspective.

5. "Degrees: Real talk about planet-saving careers"

   - **Focus:** Career stories in sustainability, featuring individuals finding meaning in their work.
   - **Why It Matters:** Personal narratives reduce anxiety about one's individual role, illustrating that moral purpose can be embedded in professional paths.

Online Communities and Discussion Forums

1. "r/ClimateActionPlan" on Reddit

- **Focus:** Crowdsourced strategies, solution-sharing, and incremental successes in climate action.
- **Why It Matters:** Interactive platform allowing conversation, peer support, idea exchange, and celebrating small victories together—fostering communal emotional resilience.

2. "Eco Anxiety Support" Groups on Facebook or Discord

- **Focus:** Emotional support, coping techniques, and balanced media sharing.
- **Why It Matters:** Normalizes anxiety, providing a safe space to process feelings, learn reframing strategies, and practice boundaries with sympathetic peers.

3. "The Solutions Journalism Network"

- **Focus:** Story database highlighting successful responses to social and environmental problems.
- **Why It Matters:** Provides a catalog of solution-oriented coverage, making it easier to find balanced stories to offset grim news and maintain grounded hope.

4. "We Don't Have Time" App and Online Community

- **Focus:** Climate action social network where users rate companies' climate efforts, share positive news, and support innovations.
- **Why It Matters:** Encourages shifting from anxiety-driven doomscrolling to constructive engagement, rating improvements, and discovering action opportunities.

5. Local and Regional Environmental Networks (e.g., city-based sustainability coalitions, community garden forums)

- **Focus:** Place-based action, cultural adaptation, and incremental change in your locality.
- **Why It Matters:** Connects global issues to tangible local initiatives, reducing isolation and anxiety by showing direct impacts of your efforts.

Approaches for Using These Resources Mindfully

Balancing Media Sources:

- Sample from solution-oriented podcasts or websites when feeling overwhelmed by negative stories.
- Alternate between global perspective news and local success narratives to maintain a sense of scale and controllability.

Setting Information Rhythms:

- Plan a weekly rotation: start Monday with a success-focused podcast, midweek check policy updates from a reputable NGO, and Friday read a reflective essay from a selected book. This pattern ensures a balanced emotional arc.

Community Engagement with Resources:

- Form a small reading group—pick a chapter from "All We Can Save" or a segment from "How to Save a Planet" podcast to discuss monthly, focusing on what insights help manage anxiety and inspire action.

Periodic Reassessment of Your Resource List:

- Every few months, evaluate which resources still serve your moral and emotional needs. If a podcast feels too alarmist, switch to another offering more balanced coverage. If a certain online community no longer meets your needs, find another that better aligns with your current focus.

Tables and Charts to Customize Your Resource Use

Table: Personal Resource Integration Plan

Resource Type	Resource Name	Frequency of Engagement	Intended Moral/Emotional Benefit	Evaluation After 3 Months
Book	"Braiding Sweetgrass"	2 chapters/month	Gain indigenous perspective, restore emotional grounding	Check if reading reduces eco-anxiety
Podcast	"How to Save a Planet"	1 episode/week	Stay solution-focused, balanced outlook	Assess if episodes inspire action
Online Community	Local eco-forum	Check biweekly	Peer support, incremental improvement stories	Note if group interactions lift mood
Organization Newsletter	EDF updates	Once a week	Track policy increments, celebrate small wins	Evaluate if updates maintain hope
Emotional Coping Tool	Anxiety tracking chart	5 min/day	Monitor stress triggers, refine coping toolkit	Check patterns and improvements

Rising Above: A Practical Guide to Overcoming Climate Anxiety and Finding Hope in a Changing World

Chart: Information-Action Cycle

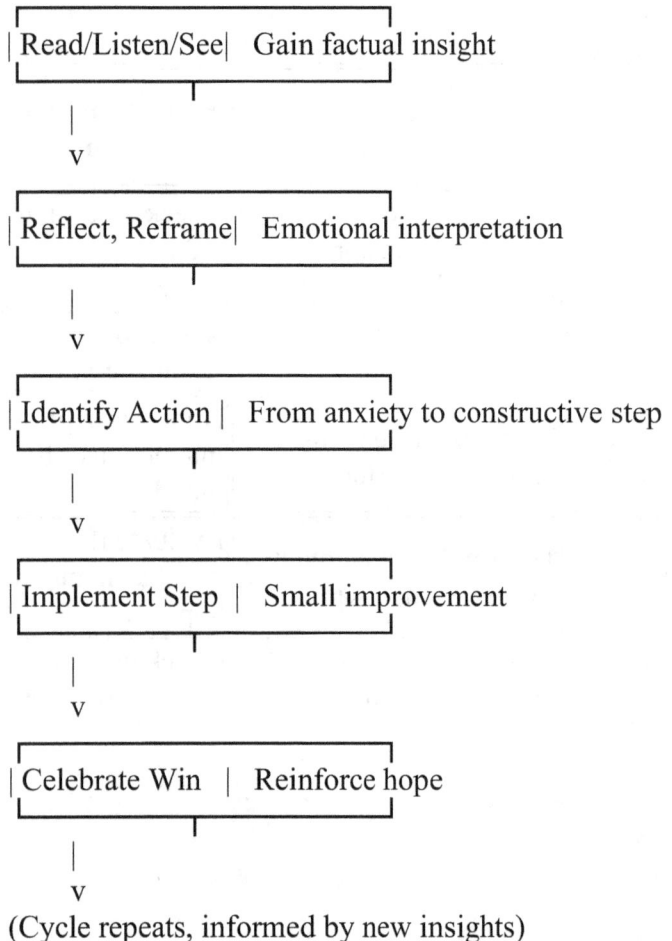

(Cycle repeats, informed by new insights)

Use this chart to visualize how media intake and resources guide you through a constructive cycle: from receiving information to framing it morally, choosing an action, executing it, and then celebrating, thus reinforcing hope and purpose.

Final Notes on Utilizing Resources

These resources are tools, not burdens. Use them strategically. If certain sources aggravate anxiety rather than informing or uplifting you, set them aside. If a podcast series you once found inspiring begins to focus too heavily on despair without offering pathways forward, feel free to switch to another that reaffirms hope and complexity. Just as you learned to adapt your coping toolkit, adapt your resource usage. The overarching aim is to maintain emotional resilience, moral clarity, and productive engagement, ensuring that as you continue your lifelong journey of caring for the planet, you remain equipped, supported, and inspired by the wealth of knowledge and experience humanity has gathered.

Glossary: Key Terms for Quick Reference

Adaptive Management:
A flexible approach to decision-making in environmental policy and practice that acknowledges uncertainty, encourages ongoing monitoring, and adjusts strategies based on new evidence and changing conditions. Rather than adhering to a fixed plan, adaptive management treats action as a continuous learning process.

All-or-Nothing Thinking:
A mindset that frames environmental solutions as either total successes or total failures. This rigid perspective often leads to despair when immediate, perfect outcomes aren't achieved. Recognizing partial progress and incremental improvements counters this extremism, allowing moral agents to value steady, cumulative gains.

Anxiety (Eco-Anxiety):
Emotional distress arising from awareness of environmental crises—such as climate change, pollution, species extinction—and uncertainty about future outcomes. While anxiety can be destabilizing, it can also prompt moral reflection, strategic thinking, and constructive engagement when channeled wisely.

Boundaries with Media Intake:
Intentional limits set on consuming environmental news and information to prevent emotional overwhelm and despair. By curating sources, timing exposure, and focusing on quality over quantity, individuals maintain emotional balance and stay sustainably informed.

Civic Participation:
Involvement in public life—voting, contacting representatives, attending hearings, joining community initiatives—to influence environmental policies and resource management. Civic participation transforms anxiety into collective moral action, demonstrating that individuals can shape larger systems through sustained engagement.

Collective Effort (Communal Moral Engagement):
The idea that environmental stewardship is not an isolated task but a shared undertaking. People, organizations, cultures, and communities collaborate across generations, combining resources, knowledge, and skills to produce more effective, enduring solutions.

Complexity Thinking:
A framework acknowledging that environmental problems, ecosystems, and societal systems are dynamic, interconnected, and non-linear. Rather than expecting straightforward solutions, complexity thinking embraces uncertainty, incremental progress, and iterative adaptation as normal features of moral life.

Creativity in Environmental Action:
The capacity to devise innovative solutions, blend diverse knowledge systems, and experiment with new strategies. Creativity helps navigate complexity, unlock fresh approaches when old methods stall, and transform anxiety about unsolved problems into a catalyst for invention.

Cultural Diversity (Moral Pluralism):
Recognizing that multiple cultural traditions, indigenous insights, philosophical schools, and spiritual beliefs contribute valuable moral perspectives. Embracing cultural diversity enriches moral frameworks, allowing varied approaches to emerge and collectively improve environmental responses.

Embrace of Complexity:
Acceptance that no single measure or viewpoint solves environmental crises. Instead of trying to eliminate uncertainty or complexity, moral agents treat them as opportunities for learning, adaptation, and creative exploration. This stance transforms anxiety from paralyzing fear into a prompt for moral innovation.

Emotional Well-Being (Emotional Resilience):
The ongoing practice of maintaining psychological health amid challenging environmental realities. Emotional well-being involves setting media boundaries, adopting mindfulness and breathwork techniques, celebrating partial successes, seeking community support, and integrating rituals that affirm moral purpose.

Feedback Loops (in Nature and Policy):
Circular processes where outputs influence inputs, shaping the trajectory of ecosystems or social systems over time. Recognizing feedback loops allows for adaptive interventions that reinforce positive trends or counter harmful cycles, reducing anxiety by highlighting opportunities to steer dynamics toward stability.

Grounded Hope:
A form of hope that acknowledges difficulties, uncertainties, and partial progress without succumbing to naive optimism. Grounded hope balances honest assessment of challenges with faith in moral agency, incremental improvements, and nature's regenerative capacities. It provides emotional stability and sustained motivation for long-term engagement.

Incremental Improvements (Partial Wins):
Small but meaningful steps that, over time, contribute to larger transformations. Recognizing incremental improvements counters all-or-nothing thinking and despair, reinforcing that moral progress often accumulates gradually. By valuing these partial gains, individuals and communities maintain momentum and build confidence.

Indigenous Wisdom:
Knowledge systems, cultural practices, and ethical frameworks developed by Indigenous communities over millennia of interacting closely with local ecosystems. Indigenous wisdom often emphasizes reciprocity, relational ethics, and long time horizons, providing moral guidance and practical strategies for sustainable resource management and ecological restoration.

Intergenerational Continuity (Generational Effort):
The understanding that environmental care spans multiple generations. Each cohort builds on past achievements, refines coping strategies, and passes moral insights forward. Embracing

intergenerational continuity relieves pressure on any single generation to "solve" everything at once and encourages moral patience and long-term perspective.

Mentoring and Paying Forward Insights:
The practice of transmitting hard-won lessons, coping strategies, moral frameworks, and effective methods to newcomers. Mentoring ensures that knowledge and emotional resilience accumulate across generations, strengthening collective moral capacity. It transforms anxiety about future challenges into confidence that others will be better equipped to continue the work.

Mindfulness (Breathwork, Grounding Techniques):
Science-backed methods—such as focusing on breath, observing sensations, or practicing the "5-4-3-2-1" sensory exercise—that help individuals remain present, calm surging anxiety, and restore emotional equilibrium. Mindfulness transforms fear spikes into opportunities to regain focus and moral clarity.

Moral Complexity:
The recognition that environmental ethics and solutions rarely lend themselves to simple moral judgments. Moral complexity acknowledges trade-offs, partial measures, incremental success, and the need to integrate multiple knowledge systems. By welcoming complexity, anxiety subsides because moral agents learn to navigate evolving landscapes with flexible, nuanced strategies.

Naïve Optimism:
A belief in easy, guaranteed solutions to environmental crises without grappling with their complexity or acknowledging potential setbacks. Naïve optimism provides short-term comfort but erodes morale when reality fails to match rose-tinted expectations. Distinguishing naïve optimism from grounded hope prevents emotional disillusionment and encourages realistic, sustained effort.

Nature's Resilience (Earth's Capacity to Rebound):
The inherent ability of ecosystems and species to recover, adapt, and reorganize after disturbances. Understanding nature's resilience fosters trust that moral actions can support regenerative processes. Recognizing this capacity counters despair, showing that healing is possible if humans reduce harm, restore habitats, and cooperate with ecological patterns.

Regenerative Practices:
Approaches that go beyond minimizing harm to actively restoring and enhancing ecosystems, soils, water systems, and cultural traditions. Regenerative agriculture, rewilding, habitat restoration, and community-based adaptations align with long-term healing. They integrate principles drawn from ecology, indigenous knowledge, and cultural values that emphasize renewal rather than mere maintenance.

Relational Ethics:
A moral perspective that centers on relationships—between humans and nature, among communities, across generations—rather than viewing individuals or ecosystems in isolation. Relational ethics encourage empathy, reciprocity, and shared responsibility, transforming anxiety

about global crises into collaborative problem-solving grounded in respect for interconnected life.

Resilience (Moral and Ecological):
The capacity to absorb disturbances, reorganize, and maintain or improve functionality. Ecologically, resilience reflects how ecosystems endure and adapt despite shocks; morally, it represents how humans maintain ethical commitments, emotional well-being, and strategic focus under uncertainty and setbacks. Both forms of resilience rely on diversity, redundancy, and adaptive learning.

Setting Boundaries (with Information and Emotional Exposure):
The practice of regulating one's consumption of environmental news, social media, and alarmist headlines to prevent emotional overload and despair. By setting boundaries—choosing reliable sources, limiting time spent on grim stories, and balancing negative content with solutions-focused coverage—individuals preserve emotional strength and maintain moral engagement over the long haul.

Sustainability vs. Regeneration:
Sustainability aims to maintain current conditions or prevent further harm, while regeneration seeks to actively improve ecological and cultural health beyond a baseline. Embracing regeneration recognizes that we can restore depleted soils, reintroduce missing species, and revitalize cultural traditions, moving from damage control to creative renewal.

Uncertainty as Possibility:
Instead of viewing uncertainty as a barrier or a source of endless anxiety, uncertainty is reframed as a space where multiple positive outcomes remain possible. This perspective encourages openness to learning, experimentation, and flexible strategies. By treating uncertainty as an opportunity for moral agility rather than a cause for paralysis, moral agents keep moving forward despite not knowing all outcomes.

Virtuous Cycle of Hope and Action:
The reciprocal reinforcement between taking meaningful steps (action) and reaffirming that change is possible (hope). Each moral endeavor, however modest, generates evidence that inspires further engagement, creating a feedback loop that sustains optimism without naivety. Anxiety diminishes as participants witness their actions feeding a cycle of continuous improvement and evolving moral clarity.

Whole-of-Life Moral Engagement:
An understanding that environmental stewardship is not a short campaign or a phase of activism, but a lifelong moral vocation. As conditions evolve, so must moral strategies and emotional coping methods. Accepting this long-term horizon transforms anxiety about immediate results into steady commitment, acknowledging that moral contributions build cumulatively across time and generations.

www.ingramcontent.com/pod-product-compliance
Lightning Source LLC
Chambersburg PA
CBHW082243220526
45469CB00009B/2864